T0320407

HPC, Big Data, and AI Convergence Towards Exascale

HPC, Big Data, and AI Convergence Towards Exascale

Challenge and Vision

Edited by
Olivier Terzo
Jan Martinovič

CRC Press is an imprint of the
Taylor & Francis Group, an **informa** business

First edition published 2022
by CRC Press
6000 Broken Sound Parkway NW, Suite 300, Boca Raton, FL 33487-2742

and by CRC Press
4 Park Square, Milton Park, Abingdon, Oxon, OX14 4RN

ISBN: 978-1-032-00984-1 (hbk)
ISBN: 978-1-032-00991-9 (pbk)
ISBN: 978-1-003-17666-4 (ebk)

DOI: 10.1201/9781003176664

Typeset in Palatino LT Std
by Newgen Publishing UK

Contents

Foreword

The convergence of HPC, Big Data, and AI represents a cornerstone of the EuroHPC JU's vision for the upcoming digital decade. HPC, in a broader interpretation, will become the backbone of future applications for numerical simulations, HPDA, and AI. With its ambition to provide a world-class pan-European HPC infrastructure, the JU will deliver state-of-the-art HPC systems across the EU in combination with a strategic research and innovation program, addressing the requirements of the interdisciplinary communities of scientists as well as public and private stakeholders. The federation of resources remains one of the priorities of the EuroHPC JU's vision, where the JU will connect the European HPC systems to a federated infrastructure and establish links to other European initiatives such as the European Open Science Cloud.

With the expected adoption of the new regulation establishing the EuroHPC JU in the years 2021–2027, substantial resources will be unlocked to address the interdisciplinary challenges associated with HPC and implement the objectives of the JU. Our activities will be centered around several pillars, including the deployment of new infrastructure, the development of new European technology, applications, skills, and education in HPC and related areas. The envisaged actions will contribute to the European autonomy and sovereignty regarding HPC technology and industrial supply chains in the context of big data, artificial intelligence, and cloud services, based on a holistic perspective on the digital ecosystem in Europe.

At this point in time, the JU, together with our participating states, has procured seven HPC systems in Europe and launched five calls for proposals, covering strategic European projects on technology, skills, and applications. Many research and innovation actions focus the development of workflows on modern heterogeneous architectures, adopt concepts from AI and ML in traditional numerical HPC applications and develop technology for the effective use and operation of a future federated HPC infrastructure. Finally, I would like to stress that establishing a dynamic, prosperous, and inclusive European HPC ecosystem is a collective challenge and I encourage the HPC community to join our efforts and contribute to the European HPC ecosystem, for example by participating in our calls or as experts in our evaluations.

Anders Dam Jensen
Executive Director of the EuroHPC Joint Undertaking

Foreword

Foreword

The convergence of HPC, big data, and AI on their trajectory toward exascale is very close to the heart of ETP4HPC, the European High-Performance Computing Technology Platform. We began work on this in 2017 by comparing the system stacks of HPC and big data and identifying the commonalities and deviations that needed to be addressed. This collaboration grew into an alliance with Europe's Big Data Value Association, which continued for a number of years and the work of which was present in two issues of our HPC Strategic Research Agenda (SRA).

At present, most of the key players of the European Digital Technology ecosystem are part of the TransContinuum Initiative (TCI), which is led by our team. It is a forum where long- and medium-term digital priorities are drafted jointly by HPC, big data, AI, cybersecurity, and mathematics experts. Platforms for big data use and associated solutions are present in this work. The convergence and integration of these technologies is one of the key objectives of TCI. Similarly to this publication, TCI analyzes use cases in order to illustrate the realm of its technologies.

I am glad to see projects, efforts, and publications that address the use of big data and HPC together with related workflows and technologies. I hope that work of this kind, often based on the ideas included in our SRA and TCI work, will continue. On the other hand, I welcome the participants and collaborators of such initiatives to join our Association, take part in writing the SRA, and thus leverage their research priorities within the European HPC ecosystem.

Jean-Pierre Panziera
Chairman of ETP4HPC

Preface

Until recently, the scientific fields of HPC, data analytics, and AI developed relatively separately and independently. Their mutual interactions were rather rare and only then in cases of mutual need. While typical HPC applications targeted computationally intensive simulations requiring extreme computing power, data and AI applications focused more on the volume of processed data. Algorithms and software tools developed within these domains were logically adapted to their requirements. However, the size of data sets produced by classic HPC simulations and the need for computational power necessary to process continuously increasing volume of data brings these primary domains of computational and data science closer. This process of so-called convergence of HPC, big data, and AI is also significantly supported by the current development of computer architectures. Thanks to the pursuit of computing power, current supercomputers rely more and more on so-called accelerated architectures, increasing the speed of interconnect and implementing high-speed data storage.

Thus, all these facts mark significant potential for a completely new approach to computing solutions for scientific and industrial problems. Multiple numerical models, fast processing of very large data sets, and the massive deployment of artificial intelligence tools, either as alternatives to numerical models or data processing tools, are the essential characteristics of upcoming applications. The complexity of such applications requires extreme computational and data performance that only a new generation of exascale computers can provide. In addition, increasing data connectivity over the internet allows users of such applications to access computing data resources as a cloud service. Modern supercomputers already have a modular structure to enable the best possible realization of such complex applications with the goal of separating users as much as possible from the technical implementation of the infrastructure.

Although we are already seeing the first applications and technical solutions combining HPC technologies with big data and AI tools available through user-friendly interfaces, we are yet on the cusp of a new era of modular exascale technologies and their strongly multidisciplinary applications. All projects contributing to this book pave the road for this new generation of technologies, support their development, and, in addition, verify them on real-world problems. I believe that the reader will be interested in the book because it brings an overview of currently available technologies that fit with the concept of unified cloud–HPC–big data–AI applications and presents examples of their actual use in scientific and industrial applications.

doc. Mgr. Vít Vondrák, PhD,
Director of IT4Innovations

Preface

Traditionally, the scientific fields of [illegible] and AI [illegible] relatively separate and independent [illegible] were [illegible] and [illegible] used. [illegible] applications [illegible] large, computationally intensive [illegible] extreme computing [illegible] data [illegible] applications [illegible] volume of processed data. Algorithms and software [illegible] these [illegible] adapted to [illegible]. However, the size of datasets produced by [illegible] HPC simulations and the need for computational power necessary to process [illegible] data [illegible] more [illegible]. This [illegible] of HPC [illegible] AI [illegible] supported by the recent development of [illegible]. Thanks to the [illegible] of computing power [illegible] more [illegible] architectures [illegible] and in [illegible].

[illegible] all these facts mark a significant [illegible] for a completely new approach [illegible] computing solutions [illegible] problems. Multiple numerical analysis, processing of large data sets, and the [illegible] development of artificial intelligence tools [illegible] numerical models and data processing tools [illegible] applications. The [illegible] computational and data performance [illegible] computers can provide. In addition, [illegible] data [illegible] the internet allows users [illegible] data sources [illegible] structure to enable the best possible [illegible] the goal of separating users as far as possible from [illegible] implementation of the [illegible].

[illegible] combining HPC technologies with [illegible] and AI [illegible] users [illegible] techniques and [illegible] applications [illegible] computing [illegible] this new generation of technologies [illegible] problems [illegible] the concept [illegible] applications [illegible] and [illegible] applications.

[illegible], PhD
[illegible]

Preface

Over the past decades, digital processing technologies have demonstrated an impressive growth in terms of performance. In this regard, it is noteworthy that the processing capability of computers has increased by a factor of 10^6, while their unitary (economic) cost has decreased by the same factor. This encapsulates the reason behind large-scale adoption of information and communication technologies (ICT): higher performance at an ever-lower cost. As such, the availability of new information and data has skyrocketed; indeed, every day our society generates more data than that was generated by human beings in a five-century time frame since Gutenberg's (mid-1400s) invention of the printing process. The consequence of this rapid and growing production of data is that it requires large and ever-more capable computing resources to be analyzed, a domain that is dominated by high-performance computing (HPC) systems. These are computers designed specifically to tackle extremely complex problems with great relevance in scientific and industrial fields.

Nowadays, supercomputers are able to compute more than 10^{15} floating-point operations per second (FLOPS) – petascale systems, with the top-10 systems in the world capable of performing more than 10^{17} FLOPS – pre-exascale systems. Next-generation supercomputers (exascale systems) will go even further by being able to reach the threshold of 10^{18} FLOPS, i.e. the astonishing number of a billion of billions of operations per second that could be theoretically reached by aggregating the computing power of all the smartphones in the EU. This incredible computing power is necessary to solve most of the problems in strategic sectors, as the creation of accurate and possibly long-term weather forecasts and climate models, the boosting of computational fluid dynamic simulations, the design of innovative chemical compounds and drugs, for analysis of massive data generated by high-energy physics (HEP) experiments. The European Commission has foreseen a major role for European industrial and academic actors, as well as private and public research bodies in this; thus, it decided to promote large investments to install (pre-)exascale machines, along with boosting other forefront fields like artificial intelligence (AI) and quantum computing and communication (QCC) technologies. Nevertheless, all these digital innovations are at the basis of important worldwide socioeconomic consequences, thus helping the EU to rival again the USA and Asian countries (indeed, only two European supercomputers are in the top-10 positions of the top 500 rank as of today: www.top500.org).

In this book, key technologies and components of the next generation of HPC systems are illustrated, with a focus on relevant scientific and industrial real-use cases. The challenges behind the creation of these new incredible machines are still many and touch different aspects: from the design and integration of heterogeneous computing devices (e.g. chips designed to accelerate deep learning algorithms), to the orchestration of complex applications in extremely parallel and distributed execution environments, to the design and implementation of middleware for managing gigantic data sets. All this represents a big challenge, but also a great opportunity to make important leaps forward; this book maps a journey into this exciting world of innovators and novel technologies.

Prof. Marco Mezzalama
President, LINKS Foundation

Acknowledgments

We would like to express gratitude to all our authors, coauthors, coordinators from LEXIS, ACROSS, EVOLVE, DeepHealth, and CYBELE European projects, and also to the European Processor Initiative (EPI) and TransContinuum Initiative for their contributions to this book. Our thanks also belong to those who provided support, assistance, collaboration, reviews, and fruitful discussion about book content together with essential comments.

It would not be possible to write a book without input from authors and their organizations that allowed the sharing of relevant studies of scientific and industrial applications, use cases in the field of high-performance computing (HPC), big data, and artificial intelligence convergence. Thanks to all.

A special thanks go to the European Technology Platform for High Performance Computing (ETP4HPC) initiative facilitating the provision of innovative solutions to tackle grand societal challenges in Europe and the EuroHPC initiative for building a globally competitive HPC technology value chain.

We would like to thank our institutions: LINKS Foundation – Leading Innovation and Knowledge for Society, IT4Innovations National Supercomputing Center, which enabled us to bring together our knowledge, experience, and the results from the H2020 and EuroHPC European projects.

Dr. Olivier Terzo wants to thank the LINKS Foundation president, Prof. Marco Mezzalama, the director, Dr. Stefano Buscaglia, and researchers and colleagues for their precious suggestions and comments.

Dr. Jan Martinovič wants to thank the IT4Innovations director, doc. Mgr. Vondrák Vít, PhD, for his long-term support, also to all research colleagues for their generosity, expertise, and feedback, which helped us to bring together this book.

A special thanks to our publishers, Nora Konopka and Prachi Mishra, for allowing this book to be published

We want to offer our sincere thanks to all readers and all those who promote this book.

Editors

Olivier Terzo, PhD, in Electronic Engineering and Communications, MSc Degree in Computer Engineering from the Politecnico di Torino (Italy). He also has a university degree in Electrical Engineering Technology and Industrial Informatics from University Institute of Nancy (France). He is currently head of the ACA Advance Computing and Applications Research Area with a staff of 18 researchers at the LINKS Foundation Applied Research Center (Italy). From 2013 to 2018, he was head of Research Area ACE (Advanced Computing & Electromagnetics) at the ISMB Research Center (Italy) dedicated to the study and implementation of computing infrastructure based on technologies of cloud computing. His main fields of research are on resources and applications orchestration on distributed infrastructure, machine learning application development on FPGA accelerators, cloud and IoT convergence on ultra-low-power devices, low-power computing and communication. He was technical and scientific coordinator of the OPERA H2020 project (ICT4 – 2015 call). Olivier is currently coordinator of the ACROSS [HPC, Big Data, Artificial intelligence Cross-Stack Platform toward Exascale] EuroHPC project and is codesign and dissemination manager of the LEXIS project ICT11-2018 Innovation Action on HPC and Big Data-enabled Large-scale Testbeds and Applications; he is also technical and scientific coordinator of the Plastic & Rubber Regional project focused on Industries 4.0 topic. From 2010 to 2013 he was head of the Research Unit IS4AC (Infrastructure Systems for Advanced Computing), in which research related to cloud computing systems infrastructures. He worked as a researcher in the e-security laboratory, mainly with a focus on P2P protocols, encryption on embedded devices, security of routing protocols, and activities on grid computing infrastructures (2004–2009). He is an active member of the HiPEAC community with the co-organization of the HelpDC workshop (Heterogeneous and Low Power Data Center Technologies) on HiPEAC. He is also a workshop organizer for conferences, member and associate editor of the *International Journal of Grid and Utility Computing*, IPC member of the International Workshop on Scalable Optimization and Intelligent Networking and peer reviewer in ICNS and CISIS conferences.

Jan Martinovič, PhD, is currently head of the Advanced Data Analysis and Simulations Lab at IT4Innovations National Supercomputing Center. He has extensive experience leading substantial R&D activities. His research activities focus on information retrieval, data processing and analysis, experimental orchestration platforms for HPC/cloud/bigdata, and traffic management. His activities also cover the development of HPC-as-a-service solutions, which enables remote use of HPC infrastructure. He is coordinator of the

ICT-11 project LEXIS (https://lexis-project.eu). He has previous experience in coordination of different contracted research activities with international and national companies. Currently, he is participating as innovation manager in the ICT-51 project EVEREST (https://everest-h2020.eu), as a work package leader in the EuroHPC project LIGATE (www.ligateproject.eu), and as codesign manager in the EuroHPC project ACROSS (www.acrossproject.eu). He is also a leading researcher at IT4I for the EuroHPC project IO-SEA. Previously, he was a leading researcher at IT4I of the two H2020-FETHPC-2014 projects ANTAREX (AutoTuning and Adaptivity appRoach for Energy efficient eXascale HPC systems, http://specs.fe.up.pt/antarex) and ExCAPE (Exascale Compound Activity Prediction). He is also responsible for the research and development team of the FLOREON+ system (http://floreon.eu) – a system for disaster management support and viaRODOS system (http://viarodos.cz/) for monitoring of the Czechia highway network. He has published more than 100 papers in international journals and conferences; Jan has been a member of HiPEAC since January 2020.

Contributors

Jean-Thomas Acquaviva
Data Direct Networks
Paris, France

Cristina Muñoz Alcalde
Everis Spain SLU
Madrid, Spain

Marco Aldinucci
University of Turin
Computer Science Dept
Turin, Italy

Emmanouil Alexakis
University of Piraeus Research Center
Piraeus, Greece

Andrea Ajmar
ITHACA-POLITO
Torino, Italy

Ivan Andonovic
University of Strathclyde
Glasgow, Scotland, United Kingdom

Antonio Andreini
University of Florence
Florence, Italy

Fabrice Auzanneau
CEA
List – Université Paris-Saclay
Paris, France

Asaf Badouh
Barcelona Supercomputing Center (BSC)
Barcelona, Spain

Frank Badstuebner
Infineon Technologies AG
Neubiberg, Germany

Angelos Bilas
Foundation for Research and Technology – Hellas
Thessaloniki, Greece

Federico Bolelli
Università degli Studi di Modena e Reggio Emilia
Modena, Italy

Lorenza Bovio
ITHACA- POLITO
Torino, Italy

Mario Brčić
University of Zagreb
Faculty of Electrical Engineering and Computing
Zagreb, Croatia

Monica Caballero
Everis Spain SLU
Madrid, Spain

Laura Canalini
Università degli Studi di Modena e Reggio Emilia
Modena, Italy

Michele Cancilla
Università degli Studi di Modena e Reggio Emilia
Modena, Italy

Barbara Cantalupo
University of Turin
Computer Science Dept
Turin, Italy

Franco Alberto Cardillo
Institute for Computational Linguistics
National Research Council
Pisa, Italy

Javier Cardona
University of Strathclyde
Glasgow, Scotland, United Kingdom

Izan Catalán
Universitat Politècnica de València
Parallel Architectures Research Group
Valencia, Spain

Ricardo Chaves
INESC- ID/ IST
University of Lisbon
Lisbon, Portugal

Antony Chazapis
Foundation for Research and Technology – Hellas
Thessaloniki, Greece

Iacopo Colonnelli
University of Turin
Computer Science Dept
Turin, Italy

Philippe Couvee
Atos
Benzons, France

Mikolaj Czerkawski
University of Strathclyde
Glasgow, Scotland, United Kingdom

Emanuele Danovaro
European Center for Medium- Range Weather Forecasts (ECMWF)
Reading, England, United Kingdom

Jean-Marc Denis
Atos
Benzons, France

Marc Derquennes
Euraxent
Cambridge, England, United Kingdom

Zeginis Dimitris
Centre for Research and Technology Hellas (CERTH)
Thessaloniki, Greece

Benoît Dinechin
Kalray
Montbonnot-Saint-Martin, France

Frédéric Donnat
Outpost24
Antibes, France

Leon Dragić
University of Zagreb
Faculty of Electrical Engineering and Computing
Zagreb, Croatia

Alen Duspara
University of Zagreb
Faculty of Electrical Engineering and Computing
Zagreb, Croatia

Denis Dutoit
CEA
List – Université Paris- Saclay
Paris, France

Jorge Ejarque
Barcelona Supercomputing Center (BSC)
Barcelona, Spain

François Exertier
Atos
Benzons, France

Christian Feldmann
Infineon Technologies AG
Neubiberg, Germany

José Flich
Universitat Politècnica de València
Parallel Architectures Research Group
Valencia, Spain

Monica Florea
Software Imagination & Vision (SIMAVI)
Bucharest, Romania

Tommaso Fondelli
University of Florence
Florence, Italy

Eugene Frimpong
Tampere University
Tampere, Finland

François Galea
CEA
List – Université Paris- Saclay
Paris, France

Antonella Galizia
Istituto di matematica applicata e tecnologie (IMATI)
Consiglio Nazionale delle Ricerche (CNR)
Genoa, Italy

Laurent Ganne
Atos
Benzons, France

Rubén J. García-Hernández
Leibniz Supercomputing Centre (LRZ) of the BAdW
Garching bei München
Munich, Germany

Yannis Georgiou
Ryax Technologies
Saint-Fons, France

Yiannis Gkoufas
IBM Research Europe
Dublin, Ireland

Martin Golasowski
IT4Innovations
VSB – Technical University of Ostrava
Ostrava, Czech Republic

Jon Ander Gómez
Universitat Politecnica de Valencia
Valencia, Spain

David González
Tree Technology SA
Madrid, Spain

Thierry Goubier
CEA
List – Université Paris- Saclay
Paris, France

Costantino Grana
Università degli Studi di Modena e Reggio Emilia
Modena, Italy

Marco Grangetto
University of Turin
Computer Science Dept
Turin, Italy

Stephan Hachinger
Leibniz Supercomputing Centre (LRZ) of the BAdW
Garching bei München
Munich, Germany

Sven Harig
Helmholtz Centre for Polar and Marine Research
Alfred Wegener Institute
Klußmannstr, Germany

Piyush Harsh
Cyclops Labs GmbH
Zurich, Switzerland

Mohamad Hayek
Leibniz Supercomputing Centre (LRZ) of the BAdW
Garching bei Munchen
Munich, Germany

Carles Hernández
Universitat Politècnica de València
Parallel Architectures Research Group
Valencia, Spain

José Ramón Hervás
Tree Technology SA
Madrid, Spain

Daniel Hofman
University of Zagreb
Faculty of Electrical Engineering and Computing
Zagreb, Croatia

Dennis Hoppe
High Performance Computing Center Stuttgart (HLRS)
Stuttgart, Germany

Sophia Karagiorgou
Ubitech Ltd
Athens, Greece

Josip Knezović
University of Zagreb
Faculty of Electrical Engineering and Computing
Zagreb, Croatia

Panos Koutsovasilis
IBM Research Europe
Dublin, Ireland

Cédric Koch-Hofer
Atos
Benzons, France

Mate Kovač
University of Zagreb
Faculty of Electrical Engineering and Computing
Zagreb, Croatia

Mario Kovač
University of Zagreb
Faculty of Electrical Engineering and Computing
Zagreb, Croatia

Christos Kozanitis
Foundation for Research and Technology – Hellas
Thessaloniki, Greece

Jan Křenek
IT4Innovations
VSB – Technical University of Ostrava
Ostrava, Czech Republic

Dimosthenis Kyriazis
University of Piraeus Research Center
Piraeus, Greece

Marc Levrier
Atos
Benzons, France

Stephane Louise
CEA
List – Université Paris- Saclay
Paris, France

Donato Magarielli
GE Avio Srl
Rivalta Di Torino, Italy

Fabrizio Magugliani
E4 Computer Engineering SpA
Scandiano, Italy

Branimir Malnar
University of Zagreb
Faculty of Electrical Engineering and Computing
Zagreb, Croatia

Santiago Marco
Barcelona Supercomputing Center (BSC)
Barcelona, Spain

Michele Marconcini
University of Florence
Florence, Italy

Oskar Marko
BioSense Institute
University of Novi Sad
Novi Sad, Serbia

Jose Maria Martínez
Universitat Politècnica de València
Parallel Architectures Research Group
Valencia, Spain

Jan Martinović
IT4Innovations
VSB – Technical University of Ostrava
Ostrava, Czech Republic

Tomáš Martinovič
IT4Innovations
VSB – Technical University of Ostrava
Ostrava, Czech Republic

Philip Mavrepis
The University of Piraeus Research Center
Piraeus, Greece

Paola Mazzoglio
Department of Environment
Land and Infrastructure Engineering (DIATI) Politecnico di Turin,
Turin, Italy

Craig Michie
University of Strathclyde
Glasgow, Scotland, United Kingdom

Spiros Mouzakitis
National Technical University of Athens
Athens, Greece

Johannes Munke
Leibniz Supercomputing Centre (LRZ) of the BAdW
Garching bei München
Munich, Germany

Philippe Notton
SiPearl
Maisons-Laffitte, France

Dana Oniga
Software Imagination & Vision (SIMAVI)
Bucharest, Romania

Daniele Pampaloni
University of Florence
Florence, Italy

Aikaterini Papapostolou
National Technical University of Athens
Athens, Greece

Antonio Parodi
CIMA Research Foundation
Savona, Italy

Andrea Parodi
CIMA Research Foundation
Savona, Italy

Branislav Pejak
BioSense Institute
University of Novi Sad
Novi Sad, Serbia

Jesus Perales
Barcelona Supercomputing Center (BSC)
Barcelona, Spain

Daniele Perlo
University of Turin
Computer Science Dept
Turin, Italy

Igor Piljić
University of Zagreb
Faculty of Electrical Engineering and Computing
Zagreb, Croatia

Christian Pinto
IBM Research Europe
Dublin, Ireland

Francesco Poli
University of Florence
Florence, Italy

Federico Pollastri
Università degli Studi di Modena e Reggio Emilia
Modena, Italy

Marcin Pospieszny
Institute of Bioorganic Chemistry of the Polish Academy of Sciences
Poznan Supercomputing and Networking Center (PSNC)
Poznan, Poland

Eduardo Quiñones
Barcelona Supercomputing Center (BSC)
Barcelona, Spain

Natalja Rakowsky
Helmholtz Centre for Polar and Marine Research
Alfred Wegener Institute
Klußmannstr, Germany

David Rodriguez
Universitat Politècnica de València
Parallel Architectures Research Group
Valencia, Spain

Agneza Šandić
University of Zagreb
Faculty of Electrical Engineering and Computing
Zagreb, Croatia

Paolo Savio
LINKS Foundation
Turin, Italy

Danijel Schorlemmer
GFZ German Research Centre for Geosciences
Potsdam, Germany

Alberto Scionti
LINKS Foundation
Turin
Turin, Italy

Tatiana Silva
Tree Technology SA
Madrid, Spain

Kateřina Slaninová
IT4Innovations
VSB – Technical University of Ostrava
Ostrava, Czech Republic

Ennio Spano
GE Avio Srl
Rivalta di Torino, Italy

Renaud Stevens
Kalray, France

Stephan Stilkerich
Infineon Technologies AG
Neubiberg, Germany

Václav Svatoň
IT4Innovations
VSB – Technical University of Ostrava
Ostrava, Czech Republic

Christos Tachtatzis
University of Strathclyde
Glasgow, Scotland, United Kingdom

Konstantinos Tarabanis
Centre for Research and Technology Hellas (CERTH)
Thessaloniki, Greece

Enzo Tartaglione
University of Turin
Computer Science Dept
Turin, Italy

Olivier Terzo
LINKS Foundation
Turin, Italy

Rafael Tornero
Universitat Politècnica de València
Parallel Architectures Research Group
Valencia, Spain

Ioannis Tsapelas
National Technical University of Athens
Athens, Greece

Srikumar Venugopal
IBM Research Europe
Dublin, Ireland

Chiara Vercellino
LINKS Foundation
Turin, Italy

Giacomo Vitali
LINKS Foundation
Turin, Italy

Paolo Viviani
LINKS Foundation
Turin, Italy

Lukáš Vojacek
IT4Innovations
VSB – Technical University of Ostrava
Ostrava, Czech Republic

Katarina Vukušić
University of Zagreb
Faculty of Electrical Engineering and Computing
Zagreb, Croatia

Etienne Walter
Atos
Benzons, France

Li Zhong
High Performance Computing Center Stuttgart (HLRS)
Stuttgart, Germany

Naweiluo Zhou
High Performance Computing Center Stuttgart (HLRS)
Stuttgart, Germany

1

Toward the Convergence of High-Performance Computing, Cloud, and Big Data Domains

Martin Golasowski, Jan Martinovič, Marc Levrier, Stephan Hachinger, Sophia Karagiorgou, Aikaterini Papapostolou, Spiros Mouzakitis, Ioannis Tsapelas, Monica Caballero, Marco Aldinucci, Jon Ander Gómez, Antony Chazapis, and Jean-Thomas Acquaviva

CONTENTS

1.1 Introduction

Massive increase of demand for the most powerful computing resources is determined by the steep increase of data available for processing and analysis. Many scientific and industrial advances have been driven by the capability to acquire insights and make precise decisions based on the data analysis outputs. International Data Corporation (IDC) predicts that the total volume of data available globally will exceed 175×10^{21} bytes, making it five times larger than the estimated 33×10^{21} bytes in 2018 [1]. The ability to process

DOI: 10.1201/9781003176664-1

the data and infer valuable information from them efficiently will be a key driving factor for sustainable economic growth in the future.

In this chapter, we provide an overview of the current state of the art in the high-performance computing (HPC), big data, and cloud domains and how different fields like agriculture, healthcare, or mechanical engineering can benefit from their convergence. The overview is provided from the point of view of four European projects funded by the Horizon 2020 Programme – CYBELE [2,3], EVOLVE [4], and LEXIS [5].

1.1.1 History of Cloud Computing

A common trait of cloud computing is easy access to data center-class computing resources without the requirement to buy any physical hardware [6]. In the cloud, a user can allocate a machine with a large amount of CPU cores and RAM by using a web-based user interface or a REST API. The first commercial offering of cloud machines was done by Amazon (Elastic Cloud 2) in 2006 [7], followed by Google (Compute Cloud) [8], and Microsoft (Azure) [9].

The main expectations about such a machine are that it will be reliable, fast, and secure and that the procurement of such service is done in *on-demand self-service* manner [6]. These properties are usually provided by hardware and network virtualization, which separates the machine from physical hardware, while also isolating the individual users sharing the same server or network infrastructure. Storage is usually encrypted as well as the application programming interface (API) with the use of Transport Layer Security (TLS).

The definition of the speed of the machine is trickier. Since cloud is usually used for fast scaling of microservices, users of such systems prefer a fast response of the API as well as the possibility to select a preference for some geographical location. For example, allocating a machine in the US–west zone means that the machine will be allocated on a physical hardware located on the West Coast of the United States, placing it closer to users coming from that area without specifying where the data are located. This property enables a truly global and cost-effective deployment of an application that must deal with a highly dynamic load. To make the cloud a highly available and easy to use service, it needs to hide the details about the physical infrastructure used to host the user machines. This fact drives some users with highly sensitive workloads, who want to use the benefits of the cloud, to deploy on-premises cloud often based on technologies like OpenStack [10] or VMware [11].

1.1.2 History of HPC

A supercomputer is a very powerful scientific computing system with computing power that could be in (or approach) the TOP500 [12] list which ranks the systems according to their performance in floating point operations per second (FLOP/s). Origins of supercomputing can be traced to the 1940s where the first machines were used to crack encryption codes or to estimate the yield of an atomic explosion [13,14].

Later, companies like PDP, IBM, and mainly Cray built large monolithic machines with powerful vector CPUs. These machines were quite expensive and only large companies or national-funded laboratories were able to obtain them. Based on that fact, access to such machines has been usually restricted only to a privileged group of scientists and engineers. These people were usually highly trained and had intimate knowledge of the system architecture, and were accustomed to UNIX-like operating systems, terminals, and command line tools [15].

In the early 1990s, the manufacturers started to build supercomputers from more common hardware as it became more powerful and much cheaper than using highly proprietary architectures, specialized, and monolithic machines. At that point, supercomputers became clusters of powerful servers interconnected with a network with very dense topology. The current standards remain very similar to this architecture. However, highly specialized chips and accelerators like GPUs, TPUs, or FPGAs are again taking place next to the general-purpose CPUs more and more often [16,17].

Despite incredible advances in the available computing power, the interface of common HPC clusters remains mostly unchanged. Jobs must be submitted to a batch scheduler which runs the individual applications and shares the workload among the compute nodes. Users usually use SSH to login to a frontend node of an HPC cluster where they prepare the data and compile their applications through a command line interface (CLI). As clusters today almost exclusively run Linux operating systems, users are required to have extensive knowledge about shell scripting, batch schedulers interfaces, and even compiler tool chains and library linking to efficiently utilize such systems. Eliminating this obstacle by creating user-friendly interfaces can open doors to HPC for a wide area of domain experts who do not have to acquire the specialized knowledge about HPC systems.

In comparison with the cloud, access to the largest HPC systems is usually granted only to selected use cases (projects) based on excellence and performance review to avoid wasting of precious natural resources as the most powerful systems can consume several MW of power when fully utilized [12]. Automation and normalization of this formal review process can open doors for wider areas of applications while keeping the requirement for efficient usage.

Such systems are traditionally used for solvers of methods coming from linear algebra like finite element method (FEM) [18], which is often used to simulate various physical processes like deformations, heat transfer, or fluid flow. The solvers usually work using matrix sizes that significantly exceed the computing power available on high-end scientific desktop workstations.

Nowadays, the typical application domains utilizing HPC power can be derived while following centers of excellence [34] supported by the European Union (EU) to strengthen Europe's leadership in HPC applications: not only engineering and industry applications, but also other domains like simulations of weather and climate, urgent computing to support crisis management, biomolecular research, personalized medicine and drug discovery, and new materials design.

1.1.3 Evolution of Big Data

Making educated decisions by interpreting available data is one of the founding principles of modern civilization. Stone tablets were used for tracking crop yields a thousand years ago. Rapid evolution of mathematics and statistics in the 16th century created the foundations for modern data science. With the onset of electronic computers in the 20th century, large amounts of data suddenly became available [19].

At the beginning, the data were stored on magnetic tape and processing was done using large mainframe computers controlled by punch cards. As these machines were expensive, they were used almost exclusively by large companies or financial institutions, mainly to automate tax and payroll processing [19].

Later, when personal computers appeared, data storage and processing became increasingly available for the general population. The Excel spreadsheets are the most common tool for basic data analysis widely available. At the same time, industrial automation started to gain traction and it was now possible to implement complex enterprise resource planning (ERP) systems to automate and optimize the logistics of very complex manufacturing processes, thus creating the business intelligence field (BI) [19].

However, the biggest explosion of data occurred with the onset of the internet and the social network revolution of the 21st century. Currently, the world's biggest companies are those which specialize in data processing, mainly for advertising and marketing purposes. This revolution not only allowed instant and free access to vast human knowledge using small handheld devices but also invented novel and advanced data analysis and machine learning (ML) techniques, which was followed by renewed interest in neural networks.

Development related to artificial intelligence (AI) techniques and the Internet of things (IoT) is now the cutting edge of data processing. This allows us to integrate the ability to communicate with services on the internet into many appliances. Such advances also allow to move part of the data processing power out of the data center closer to the places where data are generated. This approach is generally called Edge computing and contributes to the wide range of computing resources currently available [20].

1.1.4 Evolution of Big Data Storage and Tools

It would be futile to try to formalize the term big data as it is commonly used in multiple ways. First used by Roger Mougalas in 2005 to describe data "So big that it is almost impossible to manage and process using traditional business intelligence tools." Another definition describes them as data so big that it is faster to physically move the hard drives with them than to transfer them over the network. Those descriptions point to the fact that powerful computing resources are needed to make use of them [21].

Data lake is another informal term often used in this context. It relates to the storage and processing of massive amounts of unstructured and heterogeneous data only linked by their context [22]. Object storage becomes de facto a standard way to store such data as they use a more practical flat structure (with buckets of unique identifiers) than the tree structures of traditional file systems. Cloud providers started to offer HTTP-based APIs to use such object storages to follow the established principles. Technologies like CEPH [24] or GlusterFS [25] allow the deployment of such object storage on common hardware on premises with comparable performance and resilience to proprietary storage appliances.

Tools used to analyze the data also have undergone rapid evolution. In the 1990s, the data analysis was done on a single and powerful local workstation using various proprietary tools and graphical user interfaces. Soon, the size of a single data set exceeded the local workstation, and it was necessary to use a distributed and parallel approach. In 2004, Google published their paper on the Map Reduce paradigm which jump-started the development of Hadoop and their distributed file system (HDFS) [23]. It quickly started to replace custom solutions or SQL-based databases unsuitable for storing large amounts of unstructured data.

Hadoop quickly became the tool of choice for many subjects who needed to analyze their data on common server hardware in a distributed manner. Over time, projects such as Apache Spark were developed to overcome the limitations of the original Map Reduce paradigm and implemented high-level and distributed interfaces for streaming data processing, specialized ML algorithms, or possibility to query the data using SQL [24].

Recent years, R programming language gained traction mainly for its data-centric design and the growing number of packages on its CRAN repository. Interfaces to Spark are available as well as low-level libraries for MPI or creating web-based user interfaces for data analysis [27].

Boom of these tools contributed also to increased interest in the neural networks, which are now state of the art in the AI field and laid the foundation of deep learning.

Standardized ways how to describe the neural networks appeared as well as highly optimized libraries for their training. Specialized hardware accelerators such as GPUs or TPUs started to appear in both cloud and HPC environments and are used regularly for training and highly power-efficient devices for inference in power-constrained environments [28].

1.2 Exploiting Convergence

This section provides different points of view on convergence of the three fields described in the first part of this chapter – HPC, cloud, and big data.

1.2.1 CYBELE Project

Precision agriculture (PA) is vitally important for the future and can significantly contribute to food security and safety [29]. Site-specific management (precision or livestock farming) has the potential to nourish the world while increasing profitability under constrained resource conditions. Raw and semi processed agricultural and aquacultural data are usually collected through various sources, such as: IoT devices, sensors, satellites, weather stations, robots, and farm or sea food monitoring equipment. Besides, these datasets are large, complex, unstructured, heterogeneous, nonstandardized, and inconsistent; therefore, any data-mining application is considered a big data application in terms of volume, variety, velocity, and veracity. Such applications require new approaches in data collection, storage, processing, and knowledge extraction.

The CYBELE approach concentrates on the computing capacity and efficiency delivered by HPC e-infrastructures and HPC-empowered data services [31]. These technologies enable the processing of large amounts of heterogeneous data and boost contemporary scientific discoveries in the domains of PA and precision livestock farming (PLF) in the direction of efficient and fast-solving statistical and mathematical algorithms, as well as compute-intensive simulations. Within the context of CYBELE, the delivery of HPC industrial testbeds fosters the efficient utilization of diverse types of big data using parallel and distributed processing modalities for training, inference, and hyperparameter tuning [30]. The CYBELE technical solution grounds diverse business requirements and facilitates the aggregation of very large-scale datasets along with their metadata and of diverse types (sensor data, textual data, satellite, aerial image data, etc.) from a multitude of distributed data sources. Then, these datasets are semantically aligned to a common data model, that is, the CYBELE Data Model. The semantically enriched and harmonized information is used to explore and combine the underlying datasets and create complex queries on top of them to retrieve data which are then analyzed through distributed or parallelized deep learning and ML algorithms. The simulation and analysis results are further feeding intuitive and adaptive data visualization services, empowering the CYBELE end users with a more understandable and expressive interface with insights toward more informed decision-making.

The CYBELE technical solution is delivered as a collection of HPC, big data, and AI services, which facilitate diverse data interoperability and data sharing, data anonymity and encryption capabilities, and the extraction of valuable insights. The data models and the several methods are containerized and delivered as microservices on top of the CYBELE virtual infrastructure, that is, easing the tasks lifecycle management, the transparency of the resource's management, tasks scheduling and execution, to both research organizations but, mainly, to industrial communities. Special focus is given to SMEs lacking access to HPC infrastructures and the necessary competences to fully exploit them, building on actual end-user requirements, stemming from highly demanding data use cases, facilitating the elicitation of knowledge from big agri- and aqua-food-related data.

The CYBELE project currently pursues further exploitation and commercialization activities, either through the creation of new products, that is, the incorporation of new services in food safety with respect to the prediction of food recalls, or through high-end research activities. The research institutions and academic partners have strengthened their positioning and advantage in HPC, big data, and scalable AI models, while the SMEs and private organizations have developed novel products incorporating parallel or distributed processing capabilities.

1.2.2 DeepHealth Project

The DeepHealth – Deep-Learning and HPC to Boost Biomedical Applications for Health – project [3] is one of the innovation actions supported by the EU to boost AI and HPC leadership and promote large-scale pilots. DeepHealth is a three-year project, kicked off in January 2019. DeepHealth aims to foster the use of technology in the healthcare sector by reducing the current gap between mature enough AI–medical imaging solutions and their deployment in real scenarios. Its main goal is to put HPC power at the service of biomedical applications that require the analysis of large and complex biomedical datasets and apply DL and computer vision (CV) techniques to support new and more efficient ways of diagnosis, monitoring, and treatment of diseases.

DeepHealth motivation originates from two different main observations: the decoupled AI and HPC worlds and the benefits these two technologies together can bring to the healthcare domain.

The DeepHealth perspective on the current relation between HPC and AI is as follows: the ability of AI-related techniques to analyze data accurately is growing at a breakneck pace. Among these techniques, DL has benefited from crucial results in ML theory and the large availability of data that is intimately linked to the ability to generalize this data and transform it into useful knowledge. The accuracy of the process is crucially related to the quality and quantity of data and the computing power needed to digest the data. For this HPC is an enabling platform for AI. On the other hand, supercomputers are shifting to GPUs because of their better energy efficiency and need for more and more GPU-enabled workloads, such as DL.

Despite this potential, supercomputers are rarely used for AI. They are not yet equipped to effectively support specific AI software tools nor to acquire large amounts of data securely as required from medical field applications. Additionally, AI researchers are not used to the batch execution model used in supercomputers. Notwithstanding, the two communities need each other and are fated to meet; both technologies are at the crossroads of the European digital sovereignty challenge, which is one of the key items on the EU Commission's agenda.

With regard to the healthcare domain, this is one of the key sectors of the global economy, making any improvement in healthcare systems to have a high impact on the welfare of society. European public health systems are generating large datasets of biomedical data, in particular images that constitute a large unexploited knowledge database, since most of its value comes from the interpretations of the experts. Nowadays, this process is still performed manually in most cases. In the context of automating and accelerating the analysis of the health data and processes, moving toward the so-called "fourth paradigm of science," unifying the traditionally separate environments of HPC, big data analytics, and AI can overcome current issues and foster innovative solutions, in a clear path to more efficient healthcare, benefiting people and public budgets.

In this context, following its main goal, the DeepHealth project proposes a unified European framework that offers DL and CV capabilities, wholly adapted to exploit underlying heterogeneous HPC and cloud architectures by taking advantage of big data software tools to distribute computing workloads.

DeepHealth advocates system software tools and developer libraries as the cornerstones of the European value chain of digital technologies and of the AI–HPC convergence and its applications in several verticals, such as the medical domain.

The framework is composed, on the one hand, by the DeepHealth toolkit, a free open-source software [3] – that includes two libraries, the European Distributed Deep Learning Library (EDDLL) and the European Computer Vision Library (ECVL), both libraries are ready to be integrated into any software platform to facilitate the development and deployment of new applications for specific problems. On the other hand, DeepHealth also provides HPC and cloud infrastructure support, with a focus on usability and portability, so the procedure for training, predictive models could be efficiently distributed on hybrid and heterogeneous HPC, big data, and cloud architectures in a transparent manner. For that, it relies on the StreamFlow

Workflow Manager System [32] and the COMPs [33] parallel programming framework.

The DeepHealth framework allows data scientists and IT experts to train models over hybrid HPC, cloud and big data architectures without profound knowledge of DL, HPC, big data, distributed computing, or cloud computing and increase their productivity, while reducing the time required to do it. Additionally, DeepHealth widens the use of and facilitates the access to advanced HPC, big data, and cloud infrastructures to any company or institution. Furthermore, note that the usefulness of the DeepHealth proposed framework goes beyond the health sector, as it is applicable to any application domain or industry.

1.2.3 EVOLVE Project

The convergence of HPC, big data, and AI technologies is driven by user needs in the data economy. As more and more value is extracted from data, the race is about the ability to process the largest volume of data in the most efficient way to produce cost-effective business solutions. AI is therefore an actor in this race and a powerful accelerator of convergence: big data is needed to manipulate large amounts of data in complex workflows, while HPC is a requirement to meet the time-to-market goals. EVOLVE is a pan-European Innovation Action and frames itself in this context. EVOLVE relies on seven pilot applications all driven by industries from diverse fields. These pilot applications are enriched by additional proof-of-concept use cases.

Not only is AI right at the crossing of big data and HPC, but AI is also an extremely fast-growing market segment presenting some interesting characteristics. Highly parallel, computationally intensive with a low requirement on floating point precision, the AI workload is very specific, to a point where general-purpose CPUs are not the most suitable architectures. AI is taking advantage of accelerators; GPUs are the ubiquitous vehicle for high-performance AI. Therefore, to accommodate both general-purpose workload and AI EVOLVE, the platform hosts different kinds of processing units. CPU and GPU are power hungry, to lower the energetic footprint of the system and leverage the importance of image processing, EVOLVE also supports FPGA as an accelerator. Finally, in terms of data management, AI is also injecting innovation and new ideas: the AI workload is read driven. A massive amount of data is consumed to generate a relatively small neural network. Traditional storage architectures were built with the goal of balancing write and read; with AI this is no longer the case. Taking this into account leads to a heterogeneous storage as well. The EVOLVE platform is heterogeneous on the data processing and storage infrastructure levels.

This result-oriented approach allows EVOLVE to maintain a strong industrial focus. The EVOLVE paradigm is user-centric, therefore consequent attention is devoted to increase acceptance of the technology and to lower the entry cost.

The convergence with cloud technologies is critical to onboard SMEs. As small- and medium-sized businesses have successfully embraced the cloud, despite multiple initiatives (e.g. Fortissimo), the onboarding rate in HPC remains limited. Therefore, to shield end users from the underlying complexity, an appropriate front end is the key to foster the adoption of technology, this aspect has been too often overlooked. To address this point, EVOLVE has developed Karvdash — the EVOLVE dashboard — where users interact with the platform through Karvdash and to implement their applications in notebooks, using workflows that interface with the platform microservices and the visualization component of the software stack.

Another important gap identified to offer an easier and better user experience are the visualization services. Visualization should not solely be oriented toward system monitoring, but clearly decoupled in components for separation of concerns. Visualization should distinguish the process execution monitoring and the ability to visualize end-user results.

As a generalization of the point made above on visualization, there is a need to decouple. The general organization of the workflow advocates for an application less monolithic than what we've seen in the HPC field. A general trend observed in modern software frameworks is the packaging of containers for microservices. These microservices are then used as building blocks for workflow steps. EVOLVE offers a similar approach to microservices to handle compute-intensive tasks: data ingestion/extraction, flexible and efficient data transformation, data processing with integrated custom HPC kernels, and incremental one-pass analytics over bounded data streams.

The need to be compatible with standards (even emerging ones) is also a specific issue to deal with. Reimplementation of ad hoc solutions can prevent the perennity of software development. Special care should be taken to provide microservices that remain compatible with well-established software components such as Kafka, Spark, TensorFlow, MPI, and Dask.

To accommodate the unique characteristics of the project, pilot, and proof-of-concepts, EVOLVE extended Kubernetes with novel workload placement and scheduling features, as well as a new storage abstraction, called the unified storage layer (USL). The software effort has also led to integration of a wide selection of big data, cloud, and HPC software frameworks as microservices, including a state-of-the-art visualization service that extends to the front end. A vertical monitoring layer collects run-time statistics from hardware and software components, to assist developers in understanding the performance characteristics of their applications, and to enable automated fine-tuning of container placement and scheduling.

This crucial focus on applications allows EVOLVE to realistically set the bar for what is possible today and what should be the target, based on technology projections in the future. In addition, the project has enlisted more than a dozen external proof-of-concepts, that is, applications which were not part of the original planning and which are now running on the EVOLVE testbed for improving their productivity, dataset sizes, as well as processing time. EVOLVE aims at building an ecosystem around HPC-enabled big data processing, to disseminate knowledge and possibilities, foster innovation, but also solicit feedback and ensure continuity.

1.2.4 LEXIS Project

The LEXIS (Large-scale EXecution for Industry & Society) project takes the results of convergence between HPC, cloud computing, and big data as inspiration for addressing three major issues in data-driven research and development: (1) the need to describe and automatize computational workflows (e.g. simulations with subsequent data analysis steps) in a user-friendly way; (2) the need to manage distributed input and output data as well as intermediate results; and (3) the requirement to easily access federated computing and data systems with a single sign-on (SSO).

For this purpose, the LEXIS project is building an advanced engineering platform and portal at the confluence of HPC, cloud, and big data. The platform leverages large-scale geographically distributed HPC and cloud-computing resources, and has integrated existing systems including some of the world's top 20 supercomputers. It relies on best-in-class European data management solutions (EUDAT-B2SAFE/iRODS [35, 36]) and an advanced distributed orchestration framework (YORC/TOSCA [37, 38]). Built using a codesign approach with rich pilot use cases, LEXIS provides an optimum infrastructure for a broad range of applications in the fields of big data analytics, AI, and classical simulation. Commercial usage is taken into account for an accounting and billing concept. The LEXIS platform immerses innovative hardware such as data node/burst buffer systems with NVMe drives and Intel Optane DC NVDIMM memory, as well as GPU and FPGA accelerator cards for boosting the efficiency of its pilots and later users. Entry barriers to supercomputing and big data are lowered, stimulating the interest of European enterprises, science, and governmental institutions alike, and synchronizing their IT usage patterns.

The LEXIS Orchestration System based on YORC, as mentioned above, is the core component for executing computational workflows. For easy access, it features a front-end and graphical user interface, Alien4Cloud (A4C [39]), which allows users to devise workflows using templates. The orchestrator YORC is in charge of software provisioning, for example, virtual machines creation, containers uploading, and, once input data, output location, and processing steps are defined, YORC automatically distributes tasks over the LEXIS infrastructure. YORC cooperates with HEAppE Middleware [42, 43] to manage access to HPC/cloud infrastructures with an additional security layer. While HEAppE acts as a secure remote execution middleware and wrapper for any HPC-related functionality it still offers users basically the same amount of functionality as the direct access to an HPC infrastructure – ability to define multiple tasks within a single job, use job arrays, job dependencies, support for so-called long-term running jobs, etc. The whole LEXIS Orchestration System is designed for dynamic orchestration, taking into account in particular compute-data proximity aspects, computing time cost aspects, and live system availability.

Within workflows, data are automatically handled by the distributed data infrastructure (DDI), built upon iRODS and EUDAT-B2SAFE technologies. Federating with European research data infrastructure, this system provides a unified view on datasets at all LEXIS sites, as if they were stored in a single file system. Consistent access is ensured via entry points (iRODS servers) at all LEXIS sites. Data can be ingested by iRODS protocol, GridFTP, or chunked HTTP transfer. As with all other LEXIS components, the LEXIS DDI provides HTTP-REST APIs, for example, to discover and move or replicate data, trigger data encryption and compression, and to store metadata and persistent identifiers, complying with the FAIR principles [40] of data management. At the same time, proper access control is possible wherever necessary (e.g. for confidential enterprise data), leveraging the LEXIS-wide role-based access control (RBAC) model.

Alien4Cloud, workflow, and data management interfaces are conveniently integrated within the LEXIS portal, a one-stop-shop entry point for platform users, where they perform their SSO, leveraging the LEXIS authentication and authorization infrastructure (AAI). The portal features additional functions to apply for computational resources at HPC centers and view usage and billing data. User-friendliness is the essential guideline behind the portal in order to lower the threshold for industrial and academic researchers to use top-level HPC, cloud, and data systems. As an example, graphical visualization is offered as well as status-display features in order to track workflow results in near-real time.

The platform is codesigned and initially tested with three LEXIS large-scale pilot use cases, focusing on "aeronautics," "earthquakes and tsunamis," plus "weather and climate." The Aeronautics Pilot targets the optimization of turbomachinery and rotating-parts simulation workflows, by direct code acceleration, addition of capabilities to hydrodynamics codes and beyond. This pilot particularly drives LEXIS codesign by specifying requirements for check-pointing, visualization, and security.

Earthquake simulations and early tsunami warning formed the second pilot. The goal is a clear formalization and successful execution of a complex workflow, which follows an urgent computing paradigm to be implemented in the LEXIS orchestration system. At a later production stage, this will guarantee the timely delivery of results to decision makers.

Last, but not least, the Weather and Climate Pilot investigates the operational applicability of the LEXIS platform by a rich portfolio of hydrometeorological forecasting case studies, ranging from flash-flood prediction to air-quality and forest-fire risk. It requires the assimilation of in situ weather station and radar observation data on the LEXIS platform.

The LEXIS system with its use cases – as outlined above – reconciles academic and enterprise usage with a significant open-source dimension, including transparent documentation, repository usage (e.g. LEXIS group on GitHub [41]), and validation procedures. Beyond the pilot use cases, the LEXIS platform is tested and validated via use cases from an open call. This ongoing call for use cases stimulates the adoption of the project framework and also increases stakeholders' engagement with the original LEXIS Pilots. All this contributes to the LEXIS goal of increasing European competitiveness by removing obstacles for companies, society, and scientists to use Europe's most powerful HPC, cloud-computing, and storage systems for their data-driven endeavors.

Acknowledgment

This work has been created in collaboration with the following projects: Project CYBELE (www.cybele-project.eu/) received funding from the European Commission under grant agreement No. 825355; Project DeepHealth received funding from the EU's Horizon 2020 Research and Innovation Programme under grant agreement No. 825111; The LEXIS project, funded by the EU's Horizon 2020 Research and Innovation Programme (2014–2020) under grant agreement No 825532. The EVOLVE project, funded by the EU's Horizon 2020 Research and Innovation Programme (2014–2020) under grant agreement No. 825061.

References

[1] Rydning, David Reinsel, and Gantz, J. (2018). "The digitization of the world from edge to core." Framingham, MA: International Data Corporation.

[2] CYBELE. "Fostering precision agriculture and livestock farming through secure access to large-scale HPC-enabled virtual industrial experimentation environment empowering scalable big data analytics." www.cybele-project.eu/ (accessed Apr. 19, 2021).

[3] DeepHealth: "Deep-learning and HPC to boost biomedical applications for health." https://cordis.europa.eu/project/id/825111 (accessed Apr. 19, 2021).

[4] EVOLVE. "HPC and cloud-enhanced testbed for extracting value from diverse data at large scale." https://cordis.europa.eu/project/id/825061 (accessed Apr. 19, 2021).
[5] LEXIS. "Large-scale EXecution for industry & society." https://cordis.europa.eu/project/id/825532 (accessed Apr. 19, 2021).
[6] Mell, P., and Grance, T. (2011). "The NIST definition of cloud computing." https://doi.org/10.6028/NIST.SP.800-145
[7] Amazon Elastic Cloud 2. Amazon Web Services. https://aws.amazon.com/?nc2=h_lg (accessed Apr. 2, 2021).
[8] Google, Google Compute Cloud documentation. https://cloud.google.com/ (accessed Apr. 2, 2021).
[9] Microsoft, Microsoft Azure. https://azure.microsoft.com (accessed Apr. 2, 2021).
[10] OpenStack Cloud documentation. https://docs.openstack.org (accessed Apr. 19, 2021).
[11] VMWare, VMWare vSphere hypervisor. www.vmware.com/products/vsphere-hypervisor.html (accessed Apr. 19, 2021).
[12] Strohmaier, E., Dongarra, J., Horst, S., and Meuer, M. "The TOP500 list." https://top500.org (accessed Apr. 19, 2021).
[13] Ulam, S., and Metropolis, N. (1949). "The Monte Carlo method." *Journal of the American Statistical Association* 44: 335–341.
[14] Eckhardt R. (1987). "Stan Ulam, John von Neumann, and the Monte Carlo Method." *Los Alamos Science* 15: 131–136.
[15] Murray, C. J. (1997). *The Supermen: The Story of Seymour Cray and the Technical Wizards behind the Supercomputer*. New York: Wiley.
[16] Strohmaier, E., Dongarra, J., Meuer, H. W., and Simon, H. D. (2005). "Recent trends in the marketplace of high performance computing." *Parallel Computing* 31(3–4): 261–273.
[17] Pilato, C., Bohm, S., Brocheton, F., et al. (2021). "EVEREST: A design environment for extreme-scale big data analytics on heterogeneous platforms." arXiv preprint arXiv:2103.04185.
[18] Reddy, J. N. (2010). *An Introduction to the Finite Element Method* (vol. 1221). New York: McGraw-Hill
[19] Rijmenam, M. "A short history of big-data." https://datafloq.com/read/big-data-history/239 (accessed Apr. 2, 2021).
[20] S. B. Calo, Touna, M., Verma, D. C., and Cullen, A. (2017). "Edge computing architecture for applying AI to IoT." In 2017 IEEE *International Conference on Big Data* (pp. 3012–3016), Boston, MA, USA, doi: 10.1109/BigData.2017.8258272.
[21] Emani, C. K., Cullot, N., and Nicolle, C. (2015): "Understandable big data: a survey." *Computer Science Review* 17: 70–81.
[22] Sawadogo, P., and Darmont, J. (2021). "On data lake architectures and metadata management." *Journal of Intelligent Information Systems* 56(1): 97–120.
[23] Dean, J., and Ghemawat, S. (2004). "MapReduce: Simplified data processing on large clusters." OSDI 04 paper, www.usenix.org/legacy/event/osdi04/tech/full_papers/dean/dean_html (accessed Apr. 13, 2020).

[24] Salloum, S., Dautov, R., Chen, X., Peng, P. X., and Huang, J. Z. (2016). "Big data analytics on Apache Spark." *International Journal of Data Science and Analytics* 1(3): 145–164.

[25] Aghayev, A., Weil, S., Kuchnik, M., Nelson, M., Ganger, G. R., and Amvrosiadis, G. (2019, October). "File systems unfit as distributed storage backends: lessons from 10 years of Ceph evolution." In *Proceedings of the 27th ACM Symposium on Operating Systems Principles* (pp. 353–369).

[26] Boyer, E. B., Broomfield, M. C., and Perrotti, T. A. (2012). "Glusterfs one storage server to rule them all (No. LA-UR-12-23586)." Los Alamos National Lab.(LANL), Los Alamos, NM, USA.

[27] Team, R. C. (2018). "R: a language and environment for statistical computing." R Foundation for Statistical Computing, Vienna. www.r-project.org.

[28] Ertam, F., and Aydın, G. (2017, October). "Data classification with deep learning using Tensorflow." In *2017 International Conference on Computer Science and Engineering* (UBMK) (pp. 755–758). IEEE.

[29] Schrijver, R. et al. (2016). "Precision agriculture and the future of farming in Europe." Report of STOA, Science Foresight Unit, European Union.

[30] Schokker, D., Athanasiadis, I. N., Visser, B., Veerkamp, R. F., and Kamphuis, C. (2020). "Storing, combining and analysing turkey experimental data in the big data era." *Animal* 14(11): 2397–2403, ISSN 1751-7311, https://doi.org/10.1017/S175173112000155X.

[31] Georgiou, Y., Zhou, N., Zhong, L., et al. (2020, June). "Converging HPC, big data and cloud technologies for precision agriculture data analytics on supercomputers." In *International Conference on High Performance Computing* (pp. 368–379). Springer, Cham.

[32] Colonnelli, I., Cantalupo, B., Merelli, I., and Aldinucci, M. (2020). "StreamFlow: cross-breeding cloud with HPC." *IEEE Transactions on Emerging Topics in Computing.*

[33] Lordan, F., Tejedor, E., Ejarque, J., et al. (2014). "ServiceSs: an interoperable programming framework for the Cloud." *Journal of Grid Computing* 12(1): 67–91.

[34] Centres of Excellence for HPC Applications. www.hpccoe.eu/eu-hpc-centres-of-excellence2/ (accessed Apr. 1, 2021).

[35] EUDAT Ltd. (2020). B2SAFE – EUDAT. www.eudat.eu/services/b2safe (accessed Apr. 13, 2020).

[36] Xu, H., Russell, T., Coposky, J., et al. (2017). *iRODS Primer 2: Integrated Rule-Oriented Data System*. Williston: Morgan & Claypool. https://doi.org/10.2200/S00760ED1V01Y201702ICR057.

[37] Bull Atos. (2021). Ystia Suite. https://ystia.github.io (accessed Apr. 6, 2021).

[38] OASIS Open. (2013). Topology and Orchestration Specification for Cloud Applications Version 1.0 – OASIS Standard. http://docs.oasis-open.org/tosca/TOSCA/v1.0/os/TOSCA-v1.0-os.html (accessed Apr. 27, 2020).

[39] Bull Atos. (2021). Alien 4 Cloud. http://alien4cloud.github.io/ (accessed Apr. 6, 2021).

[40] Wilkinson, M. D., Dumontier, M., Aalbersberg, I. J., et al. (2019). "The FAIR guiding principles for scientific data management and stewardship." *Scientific Data* 3: 160018. https://doi.org/10.1038/sdata.2016.18.

[41] LEXIS GitHub group: https://github.com/lexis-project (accessed Apr. 19, 2021).

[42] Svaton, V., Martinovic, J., Krenek, J., Esch, T., and Tomancak, P. (2019, July). "HPC-as-a-service via HEAppE Platform." In *Conference on Complex, Intelligent, and Software Intensive Systems* (pp. 280–293). Springer, Cham.

[43] LEXIS Project deliverable 4.2: Design and Implementation of the HPC-Federated Orchestration System – Intermediate. https://lexis-project.eu/web/wp-content/uploads/2020/08/LEXIS_Deliverable_D4.2.pdf (accessed Apr. 2, 2021).

2

The LEXIS Platform for Distributed Workflow Execution and Data Management

Martin Golasowski, Jan Martinovič, Jan Křenek, Kateřina Slaninová, Marc Levrier, Piyush Harsh, Marc Derquennes, Frédéric Donnat, and Olivier Terzo

CONTENTS

2.1 Motivation

Many scientific and industrial fields can benefit from easier access to computing resources whose power rapidly increases. The LEXIS platform aims to overcome the shortcomings of complicated access to a high-performance computing (HPC) system and management of terabyte-scale datasets by providing a clean and concise interface which will abstract both HPC and cloud resources using workflow and data management orchestration [1]. This chapter provides an overview of the LEXIS platform with the focus on the overall architecture and the most important concepts the platform implements from the security, accounting, and usability points of view. A market analysis is also provided to lay ground for a business case and future exploitation of the solution.

DOI: 10.1201/9781003176664-2

The platform runs workflows comprising many dependent tasks which can be executed simultaneously on various computing resources such as cloud systems or geographically distant HPC clusters. This way, the platform can make use of cloud systems deployed in the same data center as HPC cluster to significantly decrease the time required for data staging and migration while leveraging the best features provided by the cloud systems. The implemented orchestration solution is described in Chapter 5.

The platform implements a distributed data interface (DDI) [2] which leverages EUDAT [3] and iRODS technologies [4] to enable seamless and secure ingestion, staging, curation, and publication of data. The DDI also offers in-place encryption of the data to provide the highest level of security for the most sensitive data. This interface is used by the platform users to upload their datasets and download their results. It can also be used to move data between different computing centers connected to the platform, essentially moving terabytes of data by a single click. The interface is used by the orchestrator as well, to automatically move the data to the target computing resource. Detailed description of the DDI is in Chapter 4.

All APIs implemented by the LEXIS platform are based on the REST HTTP protocol making it easy to integrate it to existing systems. Using API to allocate and use a computing resource is an inherent property of cloud computing. The LEXIS platform uses this fact as inspiration to bring the similar level of usability also to the world of HPC and Big Data. Part of the platform is a web-based graphical user interface – the LEXIS Portal, which uses the APIs to expose the platform in a user-friendly way. An important part of the platform is also an accounting and billing system which is used to track resources consumed by the platform users in each connected center.

The platform provides its own authentication and authorization interface (AAI) which uses common technologies such as OpenID and JWT tokens [5]. HPC centers and operators offer hosts their own identities used to access their system. The LEXIS platform has a solution to overcome this restriction by using the HEAppE middleware [6], which is deployed in each center which provides its resources for the platform. The HEAppE middleware implements an API for job submission and control through a predefined command template. The API can be used only by authenticated LEXIS users and the middleware uses local service accounts of the HPC center to communicate with the cluster job schedulers, thus making the local HPC AAI opaque to the LEXIS Users.

2.2 Architecture (Codesign) and Interfaces

LEXIS architecture is a complex system formed from small components (blocks) and relations between them. The design approach provides loosely coupled components, which makes them easy to modify, replace, or extend without breaking the entire system.

Component versioning is necessary to maintain consistency of the component APIs. It provides control of their dependencies, where two components must use the same API in compatible versions in order to work correctly. The system defines the versioning scheme and version compatibility.

Exposing LEXIS platform as a set of APIs allows us to create for example mobile applications (Android, iOS) or integrate the platform in existing enterprise resource planning systems. These applications will use the same LEXIS back end which is currently used by the LEXIS portal user interface.

The LEXIS platform has a three-layer architecture where each of them contains its own software components. A top-level view of the architecture is in Figure 2.1. The three layers are described below.

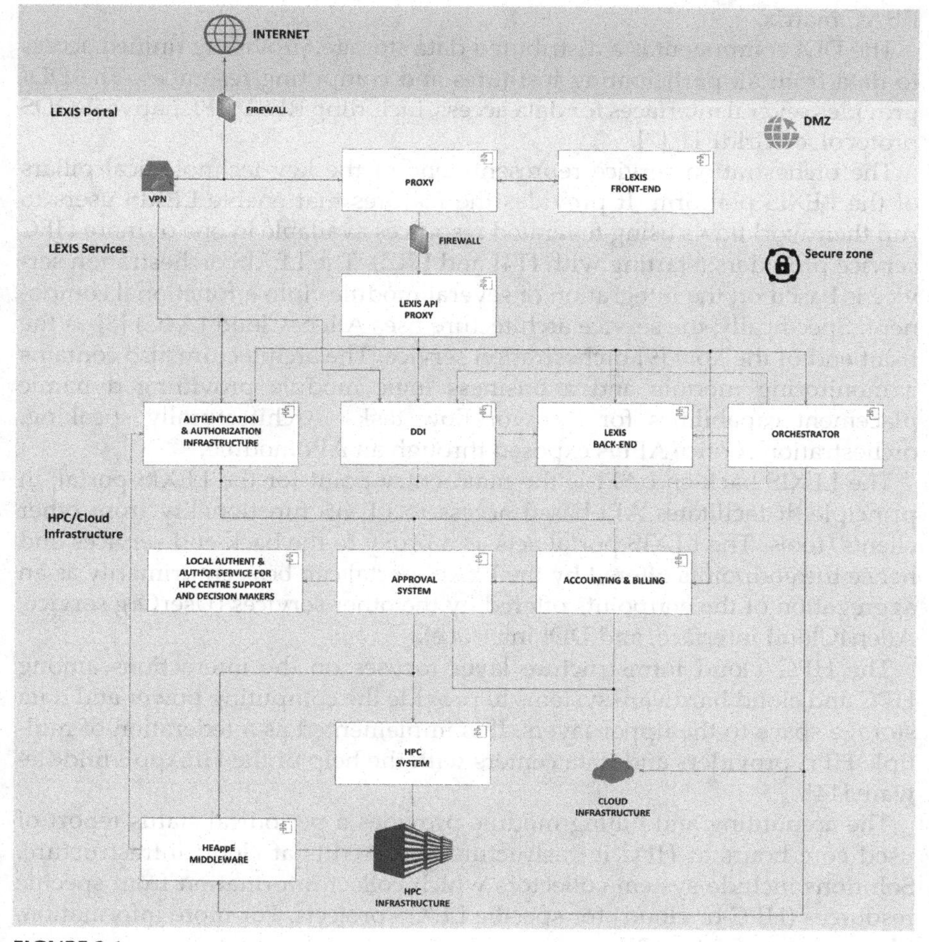

FIGURE 2.1
High-level architecture of the LEXIS platform with three-layer separation.

The LEXIS Portal Layer, providing easy access to the LEXIS platform for pilots and possible external users, contains the front-end (portal) which is the main entry point to the LEXIS platform. It is designed primarily for the users who do not necessarily have deep experience with HPC or cloud environments, so they can have an easy way to access the most advanced capabilities and features of the platform.

The LEXIS services layer lies in the middle between the portal and infrastructure. It includes federated security infrastructure (authentication and authorization infrastructure), data management (distributed data infrastructure (DDI)), and orchestration services (orchestrator).

Authentication and authorization infrastructure (AAI) is based on Keycloak which is an open-source solution. This specific core component of the architecture is in charge of handling the authentication of the LEXIS users or processes and of the authorization according to the role assignment and RBAC matrix.

The DDI component is a distributed data storage, providing unified access to data from all participating institutes and computing resources. The DDI provides several interfaces for data access, including REST API, native iRODS protocol, or GridFTP [7].

The orchestration service represents one of the key technological pillars of the LEXIS platform. It provides the features that enable LEXIS users to run their workflows using federated resources available in one or more HPC service providers (starting with IT4I and LRZ). The LEXIS orchestration service is based on the integration of several modules into a functional component. Specifically, the service architecture uses Alien4Cloud (A4C) [8] as the front end of the Yorc [9] orchestration service. The architecture also contains a monitoring module and a business logic module providing dynamic placement capabilities for the workflow tasks. Architecturally speaking, orchestration service API is exposed through an API module.

The LEXIS back-end API is the main entry point for the LEXIS portal; in principle, it facilitates API-based access to LEXIS functionality from other clients/tools. The LEXIS portal acts as a proxy to the back-end services and hence the endpoints offered by the LEXIS portal can be seen primarily as an aggregation of the endpoints offered by the other services (UserOrg service, Alien4Cloud interface, and DDI interface).

The HPC/cloud infrastructure layer focuses on the interactions among HPC and cloud hardware systems to provide the computing power and data storage space to the upper layers. It is implemented as a federation of multiple HPC providers and data centers with the help of the HEAppE middleware [14].

The accounting and billing module provides a periodical status report of used core hours at HPC infrastructure and credits at cloud infrastructure. Solutions include system collectors which collect information from specific resources (HPC or cloud) for specific LEXIS projects. For more information, please refer to Section 2.4.

2.3 Security

Since the very beginning and the design phase of the LEXIS project, security has been more than an important topic for the following two main reasons. The first one is the rise of cybersecurity incidents during the past decade, mainly due to a growing hacking activity for profit. The second one is the importance of intellectual property, data, and results for all private companies (SMEs to big enterprises) that are the targeted customer for the LEXIS platform.

It is worth mentioning that it is quite impossible nowadays to provide or add security at the end of a project due to increased complexity in projects and all technologies that may be used. Including security during the design phase and all along the project is not optional.

Due to the high security standards of all LEXIS partners, we have targeted a high level of security by aiming at the most advanced security concept and principle (at the time the project started) such as a "no trust" concept and "security by design" principle [12].

In a few words, the security concepts and principles can be summarized in the following actions:

- limiting the features and applications running in the platform to only what is required and restricting access to them for all users and processes to minimize the attack surface area;
- putting in place proper identity and access control mechanisms such as reducing privileges and permissions to bare-minimum mandatory access to follow least-privileges principles;
- validating identity and access in all layers and components of the LEXIS platform, adding auditing capabilities and a proper segregation of duties to follow a defensive approach to security; and
- putting in place proper default configurations that do not expose applications or data and do not grant access to resources to follow secure defaults.

Starting from the infrastructure codesign where several layers have been created to properly describe security zones for the segregation of duties concept, both HPC centers (IT4I and LRZ) have been interconnected using site-to-site VPN providing a default secure communication channel between them as shown in Figure 2.2.

Once the main HPC centers are interconnected, each HPC center is an equal deployment of the LEXIS platform to make handling security easier and avoid having different security policy according to the HPC center. A high-level view of the LEXIS platform in each HPC center is in Figure 2.3. The different security zones described in the diagram are a *one-to-one* mapping

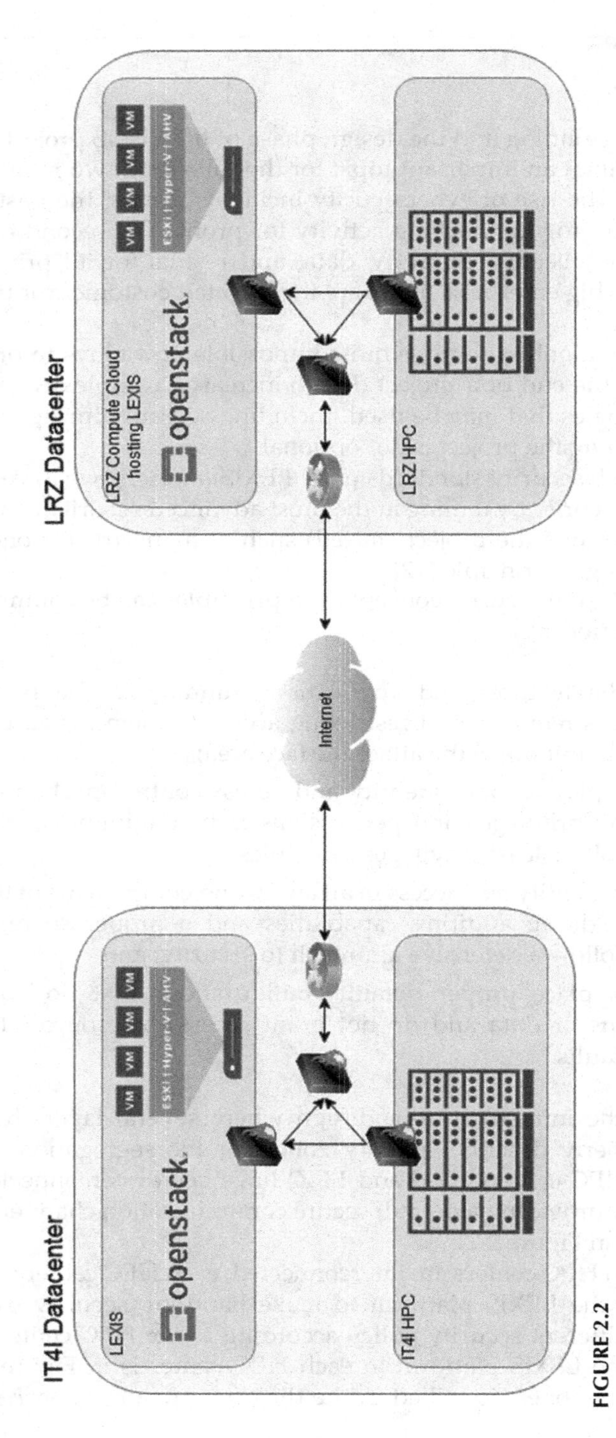

FIGURE 2.2
Federation between data centers with separation of concerns and firewalls.

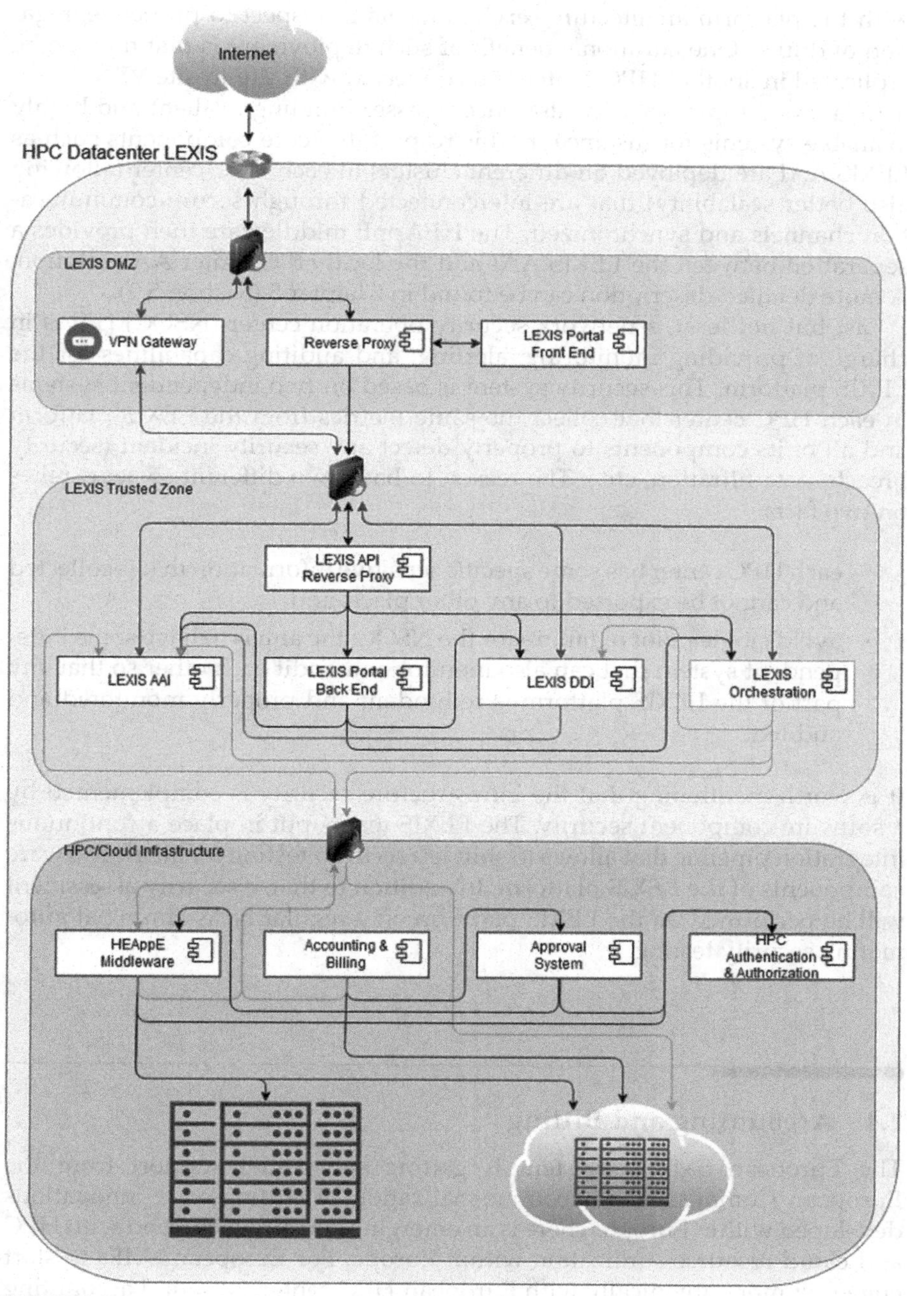

FIGURE 2.3
Security view on high-level architecture of the LEXIS platform.

with the platform architecture services layers in respect to proper segregation of duties. One additional benefit of such deployment is that this can be replicated in another HPC center and connected with site-to-site VPN.

As a broad topic, security also encompasses building resilient and highly available systems for instance. In this respect, the core components such as LEXIS AAI are deployed on different clusters in each HPC center (allowing also better scalability) that are interconnected through secure communication channels and synchronized. The HEAppE middleware then provides a separation between the LEXIS AAI and the local HPC center AAI solution. A more detailed description can be found in Chapter 5 (Section 5.2).

Last but not least, a network security operation center (NSOC) [15] is in charge of providing monitoring, alerting, and auditing capabilities for the LEXIS platform. This security system is based on two independent systems in each HPC center that collect the same metrics from the LEXIS platform and all of its components to properly detect any security incident (security breach, data filtration, etc.). The reason to have two different systems relies on two facts:

- each HPC center has some specific sensitive information that is collected and cannot be exported to any other place; and
- avoid single point of failure for the NSOC: the aim is to have some independent system that can also monitor and audit each other so that any part of the LEXIS platform is redundant and properly monitored and audited.

It is worth mentioning that the infrastructure security is complemented by a software component security. The LEXIS teams put in place a continuous integration pipeline that allows to shift left security testing for all the software components of the LEXIS platform. In addition to that, a security assessment will be performed on the LEXIS platform on a regular basis aiming at automating security testing.

2.4 Accounting and Billing

The European SME ecosystem is getting substantial support from the European Commission for commercialization of cutting-edge innovations developed within Europe. There is an emerging convergence trend with HPC and cloud resources utilization within Europe. For European SMEs to start engaging more organically with European HPC centers, a sound accounting and billing best practices are a must. The accounting and billing capabilities being developed as a part of the LEXIS project is critical to ensure the long-term financial viability of the LEXIS platform. The main requirements of the LEXIS accounting and invoicing engine are:

- ability to track resources consumed from both traditional HPC as well as cloud services offered by an HPC operator;
- ability to support a per account depleting virtual credit pool;
- ability to support separate pricing plans to offer differentiated pricing to different class of LEXIS users;
- ability to support on-demand running cost and usage consumption reporting to enable users to track their cost and resource utilizations; and
- extensibility – the ability of the accounting tool to support new products and services which can be created in the future by the HPC operator.

Figure 2.4 shows the overall architecture of the Cyclops accounting and billing engine [10] that satisfies all the above-mentioned requirements. A brief description of key components includes:

- Collector processes: periodic agents which track various resources, either HPC or cloud, and report consumed metrics to the Cyclops core, if a new service is offered by the HPC operator in future, a detached collection strategy enables the operator to develop a new collector process to bring the new service within the ambit of the Cyclops accounting and billing process.
- Plan management service: allows for registration and management of stock-keeping units (SKUs) as well as association of unit pricing rules with the defined SKU element. This service also allows grouping of various pricing rules, together with discounting criteria under one of more plans. This service enables an HPC operator to offer different service plans to different classes of customers.
- Customer data store: this service maintains the relationship between LEXIS customers, and the service plans with which they are associated. This service also allows association of different service instance identities with a customer account which allows Cyclops to aggregate multiple HPC/cloud service accounting data with the correct customer accounts.
- Credit management service: this module allows virtual credits to be added to the customer's account, which also allows other LEXIS services to query existing credit for a customer in order to take a decision whether a particular service is to be rendered to the customer or denied where allocated credits have run out.

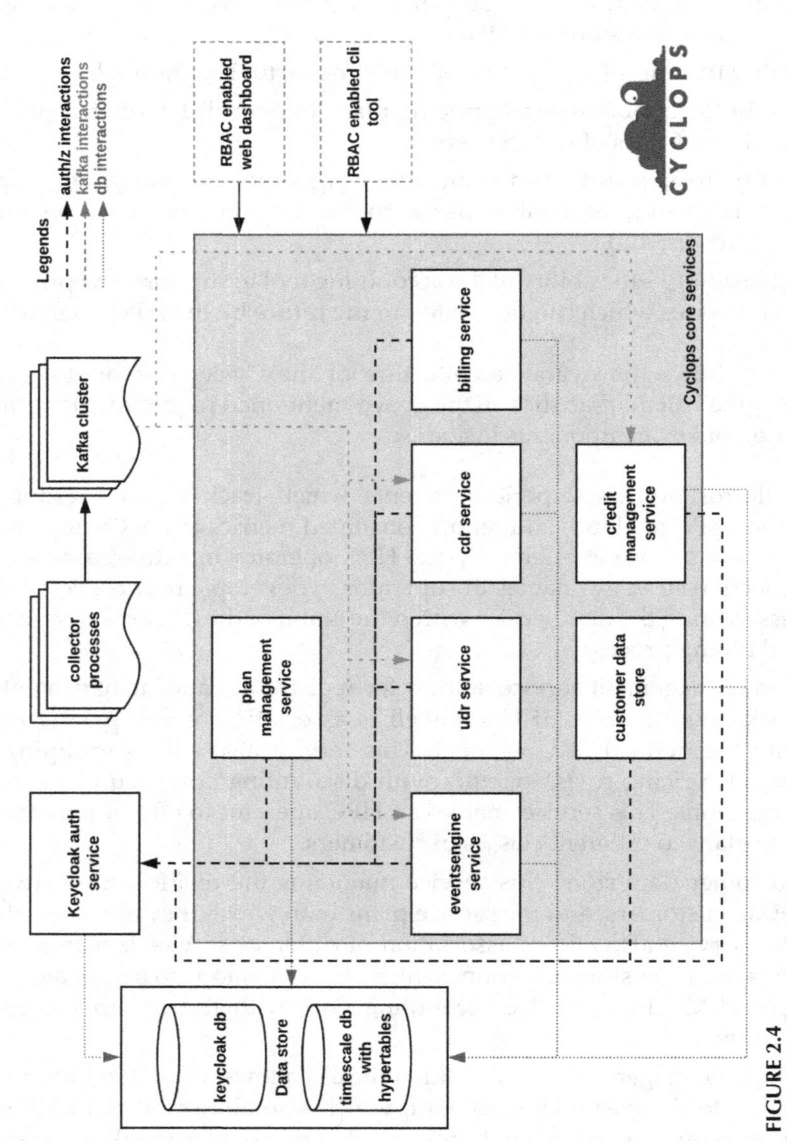

FIGURE 2.4
Architecture of the Cyclops accounting and billing system

- Events engine: this service intelligently tracks the life cycle state changes of an HPC resource. This service enables accounting of a resource based on the time interval a resource is in a given state.
- UDR, CDR services: usage record, and charge record services perform periodic aggregation of usage reports into a consolidated usage and charge reports. Conversion of usage into cost depends on the plans associated with a customer account. Using APIs of these services, the LEXIS platform displays running usage and costs information to the users.
- Billing service: depending on the customer's configured billing period (daily, weekly, monthly, quarterly, semi-annually, or annually), this service generates invoice objects while taking into account any allowed discounts depending on the linked plan with the customer account.

In order to support HPC resources accounting, a dedicated HEAppE collector has been developed which reports the amount of CPU core hours consumed by an HPC scheduled task. Cyclops existing collectors for OpenStack – server collector, floating IP collector, object and block storage collectors enable tracking of both HPC and OpenStack services used by LEXIS users.

Cyclops services coordinate with LEXIS back-end services (see architecture diagram in Figure 2.1) an establishment of linkages between HPC and cloud project identities and customer accounts.

The entire LEXIS accounting and billing workflow supported by Cyclops is aided by the RESTful interface of the framework, enabling fine-grained access control, as well as ease of integration with the rest of LEXIS services.

Flexible collector development aided by availability of a collector template enables future readiness of the LEXIS billing systems. Billing scenarios to enable invoicing of various levels of technical support hours, future services such as container run times, managed application runtime environment (like PaaS for HPC) and similar forward-looking services can be easily supported by Cyclops accounting and billing platform – as long as a sensible SKU nomenclature, linked pricing rule, and respective collector services can be created.

2.5 Easy Access to HPC/Cloud through a Specialized Web Portal

Access to HPC resources and services has been predominantly used by research centers and academic institutions. The LEXIS portal makes for easy,

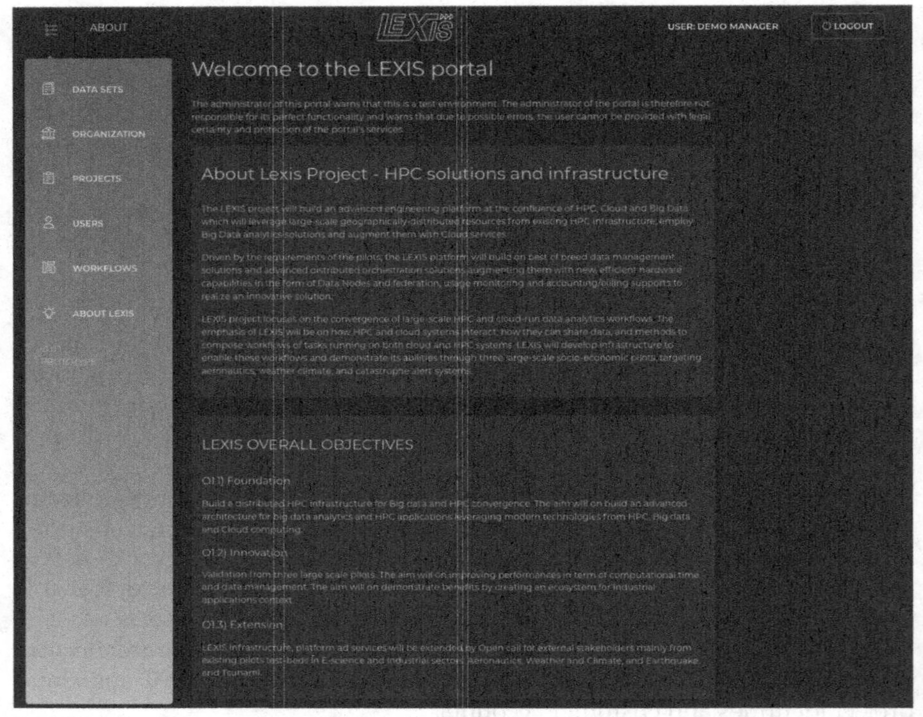

FIGURE 2.5
Screenshot of the LEXIS portal landing page

streamlined access to European SMEs removing many of the older process friction points. The design of the portal takes concrete steps to specifically address the needs of a European SME entity:

- self on-boarding by organizations and user management;
- simplified resource requisition and approval process;
- access to large inventory of popular HPC workflow templates;
- intuitive workflow execution management, post-execution results access;
- real-time usage and cost tracking; and
- fine-grained access control capabilities in line with needs of privacy and data safety best practices

Figure 2.5 shows the LEXIS portal components. The portal allows an organization to access a rich set of public datasets, as well as manage their internal datasets, stored fully encrypted for additional safety, if needed. The datasets

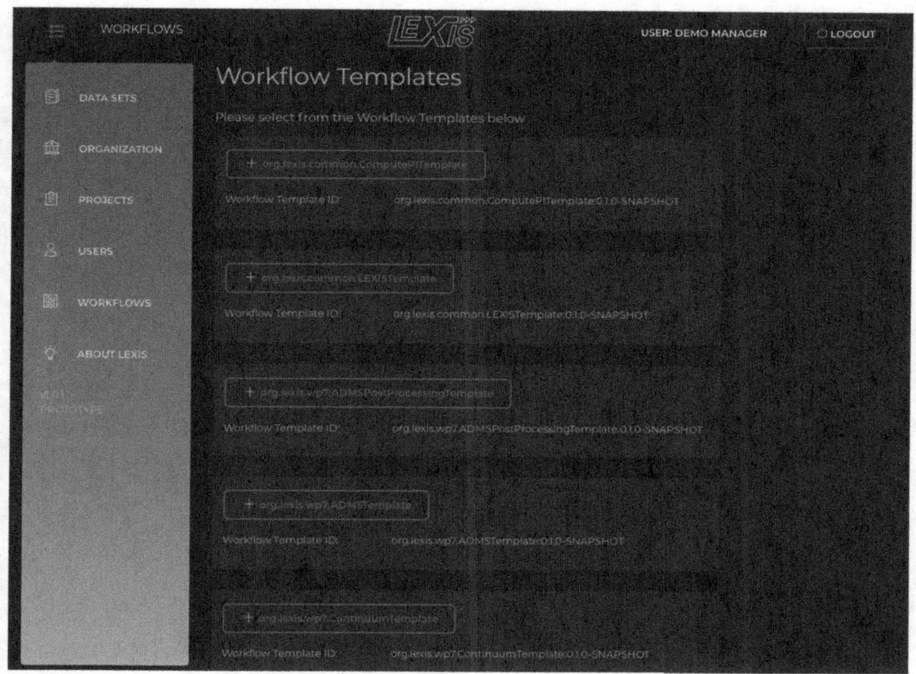

FIGURE 2.6
Screenshot of the LEXIS portal with workflow templates listing

form an essential component in workflow execution. The following access levels are configurable for every dataset in LEXIS:

- user-only – access is limited to the dataset creator only;
- project-wide – access is allowed to all members of a project; and
- public-access – the access is allowed to everyone with valid LEXIS portal credentials.

Figure 2.6 shows the LEXIS workflow template registry page. LEXIS community and participating HPC centers continually refreshes the registry adding more workflow templates covering a wider set of HPC workload use cases. A user, with the correct access rights, can create a workflow instance from the template, or may even define their own workflow if none of the available templates fits the use case. During workflow instantiation, if needed, appropriate datasets can be linked as input or output targets of the workflow.

The LEXIS portal allows the user to see visually at what stage of execution the processing is as presented in Figure 2.7. Depending on how the workflow stages are defined, the tooltip can show the user workflow stage relevant information.

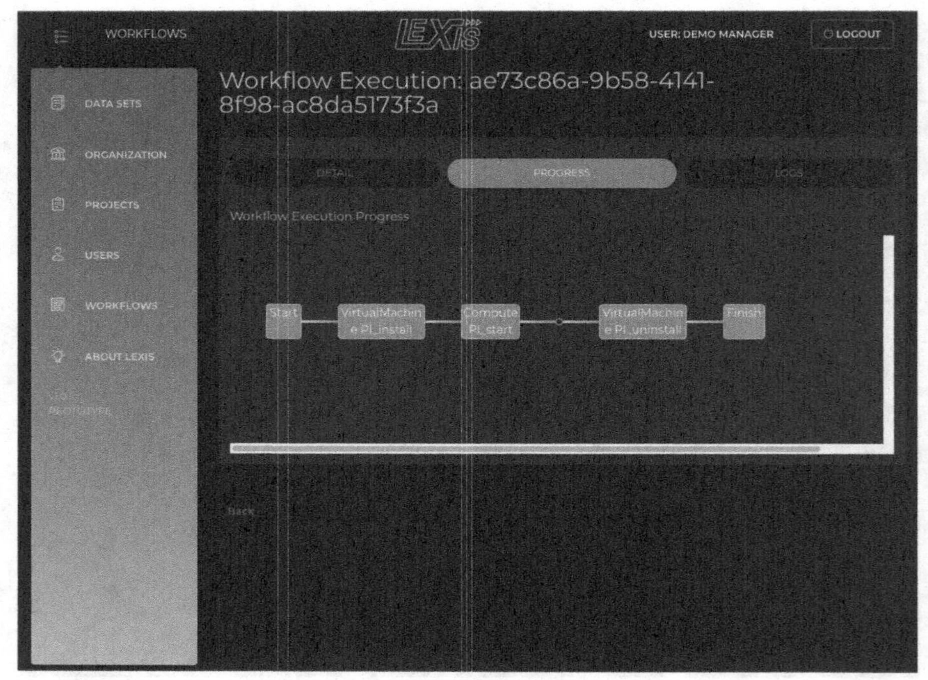

FIGURE 2.7
Screenshot of the LEXIS portal page with detail of the workflow execution

The portal also enables the user to request allocation of relevant HPC/cloud resources to a project at participating HPC centers. Once made, the request is sent to the approval subsystem of the target HPC operator. The HPC operator follows their center specific process for approval, but using a unified interface, notifies the LEXIS user of the status of the request.

The portal supports a role-based feature access by its users. The following roles are supported by the portal: LEXIS administrator; organization manager; organization user; LEXIS support user; organization financial manager; organization license manager; LEXIS project manager.

For detailed description of access rights available with the above-mentioned roles, see Section 2.3.

2.6 Market Analysis

The rapid evolving market targeted by the LEXIS project and the emergence of new critical factors such as the new EU strategic plan for digital sovereignty,

GAIA-X [11], the deployment of 5G, the COVID crisis, and the geopolitical battles for the supply of key technologies (i.e. China vs USA), have all impacted the way the LEXIS project has managed its market positioning and its targeted impacts. It has also reinforced the relevance of the LEXIS project in the EU landscape of today and tomorrow.

The global trend is clearly to develop a cloud-computing approach for HPC, big data, HPDA, AI integrating de facto the IoT and edge growing ecosystems as part of an end-to-end computing continuum. In addition, the EU wants to push organizations of all kinds to adopt practices, values, and standards aligned with its strategic goals of independence, security, and sovereignty.

All of this was taken in consideration and integrated when designing the LEXIS platform and its services. It was done in addition to the primary objective of the LEXIS project: to overcome the shortcomings of the excessively complicated access to HPC systems, large-scale datasets management, that appears to be a major barrier to the global adoption of these capabilities and technologies by companies (large, SMEs, start-ups) and even a very significant proportion of researchers, to the detriment of European competitiveness.

Instead of trying to directly compete with the dominant players (all from the USA or China), LEXIS has taken a different approach to the market and is building its offers by capitalizing on the best assets available today and in the future and developing complementary state-of-the art additional technologies to procure a clear and valuable differentiation. The key factors of LEXIS market positioning are as follows:

- ease of access;
- scaling of datasets;
- scaling and flexibility of computing, able to welcome additional new infrastructures including exascale-level ones;
- accompanying users to remove major hurdles for non-HPC professionals and to reduce TCOs, operations costs, reactivity;
- preserving key interests in IP, security, and confidentiality within the new framework and global interests of the EU;
- ability to interconnect with other infrastructures via APIs (i.e. AWS, Azure, Google) for flexibility and adaptability to various scenarios; and
- preparing the communities of researchers and companies to use the next level (exascale) in the near future

As a consequence, the positioning is visualized in Figure 2.8 [12].

	LEXIS PROJECT	HPC Centres in the EU (Private)	HPC Centres in the EU (Public)	HPC Intermediaries in the EU	ISVs	Global Players (AWS, ALIBABA, GOOGLE, AZURE...)
Data Protection from Non EU players and Intelligence gathering organisations (under American Law)	YES	Partially	YES	Partially	Partially	NO
HPC Computing Services	YES	YES	YES	NO	NO	Partially
Big Data Services & Optimised Data Management	YES	Partially	Partially	NO	NO	YES
Cloud services	YES	Partially	Partially	NO	NO	YES
Software Stack	YES	Partially	Partially	NO	NA	YES
Added Value Services, Consulting & Software developement	YES	Partially	Partially	Partially	Partially	Partially
Large execution Scalability and heterogeneity of architectures	YES	YES	YES	NO	NA	Partially
Optimised orchestration in a distributed environment	YES	NO	Partially	NA	NA	Partially
Linking HPC service providers with HPC customers & users	YES	NO	YES	YES	NO	NA
One Stop Shopping experience & Billing for all components of a computing project	YES	Partially	NO	NO	NO	YES
User Friendliness	YES	NO	NO	Partially	YES	YES
Flexibility	Partially	YES	YES	Partially	Partially	Partially

FIGURE 2.8
Positioning of the LEXIS platform on the market

2.6.1 LEXIS Project Impact

The impacts of the LEXIS project can be put into two categories. The first one is for infrastructure and data suppliers. Most HPC infrastructures in Europe are primarily used for academic and public research, with only 20% of their capacities available for the private sector. As an average, less than 4% among the 20% available are really used, with the exceptions of HLRS in Germany and CINECA in Italy (but the 20% of CINECA are mostly used by only one client) [16]. By removing the adoption barriers, the LEXIS project will allow the participating infrastructures to welcome many more large, small, and medium companies including start-ups, hence improving dramatically the ROI and TCO for each infrastructure, without major additional investment.

HPC centers will be able to welcome projects beyond their own computing capabilities, by using capabilities available in other participating infrastructures and still keeping the full management of the relationships they have with their clients/users.

Each HPC center will quickly demonstrate a major social impact by allowing economic actors to largely benefit from being empowered to use these capabilities.

Data providers will securely value their data much more by having new users in position to fully extract value from these data (previously they were unable to access computing resources and services in such an easy and affordable way).

The second market positioning category targets end users, researchers, and private companies, irrespective of industry or service type. Finding computing and big data resources is today limited in choice, with the very dominant overseas players managing more than 85% of the European market for cloud-based computing resources and storage. Working with these resources requires significant budgets, specialized skills, and in addition, once you have started with these providers you are trapped due to the difficulties and costs attached to any migration you would like to manage. The LEXIS project allows users and potential new users to access very easily the computing and data resources they need, in compliance with the EU's digital sovereignty vision, the required security, no strings attached, still keeping the flexibility to interconnect with other platforms if necessary.

- Data sources of all sizes can be accessed, connected, or uploaded, including future GAIA-X data spaces, opening the door to unlimited potential use cases.
- One-stop shopping experience for end users even for those with no HPC experience.
- Ability to innovate faster, better, and to increase competitiveness, decrease time to market, or increase the number of R&D cycles at a pace not seen before

It is believed that the LEXIS project's approach to the market will bring a real, valuable, and credible alternative to other solutions today dominating the market (all of which are under US and China control). This solution provided by the LEXIS platform is the start in setting up effective and impactful measures for the benefit of the European economy and research and its independence and sovereignty.

Acknowledgment

This chapter was supported by the LEXIS project, funded by the EU's Horizon 2020 Research and Innovation Programme (2014–2020) under grant agreement No. 825532.

References

[1] Scionti, A., Martinovic, J., Terzo, et al. 2019. HPC, cloud and big-data convergent architectures: The lexis approach. In *Conference on Complex, Intelligent, and Software Intensive Systems* (pp. 200–212). Springer, Cham.

[2] Hachinger, S., Golasowski, M., Martinovič, J. et al. 2021. Leveraging High-Performance-Computing and Cloud Computing with Unified Big-Data Workflows: The LEXIS Project. In E. Curry et al. (eds.) *Technologies and Applications for Big Data Value*. Springer, Cham.

[3] EUDAT Ltd. 2020. EUDAT – Collaborative Data Infrastructure. www.eudat.eu (accessed Apr. 6, 2021).

[4] Xu, H., Russell, T., Coposky, J. et al. 2017. *iRODS Primer 2: Integrated Rule-Oriented Data System*. Williston: Morgan & Claypool. https://doi.org/10.2200/S00760ED1V01Y201702ICR057.

[5] Jones, M., Bradley, J., and Sakimura, N. 2015. RFC7519–JSON Web Token (JWT). IETF.

[6] HEAppE Middleware. 2021. High-End Application Execution Middleware. https://heappe.eu (accessed Apr. 19, 2021).

[7] Allcock, W., Bresnahan, J., Kettimuthu, R., and Link., M. 2005. The Globus Striped GridFTP Framework and Server. In *SC '05: Proceedings of the 2005 ACM/IEEE Conference on Supercomputing*. Seattle. https://doi.org/10.1109/SC.2005.72

[8] Bull Atos. 2021. Alien 4 Cloud. http://alien4cloud.github.io/ (accessed Apr. 6, 2021).

[9] Bull Atos. 2021. Ystia Suite. https://ystia.github.io (accessed Apr. 6, 2021).

[10] CYCLOPS Labs, Cloud financial intelligence. www.cyclops-labs.io (accessed Apr. 2, 2021).

[11] Braud, A., Fromentoux, G., Radier, B., and Le Grand, O. 2021. The Road to European Digital Sovereignty with Gaia-X and IDSA. *IEEE Network*, *35*(2), 4–5.
[12] LEXIS Project deliverable 9.5: Market Analysis of Converged HPC, Big Data and Cloud Ecosystems in Europe. https://lexis-project.eu/web/wp-content/uploads/2020/08/LEXIS_Deliverable_D9.5.pdf (accessed Apr. 2, 2021).
[13] Ross, R. S., McEvilley, M., and Oren, J. C. 2018. Systems security engineering: Considerations for a multidisciplinary approach in the engineering of trustworthy secure systems [including updates as of Mar. 1, 2018].
[14] Svaton, V., Martinovic, J., Krenek, J., Esch, T., and Tomancak, P. 2019. HPC-as-a-Service via HEAppE Platform. In Conference on Complex, Intelligent, and Software Intensive Systems (pp. 280–293). Springer, Cham.
[15] Vielberth, M., Böhm, F., Fichtinger, I., and Pernul, G. 2020. Security Operations Center: A Systematic Study and Open Challenges. In IEEE Access, vol. 8, pp. 227756–227779, doi: 10.1109/ACCESS.2020.3045514.
[16] Gigler, B., Casorati, A., and Verbeek, A. 2018. Financing the future of supercomputing. Innovation Finance Advisory. *European Investment Bank* www.eib.org/attachments/pj/financing_the_future_of_supercomputing_en.pdf (accessed Apr. 19, 2021).

3

Enabling the HPC and Artificial Intelligence Cross-Stack Convergence at the Exascale Level

Alberto Scionti, Paolo Viviani, Giacomo Vitali, Chiara Vercellino, and Olivier Terzo

CONTENTS

Boundaries between scientific and industrial application domains (machine learning (ML) and deep learning (DL), big-data, HPC simulations) and between types of infrastructures (HPC– supercomputers, cloud computing) is becoming ever less defined. This is the result, in recent years, of new hybrid applications emerging, which mix tasks that are typical of the HPC domain (e.g. complex numerical simulations requiring large-scale, tightly coupled execution infrastructures) and tasks belonging to the ML/DL domains – e.g. training large artificial neural network models. These hybrid applications have their foundation in the capability of ML/DL tasks to boost the performance of the simulation frameworks and/or improve the quality of the simulation results without requiring a large number of runs. While mixing ML/DL and HPC tasks is beneficial, both domains still drive a strong demand in terms of performance, which is the main motivation for designing and

DOI: 10.1201/9781003176664-3

integrating novel acceleration devices into traditional HPC systems. A successful example of this integration is given by GPUs: thanks to their number-crunching capabilities, they happen to be very attractive both for boosting numerical simulations and for training large ML/DL models. On one hand, the increasing demand for training more and more large ML/DL models made the case for introducing ad hoc chips on the market, which offer a gain in terms of performance/watt/cost. On the other hand, the respective accompanying programming frameworks are quite different from each other and also with regard to the traditional HPC world; so, their integration into an easy-to-use platform is a challenge.

This chapter introduces a first example of a cross-stack convergent execution platform, as envisioned in the (H2020) EuroHPC JU ACROSS project.[1] As such, the chapter attempts to describe how modern heterogeneous hardware resources (in some cases also very exotic – neuromorphic processors) can be fruitfully exploited to efficiently execute hybrid scientific and industrial applications. To this end, innovations in the entire software stack will be applied, also by integrating ML/DL approaches: the artificial intelligence (AI) domain will be then able to close the loop by, on one hand, serving as a foundation of applications running on the ACROSS platform; on the other hand, being a key feature of the management system controlling the platform.

Since this solution is expected to scale toward an exascale-level performance demand, energy efficiency must be taken into account in all the platform levels. To this end, the combination of a simple way of describing applications as a set of computational tasks that can be conveniently executed on specific resources, and the use of less energy-hungry hardware accelerators (i.e. with high flops per watt rate) dynamically provisioned to these computational tasks, allows the ACROSS platform to fulfill energy-efficiency requirements.

3.1 Introduction

AI represents a broad research domain, spanning from the creation of new algorithms and models to the design and implementation of efficient hardware (performance/watt and easy to program) supporting their execution, to the assembly and deployment of these algorithms and models in useful applications. The AI domain includes subfields like ML/DL – that is, connectivism based on (deep) artificial neural networks, evolutionary approaches (e.g. genetic algorithms, ant colony optimization algorithms, bees algorithms, artificial immune systems), reinforcement learning techniques and Bayesian inference, to mention a few. AI is finding large consensus thanks to the successful application of its algorithms and models to a large number of real use cases and challenging tasks [1],[2]. The number of successful cases is growing day by day, leading the chip manufacturers to design more powerful devices which open the door to ever more complex and expensive models. Among the others, the connectionist approach, specifically in the form of deep artificial neural networks found to be very effective in many tasks such as classification, speech/object recognition, and object detection. More recently, they demonstrated strong capabilities in synthesizing images or application-specific input/output patterns. Behind these promising models, there is the execution of an impressive number of arithmetical operations, mainly in the form of matrix multiplications, matching with those required nowadays to simulate complex systems on traditional supercomputers. The request for computational power and the specific patterns of operations involved in DL tasks have been, in the last decade, the driving force for chip manufacturers to, on one hand, create silicon devices (focused on accelerating deep neural networks) ever more specialized and powerful. On the other hand, scaling and distributing the computing workload has become the key to allow these models to grow, far beyond what is possible with a single device. Nowadays, public cloud providers can easily execute models with billions of parameters, leveraging and strengthening the role of large arrays of high-performance GPUs, FPGAs (field programmable gate arrays), and special purpose accelerators. In this regard, hardware accelerators tailored for efficiently executing the training phase (i.e. the phase where the model learns from the input data) should be distinguished from those oriented to the inference phase (i.e. the phase where the trained model is used to provide a prediction; for instance, classifying an input pattern). This translated into different chip architectures, embedded memory sizes, and power envelopes. However, in a race of mimicking the behavior of the human brain, more sophisticated (and more exotic) devices have been designed: neuromorphic chips are expected to be the next step in the domain of AI accelerator architectures.

Not only did AI advance in the last years, the traditional HPC domain did likewise. The availability of more sophisticated and large-scale machines allows scientists to simulate complex phenomena to very fine granularity, and design engineers to achieve the desired solutions in a fraction of time. Hardware accelerators played (and still play) a fundamental role in the path toward the "exascale," i.e. the capability of a machine (supercomputer) to perform at least 10^{18} floating-point operations per second (FLOPS). To reach and surpass this threshold, fundamental improvements and innovations are needed in all the components, from the processing elements to the memory subsystem, to the networking and the overall software management stack. These kinds of infrastructures are also needed to cope with the newfound capability of generating significant amounts of data, often connected to the outside physical world (generally in the order of terabytes (TBs) or petabytes (PBs)), that need to be quickly analyzed. The software stack and the underlying hardware need to be adapted to support such an analysis.

Furthermore, there is a growing interest inside the HPC community for the integration of ML/DL models with traditional numerical simulations. Although DL algorithms have been successfully applied at scale to speed up complex simulation in, among others, the fields of weather and climate forecasts and high-energy physics (HEP) [3], a large adoption is still far away. The drive for more heterogeneity in the infrastructural layer is making more evident the need for modernizing the management software stack. While batch schedulers (e.g. SLURM, PBS, HTCondor) represent a "legacy" way of serving the computational resources to the users, cloud computing provides a more flexible and adaptable approach. To this end, a "cloudification" of the software management stack is foreseen as a viable solution for deploying and serving more heterogeneous hardware in the HPC context, and for allowing more tightly coupled interaction between application frameworks in the HPC, ML/DL, and big-data domains.

Dealing with (hybrid) applications means taking care of the way they are constructed and their specific requirements. This translates into understanding the specificities of the programming frameworks used to implement and run the different parts composing the applications and providing an effective mapping with the underlying execution resources: a convenient approach to describe applications is through the concept of *workflow*. It is a sequence of steps, including all their dependencies and requirements, that must be performed in order to achieve desired results (e.g. all the necessary steps needed to successfully complete a numerical simulation, including the preprocessing of the input files, performing the simulations, and visualizing the output result). When using the resources available on different data centers which are served as HPC and cloud systems, there is a need for a flexible and modular software, which takes the responsibility of governing the whole process of workflow definition, execution, and result visualization. Chapter 5 provides an insight into such modern workflow management systems (WMS), by describing the WMS underlying the H2020 LEXIS project. This solution allows abstracting the underlying infrastructures, by automating the process of assigning user computational tasks to the most suitable resource. While Chapter 5 remains focused on a high-level approach, this chapter will focus more on the way innovative techniques can be applied to the lower levels of the execution platform (i.e. the levels where the cluster resources are assigned to the single jobs; that is the batch-scheduler level for an HPC cluster, and the virtualization management system level for a cloud cluster). Also, the way heterogeneity is effectively exploited is part of such novel techniques.

The next sections describe some key ideas that lie behind the architecture of the EuroHPC JU ACROSS project platform. To this end, the project focuses on the integration of innovative acceleration technologies (e.g. neural network processors (NNPs) or neuromorphic processors) within existing modern supercomputing systems, as well as on the integration of innovative software solutions that will be the building blocks for a renewed management stack. Emphasis will also be on the energy efficiency as a hard constraint for building next-generation exascale machines. As such, the ACROSS platform devises a large use of (lightweight) virtualization technologies along with smart resource allocation and workflow task scheduling strategies.

3.2 The Rise of Convergent Infrastructures

Convergence of information technology (IT) infrastructures is a concept that was born in the enterprise world, where the pressure in providing infrastructural solutions that improve the productivity of users and reduces the management effort of the IT staff is always high. The idea behind the concept of convergence is that of reducing the complexity in integrating different components (hardware and software) into a unified system and reducing the effort in managing them. One of the foundations of the convergent infrastructures (CIs) is the integration of computing server, storage, and networking components into a single unit, that can be easily deployed in a data center and centrally managed. Cloud providers largely exploited this concept to profoundly customize their systems, also codesigning hardware and software solutions for an optimal integration. Indeed, hyperconverged infrastructures (HCIs) combine modular systems integrating storage, networking, and computing capabilities along with (low-level) software components that ease the scalability of the infrastructure and the remote management of the entire processing environment. The majority of the IT solution providers have HCI solutions in their catalog.

One of the key points of CI and HCI to improve productivity and manageability is given by resources virtualization. The tight integration of storage, networking, and computing capabilities allows the virtualization software to easily aggregate them into a virtual system that users can access, allowing the management system to optimize the physical resources usage. For instance, physical resources can be shared among different virtual systems in a multi-tenant environment, while the users still perceive their (virtual) systems as a pool of unshared resources. Virtualization technologies have leapt forward in the last years, also thanks to the hardware-assisted technologies (e.g. Intel VT-x and AMD-V), such as I/O MMU virtualization, the management of virtual interrupts, network functions virtualization, PCI passthrough, and single-root I/O virtualization (SR–IOV). All these features largely contributed to push the virtualization concept to an extreme: on the path of further improving HCI flexibility, virtualization started to be used to compose entire virtualized large-size environments. This software-defined concept has been successful in managing large infrastructures and led to a further increment of hardware assisting technologies. More specifically, the management layer (originally highly based on a monolithic approach) has been rethought to be built on top of many small functions, which can take advantage of dedicated hardware accelerators. This is the case for networking functionalities that found great benefits from FPGA acceleration or the use of smart NICs (network interface cards).

While hyperconverged systems provide (in most cases) the building blocks on which modern cloud-computing platforms are built on, the fast evolution of user applications in terms of performance demand and requirements is making such level of convergence, again, not enough. As mentioned in the introductory part of this chapter, scientific and industrial applications are going to mix very different domains, each with its peculiarities in the way underlying infrastructure must be configured. In this regard, low-level abstraction and converged functionalities are no longer sufficient to fulfill the requirements of these applications. Indeed, the access to HPC-like resources is a priority, as it includes access to all the various types of hardware accelerators, with the same degree of control and flexibility of a modern cloud environment. In other words, it makes the case for rethinking the traditional HPC world in a more cloudified way, to efficiently manage next-generation HPC systems [4][5].

3.3 The ACROSS Approach to the HPC, Big Data, and AI Convergence

In the context of ACROSS project, the platform has been thought in such a way to support the execution of hybrid workflows where there is a tight coupling among tasks belonging to different domains; thus, focusing on the convergence between HPC simulations, ML/DL, and HPDA (with an emphasis on big data) tasks. Furthermore, the ACROSS platform aims at sustaining the application scaling toward an exascale-level by focusing on two key architectural pillars: (1) the integration of different hardware accelerators, ranging from more traditional GPUs, to less common FPGAs and NNPs, to the more exotic neuromorphic devices; (2) the implementation of a multilevel, multi-domain *orchestrator* providing all the mechanisms to easily decompose an application into smaller units of computations that can be automatically and dynamically mapped onto the hardware resources that mostly fulfill the requirements in terms of performance and energy efficiency. For this purpose, the whole software stack needs to be revised and adapted to support such convergence. Moreover, the orchestrator has been imagined in a way that would allow the end user to explore the trade-off space between time-to-solution, energy efficiency, and cost (i.e. by selecting newer but busier resources, or choosing between exotic and costly accelerators or conventional architectures). In this regard, a specific approach for predicting queue waiting times for jobs executed on HPC resources will be integrated to support this vision without disrupting legacy HPC batch-scheduling infrastructures. Similar to this, advanced policies for providing cloud virtual resources and serving the large hardware heterogeneity will also be applied (see Section 3.3.2).

Figure 3.1 depicts the conceptual architectural schema of the ACROSS converged platform. Moving top-down, the reader can see how this multi-level management concept is applied. First, the application is partitioned (at a high level of abstraction), into its main constituents (phases), which generally comprise the preprocessing of input files, running numerical simulations, performing a training/inference phase of an ML/DL model, performing an HPDA task on a large dataset, post-processing output files and visualize the results. Furthermore, some of these phases may be repeated several times (e.g. in a closed feedback loop) over the application execution. One way of providing such a partitioning in a formal way is by describing the application workflow using a standard language: Chapter 5 provides an overview of different tools that have been proposed for this purpose. Cloud computing aims at providing a high level of process automation (i.e. the way computing, networking, and storage resources are acquired is automated by the underlying governing software stack) and several tools have been designed allowing the users to decompose their applications into sequences of steps that could be automatically managed by the cloud platform. TOSCA[2] is an open standard for accomplishing this purpose. It is a YAML-based domain-specific language (DSL) that provides a convenient way for defining all the unitary blocks (e.g. SQL database, a programming framework) of an application, their requirements in terms of computing resources or software dependencies (e.g. a SQL database may need a specific version of the Linux kernel, and may require a minimum amount of memory on the node that will run it), and to specify the sequence of steps needed to execute these unitary blocks in the correct order.

The ACROSS platform integrates (at the top level of the orchestrator) a set of tools that allows to extend these functionalities in such a way ML/DL models can be easily expressed, their training phase can be easily integrated into a more general workflow, and HPC jobs can be simply deployed on HPC resources. In this regard, the envisioned orchestrator is composed of a front-end module that provides all these extensions, which is coupled with a back-end engine that is responsible to deploy the various application unitary blocks on the appropriate underlying resources. For instance, the YSTIA suite (i.e. a software suite composed of the Alien4Cloud (A4C) front end, which provides both a REST API and a GUI, and the YORC back-end engine) described in Chapter 5 falls down in this type of solution. The efficient application execution requires a fine control over the underlying infrastructure; both HPC and cloud infrastructures have their own specific management

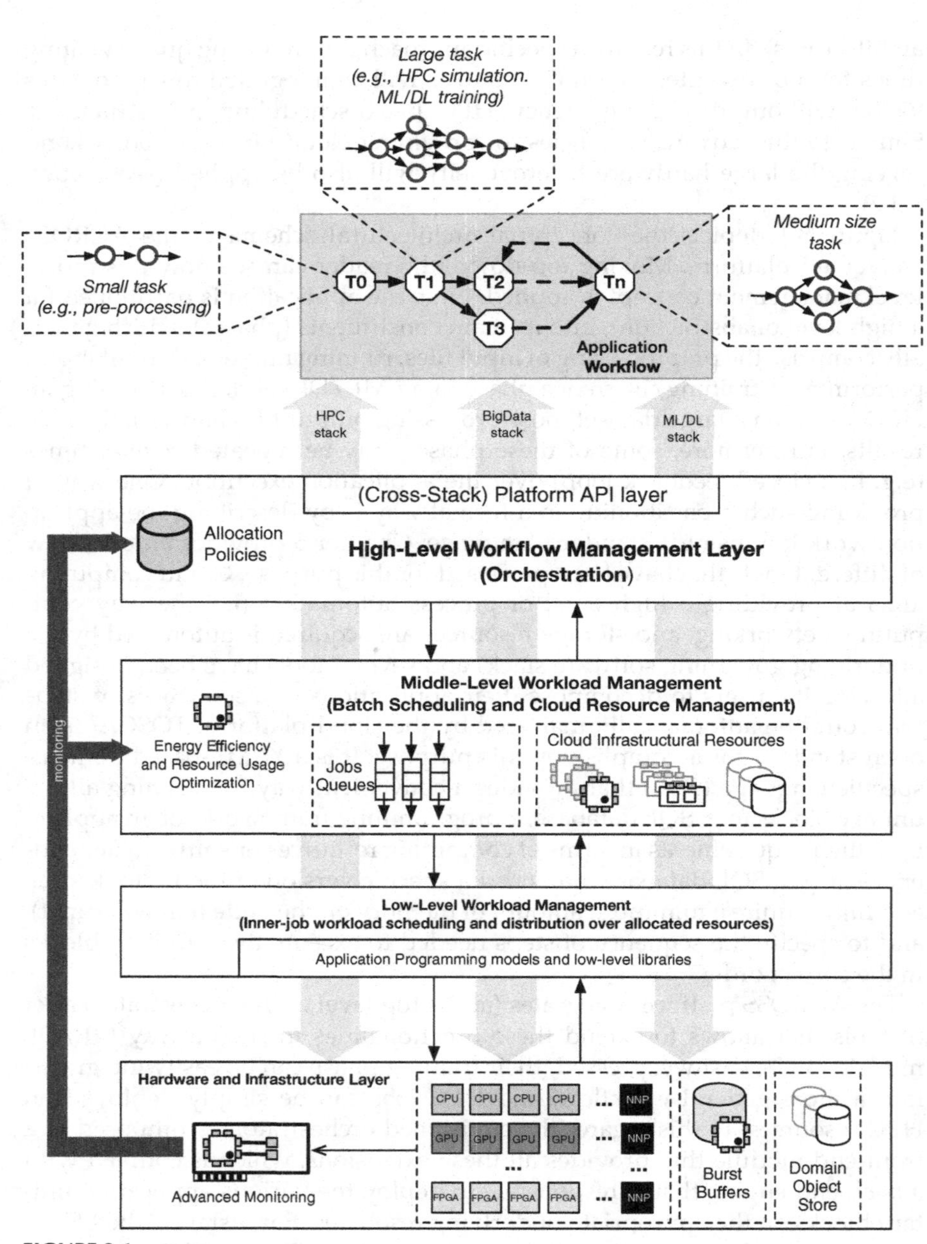

FIGURE 3.1
The overall ACROSS platform architecture with high-, middle- and low-level orchestration layers providing unification across HPC, big data, and AI stacks.

software stacks, and so, the high-level layer of the orchestrator (described above) needs to interact with them. To this end, optimizations on the HPC job scheduling policies can be applied; similarly, smart resource allocation policies can be used on the cloud. One strong limitation in doing this is that these software stacks are used for serving resources that are not part of the ACROSS platform, and thus they cannot be modified without a large (potentially negative) impact on the applications running on those resources. Also, some of these software are not designed to be customized. To circumvent all the potential limitations and restrictions, and to still move on with the goals set for the ACROSS platform, innovative techniques to analyze and predict the behavior of the jobs within the batch-scheduler queues will be put in place. Similar to this, innovative strategies to allocate virtualized resources (mostly in the form of Linux containers) and serving hardware accelerators will be applied – middle layer of the orchestrator. In this light, the batch schedulers (e.g. PBS, SLURM) and the cloud management software (e.g. OpenStack) will be used only for deploying jobs on the proper HPC systems (i.e. on the specific queue) and on the virtual composition of cloud resources (i.e. a set of Linux containers that can interact with one another). This can be enhanced by the integration of dedicated middleware that also has responsibility for security aspects (e.g. HEAppE Middleware – see Chapter 5 for a complete description of it).

Further optimization of the workload execution process can be achieved once the resources are assigned; as such, there can still be room for a smarter way of distributing the workload. This is the case of jobs whose activity can be described by a directed acyclic graph (DAG), where vertices represent the inner computational tasks to perform and the edges their temporal and data dependencies. In this regard, the ACROSS orchestrator has been devised to integrate, at the lowest level (i.e. closest to the execution resources) software elements that allow to automatically manage the distribution of the (inner-job) workload among the acquired resources. For instance, the HyperLoom [6] system provides a simple interface through which it is possible to define computational tasks and their dependencies. The framework uses a master-workers approach to distribute the computational tasks among the worker agents mapped onto the allocated resources (nodes assigned by the batch scheduler in the HPC domain, virtual machines or containers deployed by the cloud manager – e.g. OpenStack). Another example of such kind of tools is given by the COMP Superscalar (COMPSs) [7], a programming framework developed to ease the implementation of parallel applications on distributed systems spanning from HPC clusters, to clouds and computing grids. On the other hand, other middleware and dedicated (run-time) libraries can lie at the bottom of the infrastructure to ensure optimal exploitation of the managed hardware, including accelerators, e.g. Nvidia CUDA and AMD ROCm in the case of GPUs, Intel OPAE, and Xilinx XRT in case of FPGAs, to mention but a few.

While Section 3.3.1 gives an overview of the hardware heterogeneity that will be one of the key elements of the ACROSS platform, Section 3.3.2 focuses on giving the reader more insight into the way these heterogeneous resources can be effectively provisioned.

3.3.1 Heterogeneous Infrastructural Support

The first key element enabling the multi-domain execution is represented by the heterogeneous infrastructure. Hardware accelerators are good representatives of the basic concept that a more specialized architecture can do more computations in less time and with improved energy efficiency. Specialization is strictly connected to the specificity of the applications whose performance must be improved. Generally, this specialization requires analyzing the behavior of applications of interest in order to determine those portions that strongly affect the overall execution. The more specialized is the designed architecture, the more performance and energy efficiency are obtained; however, a higher specialization comes with the countervalue of a less flexibility, that translates into a more rigid programming environment and to a reduced set of applications that can be accelerated.

Over time, designers proposed acceleration architectures trying to intercept a broader set of applications, by integrating ever more functions in the same chip (system on chip (SoC)), instead of squeezing performance of homogeneous resources. A representative example of this is found in (modern) GPUs. These devices provide a huge number of simple execution cores backed by a relatively small amount of cache memory when compared to CPUs. The high number of computing units supporting double-precision arithmetic is the natural home for all the scientific applications, other than graphic applications. It also provides enough resources to concurrently execute several thousands of threads; by contrast, large server CPUs offer, at most, a few hundred concurrent threads. The way cores are hierarchically organized, along with an optimized instruction set architecture (ISA) and programming framework, allow GPUs to be very effective when performing algorithms that can be expressed as matrix multiplications, as they can be found in many scientific and graphic applications. ML/DL algorithms are characterized by the need to perform a very large number of operations, often falling down in the matrix multiplication category. ML/DL tasks also largely benefit from their tolerance to lower-precision arithmetic: in fact this fits very well with current GPU architectures, where double-precision floating-point performance is typically forgone to dedicate more silicon real estate to massive parallelism for lower-precision instructions. In the past few years this led to further optimization of GPU cores to support various floating-point data types (e.g. double, float, bfloat), as well as to embed cores better tailored for tensor operations which are very common in the ML/DL domain.

The fast pace at which ML/DL models are taken up by different application domains (including scientific ones) moved the market toward more specialized architectures, which favor training and inference performance for such (ML/DL) models (DNNs, CNNs, etc.), at the cost of a restricted domain of applications. We refer to them as NNPs. Compared to GPUs, NNPs provide a better ratio between performance, energy consumption, and cost. Many companies found that the restricted application field (neural network processing) for such specialized machines, is well counterbalanced by new AI (with a strong emphasis on deep neural networks) use cases that day by day emerge. The major efficiency of these systems come from an optimized design of the computing cores, along with the integrated interconnection and the memory subsystem. Devices tailored for supporting specifically computer vision applications are also in this category[8]. Generally, these devices are designed to scale well across multiple nodes, in order to allow processing huge models. For instance, the Habana Labs Gaudi processor embeds fast ethernet interconnection switch [9] enabling direct connections among chips and fast data movement without the intervention of any external device (support for massive scale up and scale out).

There are devices designed to mimic the way the human brain works and performs our daily (very complex) tasks, which are known as *neuromorphic* devices. Unlike any other accelerator architecture, they are designed to mimic the way neurons are activated and exchange information with the others. While an artificial neural network provides a rough approximation of the (human) brain, and so also "digital" accelerators, neuromorphic devices operate in the analog regime. Generally, the information is carried by a sequence of pulses (spikes) which enable the artificial neurons to be activated or not, depending on the cumulative number of pulses received in a certain time window. These kinds of systems offer the best approximation of the (biologic) brain of a mammalian. As such, they are gaining momentum thanks to their flexibility and plasticity in performing AI tasks. Good examples of neuromorphic chips are the Intel Loihi [10], IBM TrueNorth [11], and the SpiNNaker architecture [12]. Despite their promising capabilities, mimicking even a small fraction of a mammalian brain requires a large-scale system, where many of these devices are interconnected with each other. The programming framework is also quite different from a standard programming framework and requires great expertise to optimize the application in order to benefit from this type of device. Despite all the restrictions imposed by an extremely specialized architecture, as in the case of a neuromorphic processor, they may find applicability at a relatively smaller scale to support the execution of hybrid workflows (i.e. those mixing numerical simulation with ML/DL tasks); this is one of the objectives of the ACROSS architecture.

Historically, FPGAs found their main application field as prototyping platforms before manufacturing specific ASICs (i.e. application-specific integrated circuits). Their flexibility derives from the combination of different hardware components (flip-flops, look-up tables, memory blocks, etc.) and a mechanism to configure them in such a way that it is possible to reproduce the behavior of any digital device (the only constraint is in the limited number of these hardware components, although it becomes larger every new fabric generation). The configuration passes through a dedicated internal network-on-chip, and the use of hardware-oriented design languages (VHDL and Verilog). The increasing flexibility gained over the past years made this kind of device very attractive to many different application fields, ranging from the energy-efficient implementation of DL inference overlay architectures, to the offloading of specific functions in enterprise systems (e.g. some of the functionalities of the network stack can be easily mapped on FPGA; also network-oriented DSLs have been proposed to easily map networking tasks on them – P4[13]). The growing number of public cloud providers that use FPGAs for operating their infrastructure [14] and offer them as a virtualized resource [15], and new programming frameworks that ease their exploitation (high-level synthesis – HLS– tools allow the compilers to infer the hardware configuration from a high-level description, generally done using conventional programming languages like C/C++) make these platforms attractive for accelerating specific tasks in hybrid workflows. They also remain a good candidate to offload many platform management operations that otherwise can waste precious computing resources (e.g. in cloud environments, where the control plane software is distributed on the same nodes that execute users' virtual machines and containers).

In this regard, a converged platform as the ACROSS envisioned one, should look at these devices as a promising and flexible tool on which to leverage for easing the integration of different stacks and accelerators on a large-scale system. For instance, FPGA devices can be used to facilitate the monitoring of the running compute nodes, offloading the CPU from the task of collecting, filtering, analyzing large amounts of data. The next section focuses on the way such vast heterogeneity can be conveniently provisioned in a modern computing platform, taking into account not only the constraints imposed by the hardware specificity but also those related to scalability, sharing of resources, and networking.

3.3.2 The Management of the Convergent Platform

The second key element of the ACROSS architecture that enables the multi-domain application execution is represented by the platform management solution. This is a combination of software and infrastructural components that allow modeling the user applications, allocating required resources by picking them up from a pool of HPC and cloud clusters, controlling their execution in a shared resource environment, monitoring the status of underlying infrastructures. High-level orchestration front end and back end, as described in Section 3.3, provide all the features that allow users to describe their applications in terms of basic steps to perform, and abstract the process of mapping these steps onto the underlying infrastructure. The middle layer of the orchestrator has a fundamental role in the optimization phase of the resource allocation and mapping. Here, it is necessary to look at how the resources are organized in a platform like ACROSS.

First of all, the resources are provided by HPC data centers by getting access to HPC clusters on one hand and getting access to cloud partitions on the other hand. HPC and cloud partitions are generally physically separate and can share data through a common storage space or through a shared bus (interconnect). The HPC resources are typically organized into silos, where the nodes inside each silo are homogeneous (i.e. they have the same characteristics); a typical repartition of the resources is as follows: CPU nodes, accelerated nodes (each node is equipped with one or multiple accelerator cards, e.g. GPUs), big memory nodes (i.e. nodes with large amount of main memory). Then, each silo is served by a dedicated queue, managed through a batch scheduler which can apply specific scheduling policies (see Figure 3.2).

This situation reflects the way HPC resources were managed in the past and shows its limitation in providing enough flexibility to manage multi-domain application workflows where tasks belonging to the various domains closely interact with one another. In fact, as previously stated, this type of application requires access to different types of resources possibly at the same time, or in a fast loop. While accelerated nodes can easily support heterogeneity at the infrastructure level, there is no way of optimizing their usage over time. In fact, once the access to such resources is granted, they will consume (also energy) whether the application uses the accelerators or not. Here, the waste is both economic (the user will pay more than the actual value extracted from those nodes) and in terms of energy. Nevertheless, batch-scheduling systems will not disappear altogether, so any improvements will need to support their management.

On the other hand, cloud computing comes with the capability of dynamically allocating and scaling the resources horizontally. In a cloud environment the acquired resources can also be heterogeneous and can be added and discarded at any point in time. However, cloud environments are in general not designed to offer the same level of performance for executing those applications that have many tightly coupled processes as conventional HPC systems.

The ACROSS envisioned solution is a combination of the two approaches that we can define as a "cloudification" of the HPC systems. The basic idea is to allow the user to dynamically compose the various (heterogeneous) infrastructural pieces they need into a single logical infrastructure. This is illustrated in Figure 3.3, where different layers previously discussed are unified through an improved orchestration intermediate layer. The foundations of this improved layer are: (1) a flexible and performant virtualization technology that allows the software overhead to be minimized and which provides near bare metal

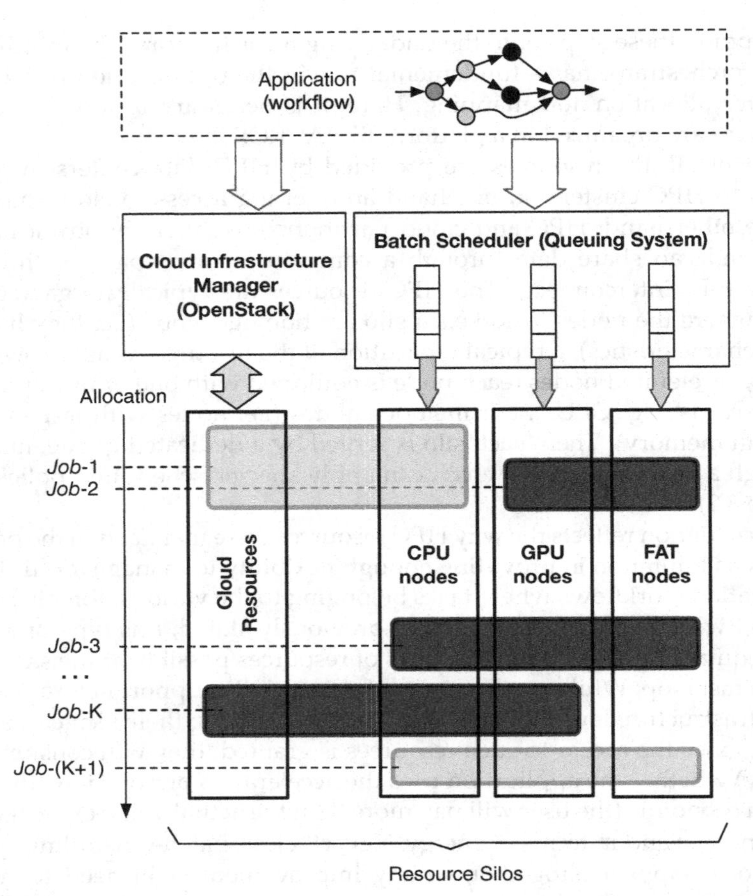

FIGURE 3.2
Flexible provisioning of HPC and cloud-computing resources across (typical) infrastructural silos.

performance; (2) a toolset that allows a virtual overlay infrastructure to be easily created, where containers can communicate with others in the same virtualized environment without any limitation imposed by the underlying infrastructure; and (3) smart allocation policies that can effectively optimize different (often contrasting) objectives – e.g. minimizing energy consumption of the whole platform, maximizing performance of the applications, reducing users' costs, maximizing the usage of physical resources. The first foundation can be fulfilled by relying on Linux containers (e.g. Docker, Singularity), which already offer a secure and mature technology. The second foundation is given by tools (e.g. Ansible, Puppet, SaltStack) that offer a large degree of control in setting up virtualized infrastructures. To overcome the limitations mentioned earlier, it is necessary to move beyond the traditional concept of

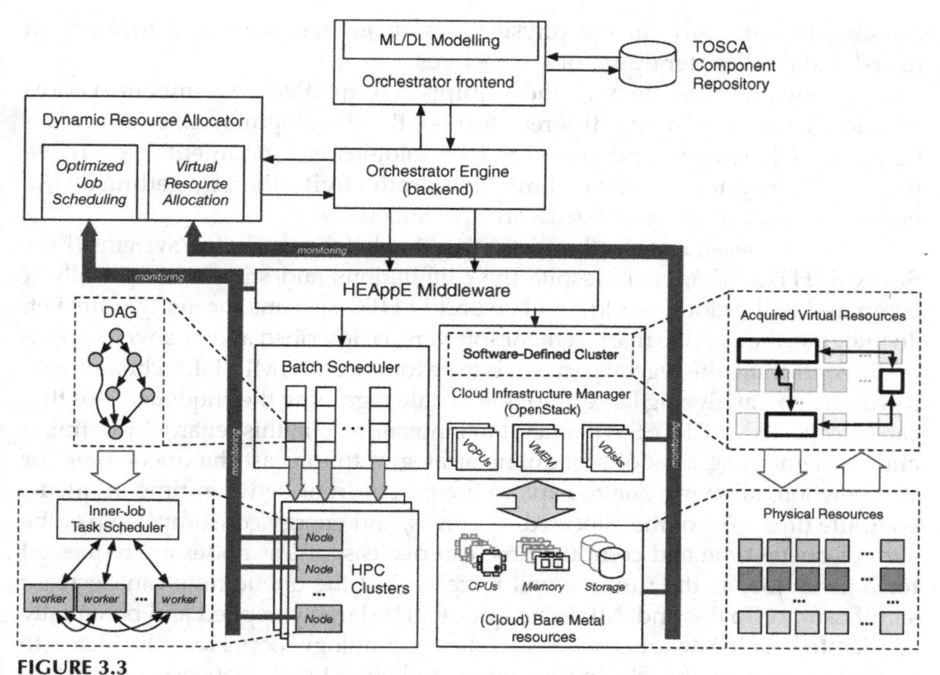

FIGURE 3.3
Overview of the multi-level ACROSS orchestration system, where innovative management strategies are applied at high-, middle- (i.e. batch-scheduling level) and low-level (in-job tasks' scheduling).

infrastructural management toward the idea of *(high-performance) resource composability*. With these terms, we refer to the capability of the user to define the virtualized infrastructure as a composition of different types of resources (including specialized accelerators) that can be dynamically acquired and released by the application, and which talk to each other via a fast interconnection (as happens on HPC cluster nodes). The core of this idea is not new, since cloud resources are already provisioned in this fashion; however, here, the intention is to stress the fact that composed resources include those generally available in a HPC cluster, and it is important that the virtualized infrastructure provides a way for running processes to communicate fast with each other, regardless of where they physically reside. A germinal concept of resource composability can be found in [16]. In this regard, the management software should allow for the flexibility and dynamicity that are typical of the software-defined systems, i.e. the software allows all the needed resources to be gathered and configured to accomplish a specific task: in our case, we refer to this as the software-defined-cluster approach. Finally, the third foundational aspect can be fulfilled by reverting the scheduling and allocation problems into combinatorial ones. By integrating an optimization box, it is possible to optimally fill the physical resources over time and to meet all users' and data center operators' objectives.

Going toward exascale machines equipped with massive numbers of cores and accelerators (of many different forms), the development of such a combination of hardware and software technologies is fundamental to enable users of such systems to not be limited in productivity, thereby getting all the benefits of the advances in hardware specialization.

Last but not least, it is worth mentioning that batch-scheduling systems (PBS, SLURM, HTCondor, etc.), despite their limitations and stiffness (especially if compared to the cloud world), will be part of HPC systems for many years yet. To integrate them in the management software as described above, a way of optimally controlling this legacy part needs to be found. Here, ML/DL techniques are a great help in analyzing large amounts of data regarding the enqueued jobs that can be collected. ACROSS will intend to innovate also in this regard, by defining and implementing a set of scheduling strategies to forecast the queue time for any new job, allowing one to explore the pareto front between time in queue, compute time, cost of the allocated resources, and energy consumption (i.e. the sum of queue time and compute time can decrease if less nodes are requested for a given job, as the runtime will increase, but the queue time can decrease significantly). To this end, ML techniques will be largely applied and potentially accelerated via hardware devices, in a close technology loop. Here, the idea is to continuously track the jobs in the queues and collect their statistics, which contain several features. Then, techniques to preprocessing these features (e.g. principal component analysis (PCA), min–max scaler) and to make predictions for new jobs (e.g. Bayesian classifier, linear regression, multilayer perceptron (MLP)) are applied. The key idea behind this approach is to learn from the historical data how to optimally map a new job to submit on one of the available systems' queues.

3.4 Related Works

The concept of *software-defined*-HPC cluster (supercomputing) is relatively new, since most of the effort in providing high-performance and heterogeneous computing resources can be found in cloud environments. In [17], the authors describe this concept applied to traditional and more monolithic HPC clusters (supercomputers). They identify the main components needed to flexibly deploy virtualized infrastructures on HPC resources and successfully tested on MPI-based applications. Having recognized the lightweight virtualization technologies like (Linux) containers as the common ground between HPC and cloud environments, other H2020-based projects built their serving platform on them. For instance, the EVOLVE project made great use of Linux containers and the Kubernetes [18],[19],[20] scheduling system to meet the demand of convergence. The H2020 SODALITE project introduced the idea of meta-orchestration [21] as a high-level management tool for orchestrating (definition, installation, launch, and deletion of workflow components in an automated way) workflows demanding different types of resources, which are provisioned on different clusters. It also explores the concept of minimalistic orchestration tool, i.e. xOPERA [22] to provide a simple and rather performance-oriented orchestration, with a minimum overhead. The Italian regional project HPC4AI [23] also explored the concept of serving HPC infrastructure through containers, departing from legacy job schedulers.

Hyperconverged infrastructures have been largely described and analyzed in literature. Most of the references are for cloud-oriented solutions [24],[25],[26].

The software-defined concept has been popularized by the large community of developers and finds common application in the management of the networking stack. In this regard, software-defined networks (SDNs) leverage on applying (many) different network functions on exchanged traffic by executing specific pieces of software, mostly on a standard CPU [27]. This strongly contrasts with the historical network processing performed on network equipment and using specialized ASICs. To overcome the limited flexibility of such ASICs and leveraging on the continuous boost in performance of modern CPUs, SDNs integrated more and more functionalities over time. However, this massive reliance on host CPUs for performing network processing (but also supporting the execution of all the tasks related to security checks, etc.) put a cap on the resources actually made available to the virtualized systems. Thanks to the ever-increasing performance of CPU microarchitectures, their energy-efficiency improvement, and the capability of integrating more and more functional blocks on silicon, led providers of network equipment to offload such virtualized functions onto dedicated acceleration boards (often referred to as smart NICs) [28],[29]. In this case, low-power multicore processors are coupled with dedicated IP cores in such a way that it is possible to perform traffic checks and manipulation at high speed (100Gbps and more), while leaving host CPUs as workhorses for user applications. Large public cloud providers also rely on such an approach to serve resources on their large-scale data centers; in this regard they have their own acceleration boards, which are highly customized for matching their data center requirements (e.g. AWS Nitro cards).[30]

Integration of specific functionalities in a modern computing architecture is the core concept behind hardware accelerators. Modern accelerators are designed with a specific application domain in mind; so, additional functions to those provided by a general-purpose computing system are melded together. This translates into strong customization of the processing core microarchitecture, the integration of special-purpose cores, the integration of a large number of processing cores served by a customized memory hierarchy and interconnection. Over the years hardware acceleration has been a synonym of GPU. With the recent revolution brought by AI, hardware accelerators started to refer to a broader set of specialized hardware architectures. Tensor Processing Unit (TPU) [31] is the full custom accelerator designed by Google to boost performance of all the AI-related tasks they have to manage daily through their cloud platform. Other AI-oriented architecture can be found in training and inference chips [8],[9],[32],[33]. Many (micro) architectures have been researched to provide the best ratio between performance (number of operations per seconds) and power consumption [34].

FPGAs represent an intriguing resource: on the one hand their reconfigurability has been explored to create flexible smart-NICs architectures and deploy them at scale, also with a relevant example given by a large cloud provider [35]; on the other hand, they have been explored to boost user applications or to accelerate services in large public cloud environments [36]. In the latter case, multiple FPGAs are used to map all the basic functions used by the Bing[3] search engine to parse user queries, processing them and providing relevant results. Specifically, the system has been designed in such a way, a portion of the FPGA is reconfigured on-the-fly, while the remainder of the fabric provides a fixed interface toward the hosting server and the other FPGA cards. FPGAs also offer an interesting ratio between performance and power consumption when applied to accelerate DL algorithms, with a focus on binarized DNNs and CNNs [37].

Neuromorphic model of computation has been developed as a way of mimicking the behavior of mammalians' brains. In the neuromorphic model, the abstract computational units are designed in such a way to resemble as closely as possible the biological neuron counterpart. Then, a large amount of such units can be arranged to form a neural network in which information is transferred in the analog form (instead of digital) [38],[39],[40]. More specifically, these artificial neurons exchange trains of analog pulses. To support this kind of computing model, a very different (analog or mixed analog–digital) chip architecture has been designed, also taking in mind the scalability of the solution at large scale. Examples of such devices can be found in the Intel Loihi [10], IBM TrueNorth [11], BrainScaleS [41], and SpiNNaker system [12]. In [42] and in [43] authors showed how the Spinnaker neuromorphic architecture, once scaled to several nodes, has been used to successfully simulate large-scale plastic neural networks, as well as mapping very large task graphs.

3.5 Conclusions

Digital technologies, in the last decades, became a precious instrument for science and industry and will continue to be an essential tool for supporting the next discoveries. As with other human invention, it is evolving to be more and more flexible and powerful. In this regard, HPC and cloud-computing paradigms drove (and still do it) this evolution process. Recently, another major driving force joined these two: AI, with a specific focus on ML/DL techniques, made possible to boost the performance of traditional numerical simulations, as well as to provide new powerful tools for large data analysis. While these technological fields remained clearly separate for a long time, the emergence of new applications strongly required them to be mixed, which again raised a historical problem: how do we better support these new applications? How do we speed up their execution? Answering these questions means rethinking the way large data center infrastructures provide their computing, storage, and networking resources. Tight coupling among numerical simulation, ML/DL training and inference tasks, data analytics on large datasets requires a new level of convergence. The resulting platforms are at the forefront of innovation and today looks like the solution for optimally provisioning all the needed resources, including hardware accelerators, in a fast, dynamic, and more flexible way.

In this chapter, we briefly reviewed this evolutionary trend, and we provided the conceptual vision for a new convergent AI, HPC, and big-data platform. To this end, the reader can grasp how the combination of virtualization technologies, tools for managing, monitoring, controlling the instantiation of virtual resources and their execution, and highly specialized hardware accelerators provide the foundations for implementing such new highly converged systems. In this regard, the chapter is mostly driven by the EuroHPC JU ACROSS project vision, where all these elements are fused together to provide a highly flexible and scalable cross-domain execution platform.

Acknowledgment

This work is funded by the EU's Horizon 2020 Research and Innovation Programme (2014–2020) under grant agreement No. 955648 (ACROSS Project – HPC, Big Data, Artificial Intelligence Cross Stack Platform Towards Exascale) and being part of the work programme European High-Performance Computing Joint Undertaking Annual Work Plan 2019 – Research and Innovation actions.

Notes

1 www.acrossproject.eu
2 www.oasis-open.org/committees/tc_home.php?wg_abbrev=tosca
3 www.bing.com

Bibliography

[1] Gibney, Elizabeth. "Google AI algorithm masters ancient game of Go." Nature News 529.7587 (2016): 445.

[2] Ferrucci, David, et al. "Watson: beyond jeopardy!" Artificial Intelligence 199 (2013): 93–105.

[3] Kurth, Thorsten, et al. "Deep learning at 15pf: supervised and semi-supervised classification for scientific data." *Proceedings of the International Conference for High Performance Computing, Networking, Storage and Analysis*. 2017.

[4] Allen, Benjamin S., et al. "Modernizing the HPC System Software Stack." arXiv preprint arXiv:2007.10290 (2020).

[5] Shabanov, Boris M., and Oleg I. Samovarov. "Building the software-defined data center." Programming and Computer Software 45.8 (2019): 458–466.

[6] Cima, Vojtěch, et al. "Hyperloom: a platform for defining and executing scientific pipelines in distributed environments." *Proceedings of the 9th Workshop and 7th Workshop on Parallel Programming and RunTime Management Techniques for Manycore Architectures and Design Tools and Architectures for Multicore Embedded Computing Platforms*. 2018.

[7] *COMP Superscalar, an interoperable programming framework, SoftwareX*, Volumes 3–4, December 2015, pp. 32–36, Badia, R. M., J. Conejero, C. Diaz, J. Ejarque, D. Lezzi, F. Lordan, C. Ramon-Cortes, and R. Sirvent, DOI: 10.1016/j.softx.2015.10.004

[8] Moloney, David, et al. "Myriad 2: Eye of the computational vision storm." *2014 IEEE Hot Chips 26 Symposium (HCS). IEEE*, 2014.

[9] Medina, Eitan, and Eran Dagan. "Habana labs purpose-built AI inference and training processor architectures: Scaling AI training systems using standard Ethernet with Gaudi processor." *IEEE Micro* 40.2 (2020): 17–24.

[10] Davies, Mike, et al. "Loihi: A neuromorphic manycore processor with on-chip learning." *IEEE Micro* 38.1 (2018): 82–99.

[11] DeBole, Michael V., et al. "TrueNorth: Accelerating from zero to 64 million neurons in 10 years." *Computer* 52.5 (2019): 20–29.

[12] Young, Aaron R., et al. "A review of spiking neuromorphic hardware communication systems." *IEEE Access* 7 (2019): 135606–135620.

[13] Bosshart, Pat, et al. "P4: Programming protocol-independent packet processors." *ACM SIGCOMM Computer Communication Review* 44.3 (2014): 87–95.

[14] Abel, Francois, et al. "An FPGA platform for hyperscalers." 2017 IEEE 25th Annual Symposium on High-Performance Interconnects (HOTI). IEEE, 2017.

[15] Amazon AWS: https://aws.amazon.com/ec2/instance-types/f1/ (2021). Accessed 19 April 2021.

[16] Colonnelli, Iacopo, et al. "StreamFlow: cross-breeding cloud with HPC." *IEEE Transactions on Emerging Topics in Computing* (2020).

[17] Lozano-Rizk, J. E., et al. "Software Defined Data Center for High Performance Computing Applications." International Conference on Supercomputing in Mexico. Springer, Cham, 2019.

[18] H2020 EVOLVE project – D5.5: Workflow resource allocation mechanisms and policies.

[19] H2020 EVOLVE project – First report on Workflow task scheduling mechanisms and policies.
[20] H2020 EVOLVE project – First Report on Data Locality and Load- balancing mechanisms and policies.
[21] Carnero, Javier, and Francisco Javier Nieto. "Running simulations in HPC and cloud resources by implementing enhanced TOSCA workflows." 2018 International Conference on High Performance Computing & Simulation (HPCS). IEEE, 2018.
[22] Carbonell, M.: xopera: an agile orchestrator. www.sodalite.eu/content/xopera-agile-orchestrator (2019). Accessed 19 April 2021.
[23] Aldinucci, Marco, et al. "HPC4AI: an AI-on-demand federated platform endeavour." *Proceedings of the 15th ACM International Conference on Computing Frontiers*. 2018.
[24] Melo, Carlos, et al. "Availability models for hyper-converged cloud computing infrastructures." 2018 Annual IEEE International Systems Conference (SysCon). IEEE, 2018.
[25] Koziris, Nectarios. "Fifty years of evolution in virtualization technologies: from the first IBM machines to modern hyperconverged infrastructures." *Proceedings of the 19th Panhellenic Conference on Informatics*. 2015.
[26] Leite, Rodrigo, Priscila Solís, and Eduardo Alchieri. "Performance Analysis of an Hyperconverged Infrastructure using Docker Containers and GlusterFS." *CLOSER*. 2019.
[27] Kirkpatrick, Keith. "Software-defined networking." *Communications of the ACM* 56.9 (2013): 16–19.
[28] Agilio CX SmartNICs. www.netronome.com/products/agilio-cx/, 2018.
[29] Liu, Ming, et al. "Offloading distributed applications onto smartnics using ipipe." *Proceedings of the ACM Special Interest Group on Data Communication*. 2019. 318–333.
[30] NITRO CARD Shalev, Leah, et al. "A Cloud-Optimized Transport Protocol for Elastic and Scalable HPC." *IEEE Micro* 40.6 (2020): 67–73.
[31] Jouppi, Norman P., et al. "In-datacenter performance analysis of a tensor processing unit." *Proceedings of the 44th annual international symposium on computer architecture*. 2017.
[32] Habana Labs, "GoyaTM Inference Platform White Paper," https://habana.ai/wp-content/uploads/pdf/habana_labs_goya_whitepaper.pdf
[33] Emani, Murali, et al. "Accelerating Scientific Applications with SambaNova Reconfigurable Dataflow Architecture." *Computing in Science & Engineering* 23.2 (2021): 114–119.
[34] Long, Yun, Xueyuan She, and Saibal Mukhopadhyay. "Design of reliable DNN accelerator with un-reliable ReRAM." *2019 Design, Automation & Test in Europe Conference & Exhibition (DATE). IEEE*, 2019.
[35] Firestone, Daniel, et al. "Azure accelerated networking: Smartnics in the public cloud." 15th USENIX Symposium on Networked Systems Design and Implementation (NSDI 18). 2018.
[36] Chiou, Derek. "The Microsoft catapult project." 2017 IEEE International Symposium on Workload Characterization (IISWC). IEEE Computer Society, 2017.

[37] Geng, Tong, et al. "FPDeep: Acceleration and load balancing of CNN training on FPGA clusters." 2018 IEEE 26th Annual International Symposium on Field-Programmable Custom Computing Machines (FCCM). IEEE, 2018.
[38] Nawrocki, Robert A., Richard M. Voyles, and Sean E. Shaheen. "A mini review of neuromorphic architectures and implementations." *IEEE Transactions on Electron Devices* 63.10 (2016): 3819–3829.
[39] K. Likharev, A. Mayr, I. Muckra, and O. Türel, "CrossNets–High- performance neuromorphic architectures for CMOL circuits," in *Molecular Electronics III*, vol. 1006, J. R. Reimers, C. A. Picconatto, J. C. Ellenbogen, and R. Shashidhar, eds. New York: New York Academy of Sciences, 2003, pp. 146–163.
[40] D. B. Strukov and K. K. Likharev, "Reconfigurable nano-crossbar architectures," in Nanoelectronics and Information Technology, R. Waser, ed., 3rd ed. New York: Wiley, 2012.
[41] J. Schemmel, D. Briiderle, A. Griibl, M. Hock, K. Meier, and S. Millner, "A wafer-scale neuromorphic hardware system for large-scale neural modeling," in *Proc. IEEE Int. Symp. Circuits Syst.*, May/June 2010, pp. 1947–1950. doi: 10.1109/ISCAS.2010.5536970.
[42] Sugiarto, Indar, et al. "Optimized task graph mapping on a many-core neuromorphic supercomputer." 2017 IEEE High Performance Extreme Computing Conference (HPEC). IEEE, 2017.
[43] Knight, James C., et al. "Large-scale simulations of plastic neural networks on neuromorphic hardware." *Frontiers in neuroanatomy* 10 (2016): 37.

4

Data System and Data Management in a Federation of HPC/Cloud Centers

Johannes Munke, Mohamad Hayek, Martin Golasowski, Rubén J. García-Hernández, Frédéric Donnat, Cédric Koch-Hofer, Philippe Couvee, Stephan Hachinger, and Jan Martinovič

CONTENTS

DOI: 10.1201/9781003176664-4

4.1 Introduction: Data Federation of European HPC/Cloud Centers

The Federation of European computing centers, for science and engineering to profit from the best-suited computing resources regardless of location, has been a hot topic for more than a decade. Many projects and initiatives have been trying to realize this by large-scale grid-computing infrastructures [1] like TeraGrid [2], WLCG [3], and EGI [4], to name but a few. Crucial to ongoing approaches (see, e.g. [4,5]), building on that work, are easy usability for multiple use cases and user-friendly authentication and authorization (e.g. [6,7]). To profit from the use of geographically distributed compute resources, it is necessary to access data in an efficient way independently of the location of the data store. To cover this demand, the EUropean DATa (EUDAT [5]) initiative offers several tools and services for collaborative and distributed data management.

This chapter describes our approach for the implementation of an EUDAT-based distributed data infrastructure (DDI) which is part of a data-driven computing platform enabling Large-Scale Execution for Industry and Society (LEXIS [8] – and see Chapter 2 in this volume). The platform executes orchestrated big data workflows on European high-performance-/cloud-computing resources in a multi-site setup. The LEXIS project is coordinated by the IT4Innovations National Supercomputing Center (IT4I, Ostrava, CZ) which also represents one compute/data site. The Leibniz Supercomputing Centre (LRZ, Garching b. München, DE) is the second founding partner of the computing and data federation, with both partners providing access to classical high-performance computing (HPC) and infrastructure-as-a-service-cloud (IaaS-cloud) resources. Both centers are representative (Tier 0/1) sites within the PRACE federation of European HPC centers. The next centers connecting to the platform are the European Center for Medium-Range Weather Forecasts (ECMWF, Reading, UK) and the Irish Centre for High-End Computing (ICHEC, Dublin, IE).

Besides the DDI, LEXIS offers advanced workflow orchestration (Bull/ATOS Ystia Orchestrator [9], employing TOSCA [10] and Alien4Cloud [11]), an accounting and billing system, and a portal for workflow and data control. The different access modalities of the compute resources are overcome by the use of the high-end application execution middleware (HEAppE [12]). The infrastructure planning and development is driven by the requirements of several pilot test cases: (1) the computational modeling of a data-intensive turbo-machinery and gearbox system in aeronautics; (2) real-time data processing and simulation of earthquakes and tsunamis; and (3) weather and climate models based on massive amounts of in situ data.

LEXIS draws advantage from its multisite architecture for computing and data handling. With respect to the DDI, obvious strong points are the redundancy and data safety aspects in a geo-distributed storage/mirroring scheme. As a core component and data middleware for the DDI, the Integrated Rule-Oriented Data System (iRODS [13]) and EUDAT-B2SAFE [14] are used and every site is represented by one iRODS zone. iRODS holds file-system-like and individual metadata for data sets in a metadata catalogue (iCAT), enabling FAIR (findable, accessible, interoperable, and reusable [15]) data management. Every zone relies on an own catalogue service provider (iCAT server), which in LEXIS is set up redundantly for eventually reaching high availability (HA). iRODS facilitates a unified view on the user's data from all sites, contributing to the unified LEXIS look and feel conveyed by the LEXIS authentication and authorization infrastructure (AAI), which provides a single sign-on (SSO) with Keycloak [16]. The DDI integrates with the other components of the LEXIS platform via several REST APIs (e.g. for data transfer). It is embedded into the European data landscape by using EUDAT tools and federates easily with further EUDAT/iRODS sites.

In Sections 4.2 and 4.3, we start out describing design requirements and middleware-based implementation of the LEXIS DDI. Section 4.4 discusses the hardware backend of the DDI, with emphasis on the conceptually new usage of buffering servers (burst buffers, data nodes) in order to accelerate data transfer and conversion. Section 4.5 describes the integration of the system with the LEXIS AAI, and Section 4.6 the REST-API-based integration with other functional units of the LEXIS platform. After elaborating on the

European (EUDAT) integration and FAIR data aspects of the DDI (Section 4.7), we conclude (Section 4.8).

4.2 Requirements on the LEXIS DDI

Within the LEXIS federation, users and orchestration systems must be able to retrieve data in a secure and efficient way independent of location. In the context of the LEXIS project, this idea resulted in the following detailed requirements to the LEXIS DDI for the design process.

4.2.1 Unified Data Access

Since user and workflow data in the distributed LEXIS platform will be used in a cross-site manner, it has to be uniformly accessible from everywhere. iRODS stores files (data objects) in folders (collections) and subfolders (subcollections) and thus provides a hierarchical structure comparable to a unified file system, with data at all sites integrated into it (see [13]).

4.2.2 Usage and Federation of Diverse Data Backend Systems

Federating different HPC/data centers means bringing storage resources of different technological nature into line. As LEXIS is aiming at a growing federation, the DDI must be versatile concerning different data backend systems. In this sense, iRODS is a good choice since it acts as a middleware and supports a variety of storage types. A description of the various hardware backend systems used at IT4I and LRZ can be found in Section 4.4.

4.2.3 Reliability and Redundancy

The LEXIS project sets up reliable services in terms of availability and data safety. High-Availability iRODS Systems (HAIRS) are described in [17] and [18]. Section 4.3 describes the LEXIS implementation of a redundant, geographically distributed iRODS system for the LEXIS DDI following this approach. This offers the possibility of data mirroring and therefore fault-tolerance concerning site-specific data loss.

4.2.4 AAI Support

For a cross-site unified user access, the DDI must be compatible with a unified, LEXIS-wide AAI system, which nowadays means supporting interfaces like OpenID Connect or SAML [19, 20]. Authentication with the Keycloak-based

LEXIS AAI is enabled using an adapted iRODS-OpenID plugin, and access rights on collections and data objects are then controlled via iRODS's built-in mechanisms (see Section 4.5).

4.2.5 APIs

The immersion of the LEXIS DDI within the LEXIS platform, that is, the connection to, for example, LEXIS Portal and LEXIS Orchestrator is implemented with several RESTful APIs. The already existing iRODS APIs and client toolkits are augmented with a set of custom-designed REST APIs specific to LEXIS (Section 4.6).

4.2.6 State-of-the-art Research Data Management

To facilitate an adequate and FAIR [15] research data management and integrate with the European data management landscape and standards, LEXIS mainly makes use of EUDAT Services, as described in Section 4.7. iRODS's capability to hold basic metadata of data sets, in combination with PID assignment by EUDAT's B2HANDLE [21] is a key component for FAIR data management.

4.3 Federation via a DDI Based on iRODS

LEXIS looked into different data management solutions that could fit our requirements described in Section 4.2. As a result, the iRODS middleware [13], also used by EUDAT [5], was finally chosen to be the core of the LEXIS data system. Data ingestion, movement, and retrieval within/from the DDI are exclusively performed via REST APIs described in Section 4.6 in order to ensure sanitized usage patterns and security (restriction of entry points) within the LEXIS ecosystem. In the following sections, we summarize some basics of iRODS and then discuss the deployment of iRODS at LRZ and IT4I with one redundantly set-up iRODS zone per center, and the federation of these zones.

4.3.1 Relevant Basic Properties of iRODS

As briefly outlined in the previous sections, iRODS uses backend file systems (or also object storage systems) to provide a unified file-system-like structure of data objects and collections (similar to files and folders). To this end, an iRODS zone – the smallest usable entity of an iRODS federation – runs one iRODS server (provider server) with metadata catalogue iCAT, and zero or more iRODS servers on machines with storage attached (consumer servers, connected to the provider). Several iRODS zones can be federated to form a large system. Beyond these federation capabilities, the middleware excels through a rule engine, running scripts on certain events (like file creation), and has the ability to store metadata with each data object or collection in an attribute–value–unit tuple (AVU – see [13]) store. Also, iRODS offers diverse client suites (command-line interface, Python bindings, etc.) which enable us to address it from our LEXIS-specific APIs (Section 4.6).

4.3.2 iRODS HA Setup

Although the iRODS zones in LEXIS are independent entities (one per computing/data center), unavailability of one of the zones could have strong consequences on workflows being executed. Such consequences can be a complete failure of the workflow, or a slower execution when the orchestrator has to get data from a different iRODS zone, far from the HPC/cloud resources used. To improve the reliability of each zone, we thus establish redundancy and a failover mechanism for the iCAT server and its database. The iRODS backend was deployed following the HAIRS concept [17,18] and the PostgreSQL database was deployed following the concept in [22], which is based on Pgpool-II [23] and repmgr [24]. Figure 4.1 depicts the setup which can reach high availability if required.

4.3.3 iRODS Zones Federation across Centers and Data Movement

An iRODS zone manages the physical storage resources at each center. It holds metadata on stored data, users, and their access rights. All this information is saved in the local iCAT database. LEXIS uses the iRODS federation mechanism to connect the iRODS zones of the centers. This allows users from a zone to access the data in a different zone while being authenticated to their home zone. Effectively, users thus have a unified access and view to data located at multiple centers. When data are only available in a remote zone, it is transparently acquired on access via the internal transfer mechanisms of iRODS. Thus, the iRODS federation covers core requirements on the LEXIS data system.

4.3.4 Storage Tiering and Underlying Data Storage

In the setup described in Section 4.3.2, the iRODS server and its iCAT database run in high-availability mode. However, if the physical storage resource is down or corrupted, data are either inaccessible or lost. While this can be avoided through geographical data mirroring via the Replication API (see Section 4.6), the iRODS storage tiering plugin [25] deployed in LEXIS provides flexible management of backend storage, including redundancy

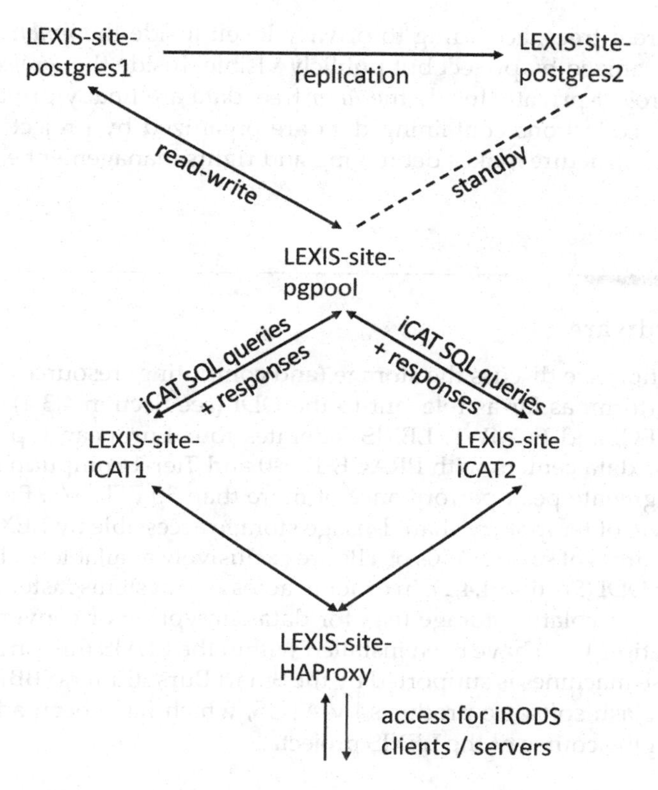

FIGURE 4.1
Redundant iCAT-PostgreSQL system following the HAIRS concept. iRODS clients and servers talk to a HAProxy which acts as a load balancer for iCAT1 and 2. The SQL queries are forwarded to the postgres1 database if it is available; otherwise failover to postgres2 is triggered. To maintain consistency, data are continuously replicated from postgres1 to 2 before failover.

when necessary. The plugin organizes multiple storage resources into so-called tiers. Once data are transferred to an iRODS zone, the data are stored automatically in the highest storage tier. The plugin sets an expiry time for the data at each tier and once this time is exceeded, the data are moved or replicated to the next tier.

4.3.5 Logical Structure of the DDI

Within the DDI, data are organized as so-called data sets (corresponding to an iRODS collection with data objects and possibly sub-collections in it). These data sets have a universally unique identifier (UUID) and can contain input and output data for a specific workflow. Internally, the DDI organizes data sets for each LEXIS computational project (see Section 4.1) and user in three different trees, according to privacy level: Inside the */public* tree, data sets are organized by project but publicly visible. Inside the */project* tree, the data are project-private. Inside the */user* tree, data are finally private to each user (here, collections containing data are organized by project and user). This logical structure makes debugging and rights management easier.

4.4 Hardware

In this section, we discuss the storage (and computing) resources behind the LEXIS platform, as far as relevant to the DDI (see Section 4.3.4). With IT4I, LRZ, ICHEC, and ECMWF, LEXIS federates four European top-level computing and data centers with PRACE Tier-0 and Tier-1 computing resources and an aggregate peak performance of more than 30 PFlop/s. Besides more than a PByte of temporary shared-usage storage accessible by LEXIS (Section 4.4.1), resources of several 100s of TBs are exclusively available and immersed within the DDI (Section 4.4.2). In order to access file systems faster and to provide very fast volatile storage (e.g. for data encryption or conversion), data nodes (Section 4.4.3) have been installed within the LEXIS infrastructure. The use of these machines is supported by the Smart Burst Buffer (SBB) and Smart Bunch of Flash software products by ATOS, which have been adapted and evolved in the course of the LEXIS project.

4.4.1 Storage Systems for HPC and Infrastructure-as-a-Service-Cloud Clusters

HPC cluster storage usually consists of several tiers, offering different economically viable combinations of space, speed, and reliability characteristics (where usually large space and high bandwidth anti-correlate with reliability and data safety). Parallel file systems such as LUSTRE [26] (at IT4I) or IBM Spectrum Scale (ex GPFS [27], at LRZ) provide the necessary I/O bandwidth for parallel applications, as they are typically executed on HPC clusters. For security reasons, access to these large (e.g. 1.4 PB on LRZ's Linux Cluster) storage systems from outside is restricted only to a handful of protocols such as SSH (i.e. SCP, SFTP) or GridFTP [28].

For applications which are not relying on distributed-memory parallelization and fast interconnects, but rather need configurability and large run time (or uptime in case of services), IaaS-clouds are an optimum environment. These environments ideally complement the HPC ecosystem, and allow for the execution of smaller applications (pre/postprocessing) and services. They can be extended by container-orchestration frameworks such as Kubernetes as well. Consequently, LEXIS immerses OpenStack- and VMWare-based

IaaS-cloud resources provided by IT4I and LRZ (with several thousand CPU cores in total), both backed by large CEPH [29] storage clusters (100 TB–1 PB, partially SSDs). The storage resources are usually addressed from within virtual machines.

4.4.2 Storage Systems Dedicated to LEXIS

IT4I's Cloud CEPH storage cluster provides 120 TB of raw HDD backed storage and 40 TB of raw SSD backed storage. Via the POSIX compatible file system CephFS, it is used as resource for IT4I's iRODS zone. In LRZ, iRODS uses a two-tier (see Section 4.3.4) backend, with the high (fast) tier being a 50 GB partition of LRZ's Data Science Storage (DSS), based on IBM Spectrum Scale and accessible via NFS. The lower tier (with slower network connection) is a 150 TB LRZ-LEXIS Experimental Storage system (legacy IBM DS3500, being replaced and extended to 300 TB), which also hosts the LEXIS Weather and Climate Data API (WCDA).

4.4.3 HPC–Cloud-Storage Interconnect and Data Node/Burst Buffer Concept

In LEXIS, practically all systems at each site are interconnected with 10 Gbit Ethernet or better (with exception of the LRZ-LEXIS Experimental Storage). This supports the LEXIS idea of running mixed data-driven workflows using HPC and IaaS-cloud computing infrastructures, demonstrating and driving the convergence of computing paradigms. To further optimize data flows and implement in-memory data encryption/decryption, (de)compression, augmentation, or preprocessing tasks, data nodes – that is, servers with TBytes of NVMe-SSD storage and Intel Optane DC NVDIMMs – were acquired. The characteristics of the systems at IT4I and LRZ are given in Table 4.1. They can

TABLE 4.1

Characteristics of data node/burst buffer servers at IT4I and LRZ

Characteristic	IT4I node 1	IT4I node 2	LRZ node 1	LRZ node 2
CPU	2 Intel Skylake Xeon Gold 6230 (2x20 cores)	Intel Skylake Xeon Gold 6230 (2x20 cores)	Intel Skylake Xeon Gold 6230 (2x16 cores)	Intel Skylake Xeon Platinum 8260M (2x24 cores)
RAM	192 GB	192 GB	384 GB	384 GB
NVMe Capacity	12.8 TB	12.8 TB	12.8 TB	—
Optane DC NVDIMM capacity	512 GB	512 GB	1.5 TB	3.0 TB
Accelerator card	NVIDIA Quadro RTX 6000	FPGA card Bittware 520N with Intel Stratix 10	NVIDIA V100	NVIDIA V100

be used as bare-metal machines for data-intensive tasks, to host a hypervisor for virtual machines, or to run software of the project partner ATOS and be used as SBB or Smart Bunch of Flash hosts. These two concepts are set out below.

4.4.3.1 SBF (Smart Bunch of Flash)

The Smart Bunch of Flash component of the ATOS Flash Accelerator software creates a persistent NVMe volume, spread over all NVMe devices, and an XFS file system in it, which is exported to computing servers using the NVMeOF protocol [30]. The allocation of these volumes can be automated via SLURM or managed via OpenStack Cinder and the LEXIS orchestration system. These fast volumes are ideal to accelerate I/O-bound tasks (simple compression, conversion, encryption) when the pure network speed to Data Node servers is superior to the write rate of (largely HDD-based) file systems. Thus, SBF volumes will, for example, be used as a backend storage for the compression/encryption API in LEXIS (see Section 4.6.5).

4.4.3.2 SBB

In a parallel I/O, HPC setting, data nodes can be used for buffering input and output between the compute nodes of a cluster and its parallel file system (see Section 4.4.1). As illustrated in Figure 4.2, the SBB component of the ATOS Flash Accelerator suite is designed to implement a transparent cache for a parallel file system such as Lustre, GPFS, or CephFS, in such use cases. To this end, it provides:

- a client-side library intercepting I/O calls to glibc and forwarding them to a *sbbd* daemon on the server side; this library (see Figure 4.2, left part) is engaged with a simple LD_PRELOAD, and works without modifying the application – at least if linked dynamically – and without root access; and
- a server-side *sbbd* daemon (Figure 4.2, lower middle part) in charge of processing the intercepted I/O calls, managing the cache, and finally asynchronously destaging the cached data to the parallel file system (Figure 4.2, right part).

Likewise, before processing, a prefetching of data into the data node can be implemented to make data available to an application (Smart Prefetch, Figure 4.2, upper middle part). As already mentioned, within the data flow the data node can be used for pre/postprocessing tasks as well.

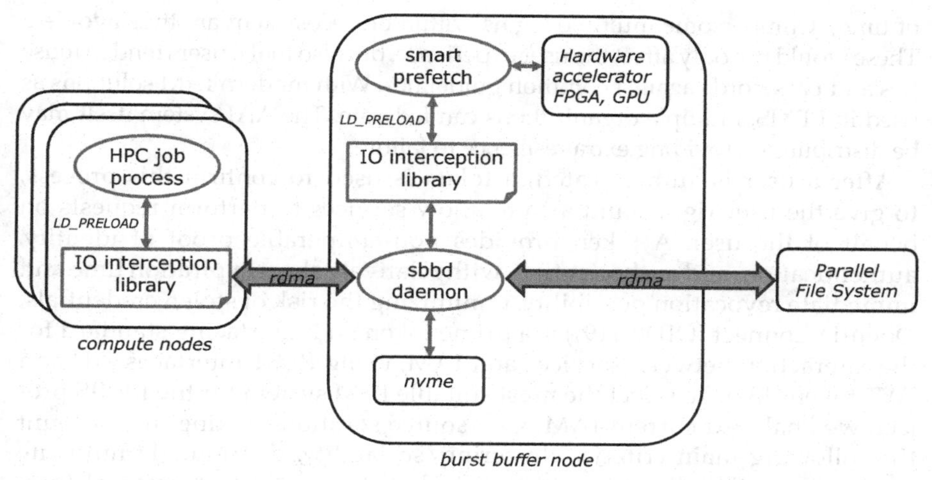

FIGURE 4.2
ATOS SBB operation concept (for an explanation of the key architectural components (I/O interception library, sbbd, prefetch mechanism), refer to the text).

4.5 Unified Access to the Platform Based on an AAI

In multiuser IT services such as the LEXIS platform and DDI, authentication of users and identity management is crucial to the service provider. The topic of authentication, authorization, and access control interplays with legal aspects (e.g. privacy, data security, and responsibilities after a hacking attempt), fundamentals of accounting (who to bill for services rendered), and informational aspects (e.g. user contact data) in a complex way. A verified identity allows the platform to identify permission classes (roles) for the users, granting them certain access rights and enabling their actions. Distributed data and computing infrastructures [1,2,3,4] have long promoted an identity management based on X.509 certificates, certification, and registration authorities. Recent approaches prefer less bureaucratic solutions based on, for example, OpenID Connect [19] or SAML [20] tokens and SSO.

4.5.1 LEXIS Identity and Access Management (IAM) Solution, SSO, and AAI

In this spirit, a LEXIS identity management and SSO system (LEXIS authorization and authentication infrastructure (AAI)) grants access to all the platforms' geographically distributed services with one login action (for a certain period of time). Cumbersome multiple logins within our ecosystem are thus avoided. These would not only affect the user experience, but also make users tend to reuse IDs and passwords against common guidelines. With modern IAM solutions as used in LEXIS, multiple organizations can federate. The AAI system itself may be distributed, providing extra resilience to failure.

After a user is authenticated, a token is used to confirm this process, to give the user rights, and also to allow services to perform requests on behalf of the user. A token provides non-tamperable proof of identity, authentication, and authorization, with relatively short expiration time and immediate revocation possibility, minimizing the risk of stolen credentials. OpenID Connect (OIDC [19]) is a prime token and interfacing standard for the interaction between services and IAM, using REST interfaces [31] and JWT tokens [32]. To select the most suitable IAM system for the LEXIS project, we analyzed current IAM open source solutions, taking into account the following main criteria: clustering/scalability, distributed/multi-site setup capabilities, disaster recovery/backup, authentication protocol, integration capabilities, and functionalities such as user/group management with role- or attribute-based access control (RBAC/ABAC). Finally, we decided to use the product Keycloak [16] which RedHat also employs in its SSO solution.

4.5.2 Platform Services vs. AAI: Separation of Concerns

The clear separation of Keycloak-based AAI and other LEXIS services (where communication happens via defined APIs) follows the security principle: "Do one thing and do it well." The user has only one identity and different access possibilities (roles) on different systems. This also enables proper auditing, allowing traceability, which is particularly important from a security perspective.

4.5.3 LEXIS DDI and IAM/AAI System

When building the LEXIS DDI, the EUDAT-iRODS system was heavily customized with an adapted plugin in order to provide proper support of Keycloak's OpenID Connect tokens. Thus, access, data privacy and data sharing permissions can be conveniently and centrally controlled. The iRODS data management software provides its own authentication and authorization solution (see [13]), and controls access with an extension to the Unix permission model (i.e. via users and groups, and permissions to own, read, and modify data in access control lists). While it allows authentication to be delegated, the iRODS user-creation process has still to be separately executed for each user in each zone's iCAT. LEXIS users are mapped to iRODS users, while LEXIS projects (see Section 4.1) are mapped to iRODS groups. When a user is created and added to a project, in each iRODS zone of LEXIS the corresponding iRODS user is created and added to the relevant groups, and directories for their private data within the projects are set up. This is at the moment a semi-manual process (with automated progress monitoring), handled via ticket creation in a LEXIS trouble-ticket system and administrators triggering appropriate API endpoints (Section 4.6.4).

When authenticating via OpenID Connect, there are limitations to the length of the OpenID tokens that iRODS will accept (approximately 1 kB, which is an issue with Keycloak's JWT tokens). We worked around this by modifying the iRODS-OpenID plugin architecture, such that tokens are prevalidated and only a token hash transverses the iRODS core [33]. This workaround, eventually to be superseded by an upstream solution within the iRODS framework, requires changes to the client programs (e.g. iRODS Python client).

Similar challenges have arisen when connecting the LEXIS orchestration system to the LEXIS AAI. Due to the usage of OpenID Connect and SAML in different orchestrator components (see Chapter 5), the LEXIS Keycloak system had to be configured to support SAML as well.

4.6 Data Management via APIs

To facilitate automated, controlled, and secure data handling in LEXIS, one central design principle of the LEXIS platform is that data handling, manipulation, discovery, upload, and download are always initiated via REST API calls. These calls can be initiated by users via the LEXIS web portal (including a data set-management interface), or by the LEXIS Orchestration System within a workflow. The portal or orchestrator authenticates on behalf of the user to the DDI using tokens from the LEXIS AAI. In order to provide all needed functionality, APIs for metadata-based search, (meta)data upload and download (Section 4.6.1), data staging and replication (Sections 4.6.2, 4.6.3), helper functions (Section 4.6.4) and finally data (de)compression and encryption/decryption (Section 4.6.5) have been programmed and deployed. Details on these APIs and their endpoints are laid out below. The LEXIS project intends to publish current versions of API documentation and public code on its github page [34] and/or zenodo community page [35].

4.6.1 Data Search, Upload, and Download APIs

These APIs with their group of endpoints (Table 4.2) allow for basic data management operations. Data sets can be searched for substrings according to name or metadata fields (Table 4.2, lower rows). In the backend, this search uses custom Python scripts executed on the metadata of the DDI, which is retrieved with the iRODS Python client.

TABLE 4.2

Data search, upload/download REST API endpoints

Endpoint	Method	Request body[1]	Response body	Comments
/staging/download	POST	source_system, source_path	<file contents>	Download from staging.
/dataset/download	POST	internalID, access, project, push_method, compress_method, path	<file contents>	Download from iRODS.
/dataset	POST	file, name, internalID, access, project, push_method, compress_method, path, metadata	internalID	Create or update a data set or sub-data set.
/dataset	DELETE	internalID, access, project, path	—	Delete data set or sub-data set.
/dataset/search/metadata	POST	<Object with any metadata>	<list of {location, metadata, eudat}>	Return metadata of matching data sets.
/dataset/search/metadata	DELETE	<Object with any metadata>	—	Delete matching data sets.
/dataset/listing	POST	internalID, access, project, path, recursive	name, type, size, create_type, checksum, contents	Contents provides a list with the metadata (name, type, …) for each file in a directory.

[1] push_method may be "empty" (empty data set), "directupload" (json-encoded in file parameter), "tus" (pre-uploaded using TUS protocol, with path stored in file parameter). compress_method: "file" or "zip." access: "user," "project," "public."

Further endpoints are offered to create, upload, download, and delete data sets, and to modify these data sets by adding and removing files. In order to make uploads and downloads convenient and feasible, file decompression/compression during upload/download (i.e. upload/download of zipped files, with the contents appearing in the DDI) is supported on user request, as well as resumable downloads using the TUS protocol [36]. In some cases, the API makes use of the staging API (Section 4.6.2) as a backend. Compression/decompression functionalities are being expanded by a compression/encryption API (see Section 4.6.5). For extremely large data sets, the LEXIS portal offers the users an upload/download link based on GridFTP/B2STAGE (see Section 4.7.3).

The API group contains further endpoints which allow platform users to add, remove, and modify additional information (metadata) regarding their data sets, as a basis for FAIR research data management. This information may be authorship (creator, contributor), publication (publisher, year), copyright (license), type (e.g. audiovisual, data set, image, interactive resource, model, service, software, sound, text, workflow), identifiers (DOIs, PIDs), among others. We use standard fields defined by DataCite [37] (schema: [38]), and we allow users to define their own additional metadata and metadata schemas, performing validation.

4.6.2 Staging API

Running workflows in the LEXIS federated environment, the orchestrator has to trigger data movement between different computing and data systems across different centers. Thus, input data are made available for computing tasks, and output data are safely stored back in the DDI for later reuse. The orchestrator performs these transfers (and necessary deletion of files) through the staging REST API (endpoints: see Table 4.3).

By default (as, e.g. in the API set discussed in Section 4.6.1), REST APIs are synchronous, which means the API call blocks until the task performed is either completed successfully or has failed. However, large data transfer between the different storage facilities requires time and the API must not wait until the transfer is completed. To allow asynchronous calls to the API, a distributed data scheduler connected to a message broker was thus introduced below the REST API. When a staging API call occurs, a data

TABLE 4.3

Staging REST API endpoints

Endpoint	Method	Request body	Response body	Comments
/stage	POST	source_system, source_path, target_system, target_path, metadata, job_id(optional), task_id(optional)	requestID	Stage data between different data sources.
/stage/<request_id>	GET	—	status, target_path	Returns the status of the data transfer.
/delete	DELETE	target_system, target_path, job_id(optional), task_id(optional)	—	Delete a data set.
/delete/<request_id>	GET	—	status	Returns the status of data deletion.

transfer or deletion task is pushed to the queue and the endpoint returns an ID for the orchestrator to track the status of the operation through helper endpoints (see Table 4.3).

Django [39] is used as REST API framework due to its robustness and scalability. Celery [40] was chosen as a task queue and RabbitMQ [41] as a message broker. The resulting status of transfers is saved in a PostgreSQL database.

4.6.3 Replication and PID Assignment API

Data transfer within the DDI, that is, between different iRODS zones in the federation, is handled differently than staging API transfers from or to external source/target systems. Data replication between iRODS zones in the context of LEXIS is needed to ensure data safety and provides flexibility for the orchestrator to choose the data source closest to the cloud or HPC resources.

To this end, we developed a replication API, based on EUDAT B2HANDLE [21] and B2SAFE [14]. As described in Section 4.7, B2SAFE provides a mechanism to replicate data across iRODS zones and keeps track of replica locations by using persistent identifiers (PIDs) assigned by B2HANDLE. The replication API calls B2SAFE rules to trigger data replication between two different iRODS zones. The implementation, similar to the staging API, allows for asynchronous execution.

To adhere to FAIR [15] data principles, PIDs should be assigned to data whether the data are replicated or not. PIDs allow long-term identification of data and thus serve a crucial role in data discovery. The replication API thus offers endpoints that trigger B2HANDLE to assign PIDs and check the assignment status.

4.6.4 Helper APIs

Several helper APIs have been programmed to address mostly administrative problems, some examples of which are mentioned here. A SSHFS API helps to export data sets via SSH to virtual machines in LRZ's IaaS Compute Cloud; and endpoints in the upload/download API (Section 4.6.1) allow for the permission management of newly created users and projects within the DDI.

Direct access to the DDI via Globus/GridFTP/B2STAGE for transferring large data sets eventually uses GridFTP for data access (see Section 4.6.1), which authenticates using certificates. A Gridmap File API allows users to map and unmap their certificate to their LEXIS/DDI user when needed. Also, we set up an API returning the sizes of LEXIS data sets, which users or the LEXIS orchestrator can use to judge whether data staging is a good option or computations should rather take place on the site where the data are stored.

4.6.5 Compression/Decompression/Encryption/Decryption API

Due to corporate usage, critical data have to be stored encrypted on the LEXIS DDI, that is, on upload or staging to the DDI, encryption must take place, with decryption being triggered on download/reverse staging (e.g. to HPC systems). In order to offer this functionality, and also compression/decompression of data sets to accelerate staging of many small files, an API is being programmed with endpoints for data encryption/decryption and (de)compression. To actually execute these processes within a workflow, the data are first staged (with the staging API) on NVMe or NVDIMM volumes of the LEXIS data nodes (Section 4.4.3). Then, by calls to the compression/decompression/encryption/decryption API, encryption/decryption and (de)compression is triggered and executed in a speed-optimized manner before the data are sent on. This follows and expands on the ideas behind the compression/decompression functions in our upload/download API endpoints.

4.7 Integration with EUDAT Services

The EUDAT Collaborative Data Infrastructure (CDI) [5] is the result of a growing group of organizations providing European-scale data management services. LEXIS aims at integrating with these services. Out of the large EUDAT service portfolio, B2HANDLE [21] was chosen for PID assignment, B2SAFE [14] for data safety, and B2STAGE [42] for data staging between iRODS and HPC-cluster file systems. The usage of further services in LEXIS is being investigated, envisaging – for example – a connection to B2FIND [43], the metadata-based data discovery portal of EUDAT. Below, we go through the EUDAT services LEXIS works with in some more detail.

4.7.1 EUDAT B2HANDLE

B2HANDLE [21] is EUDAT's PID service. By assigning a unique identifier, B2HANDLE helps in long-term identification of data and helps achieving the findability in a FAIR [15] data management system.

The B2HANDLE client is deployed as a Python library on all iRODS machines in LEXIS. B2HANDLE is based on the Handle system [44] and provides an interface to access it. Correspondingly, instances of the handle server system were deployed at LRZ and IT4I. The two instances are assigned to the same handle prefix and with one another through a built-in mechanism. EUDAT offers a reverse lookup servlet [45] that enables searching against the local B2HANDLE instance. The servlet shortens the time needed to check whether a data set has already been assigned a PID, which is a prerequisite for using B2SAFE, B2STAGE, and B2FIND with that set.

In LEXIS, iRODS collections in the federated zones are assigned PIDs. B2HANDLE assigns specific iRODS metadata to these collections, reflecting the PID assignment. The collections are then findable through their identifier.

4.7.2 EUDAT B2SAFE

B2SAFE is EUDAT's core service that ensures data safety. It provides a mechanism to replicate data between different iRODS zones deployed at different centers. B2SAFE uses core iRODS rules and provides a set of high-level iRODS rules. It also takes advantage of B2HANDLE to maintain information about the location of the replicated data. Once replication is triggered, a PID is assigned to the parent data in case of absence, and a different PID is assigned to the replicated data. B2SAFE writes iRODS metadata reflecting the replication process, such as the data set PID and the parent PID for each replicated data set. In LEXIS, B2SAFE is used as a backend to the Replication REST API (Section 4.6.3).

4.7.3 EUDAT B2STAGE

B2STAGE is EUDAT's high-performance data transfer service. It provides the ability to move data between a federated iRODS-B2SAFE infrastructure and HPC resources. B2STAGE deploys a GridFTP server on top of iRODS, allowing any client supporting GridFTP to address it. Thus, big data sets can reliably be staged to and from the LEXIS DDI using B2STAGE. On the one hand, this makes B2STAGE an ideal backend to the Staging REST API for at least part of the transfers; on the other hand, controlled data upload/download possibilities via B2STAGE can be offered to the LEXIS user for large data sets.

4.8 Conclusion

In this chapter, we presented the LEXIS DDI, based on the iRODS and EUDAT services. We laid out its significance as a data backend for the LEXIS platform, a forefront infrastructure federating European top-level HPC and cloud systems and making these landscapes converge.

The hardware backend of the DDI consists of innovative systems like CEPH, IBM Spectrum Scale and burst buffers relying on the latest Intel Optane technology. iRODS and EUDAT-B2SAFE fulfil the basic requirements to serve as a data system for LEXIS, and make for a solid federation between the data centers within the LEXIS platform. The real uniqueness of the LEXIS

DDI with respect to similar data infrastructures is however in its richness in modern interfaces.

It connects to the LEXIS ecosystem with a number of well-documented HTTP-REST APIs according to modern standards. The data search API with its rich metadata search possibilities stands out as well as the staging API, allowing for convenient, asynchronously handled data transfer requests. The APIs make interfacing and automation an easy task, enabling the LEXIS orchestration system to execute workflows in a way that optimizes data transfer times.

Seen from the outside, the LEXIS DDI uses the EUDAT kit of tools and services in order to immerse itself with pan-European data infrastructures. LEXIS data are equipped with handles from EUDAT-B2HANDLE, assigned metadata to follow FAIR standards, and replicated using EUDAT-B2SAFE. On this basis, federation with more centers and EUDAT-driven data systems has been efficiently accomplished. As a next step, we plan to expose public LEXIS data via the B2FIND data search/discovery portal.

The open and modern architecture shows its strength not only in the federation context, but also when industrial applications are executed on the LEXIS platform. The DDI has an appropriate access-rights management, interfaces to meet industry standards, and is being equipped with encryption capabilities for extra security. Service quality is ensured via the LEXIS trouble-ticket system, where also service requests (e.g. for user creation) are handled. With all these characteristics, the LEXIS DDI not only helps to accomplish the LEXIS mission of converging the HPC, cloud, and big data worlds, but also drives collaboration of industry, SMEs, and scientific research in the computing and data science sectors.

Acknowledgment

This work and all contributing authors are funded/cofunded by the EU's Horizon 2020 Research and Innovation Programme (2014–2020) under grant agreement No. 825532 (Project LEXIS – Large-scale EXecution for Industry and Society).

References

[1] Foster, I. and C. Kesselman, ed. 2004. *The Grid 2: Blueprint for a New Computing Infrastructure*. San Francisco: Morgan Kaufmann. https://doi.org/10.1016/B978-1-55860-933-4.X5000-7.

[2] Pennington, R. 2002. Terascale Clusters and the TeraGrid. In *Proceedings of the 6th International Conference on High-Performance Computing in Asia-Pacific Region (HPC Asia 2002)*, Bangalore.

[3] Shiers, J. 2007. The Worldwide LHC Computing Grid (Worldwide LCG). *Computer Physics Communications* 177, no. 1–2: 219–223. https://doi.org/10.1016/j.cpc.2007.02.021.

[4] Kranzlmüller, D., J. M. de Lucas, and P. Öster. 2010. The European Grid Initiative (EGI). In *Remote Instrumentation and Virtual Laboratories*, ed. F. Davoli, N. Meyer, R. Pugliese, and S. Zappatore, 61–66. Boston: Springer. https://doi.org/10.1007/978-1-4419-5597-5_6.

[5] EUDAT Ltd. 2020. EUDAT – Collaborative Data Infrastructure. www.eudat.eu (accessed Apr. 6, 2021).

[6] Solagna, P. 2014. EGI position paper for a European identity federation for researchers. https://documents.egi.eu/document/2049 (accessed Apr. 6, 2021).

[7] EUDAT Ltd. 2020. B2ACCESS – EUDAT. www.eudat.eu/services/b2access (accessed April 6, 2021).

[8] LEXIS consortium. 2020. LEXIS Project – High Performance Computing (HPC) in Europe. https://lexis-project.eu (accessed Apr. 6, 2021).

[9] Bull Atos. 2021. Ystia Suite. https://ystia.github.io (accessed Apr. 6, 2021).

[10] OASIS Open. 2013. Topology and Orchestration Specification for Cloud Applications Version 1.0 – OASIS Standard. http://docs.oasis-open.org/tosca/TOSCA/v1.0/os/TOSCA-v1.0-os.html (accessed Apr. 27, 2020).

[11] Bull Atos. 2021. Alien 4 Cloud. http://alien4cloud.github.io/ (accessed Apr. 6, 2021).

[12] Svatoň, V., J. Martinovič, J. Křenek, T. Esch, and P. Tomančák. 2019. HPCas-a-Service via HEAppE Platform. In *Proceedings of the 13th International Conference on Complex, Intelligent, and Software Intensive Systems (CISIS-2019)*, Sydney. https://doi.org/10.1007/978-3-030-22354-0_26.

[13] Xu, H., T. Russell and J. Coposky, et al. 2017. *iRODS Primer 2: Integrated Rule-Oriented Data System*. Williston: Morgan & Claypool. https://doi.org/10.2200/S00760ED1V01Y201702ICR057.

[14] EUDAT Ltd. 2020. B2SAFE – EUDAT. www.eudat.eu/services/b2safe (accessed Apr. 13, 2020).

[15] Wilkinson, M. D., M. Dumontier and I. J. Aalbersberg, et al. 2019. The FAIR Guiding Principles for scientific data management and stewardship. *Scientific Data* 3: 160018. https://doi.org/10.1038/sdata.2016.18.

[16] JBoss (Red Hat, Inc.). 2021. Keycloak. www.keycloak.org (accessed Apr. 6, 2021).

[17] Kawai Y. and A. Hasan. 2010. High-Availability iRODS System (HAIRS). In *Proceedings of the iRODS User Group Meeting 2010: Policy-Based Data Management, Sharing and Preservation*, Chapel Hill. ISBN: 978-1-452813-42–44, https://irods.org/uploads/2010/Kawai-HAIRS-paper.pdf.

[18] James, J. 2015. Configuring iRODS for High Availability. https://irods.org/2015/07/configuring-irods-for-high-availability (accessed Mar. 17, 2021).

[19] Sakimura, N., J. Bradley, M. B. Jones, B. de Medeiros, and C. Mortimore. 2014. OpenID Connect Core 1.0 incorporating errata set 1. https://openid.net/specs/openid-connect-core-1_0.html (accessed Nov. 6, 2020).

[20] Cantor, S., J. Kemp, R. Philpott, and E. Maler. 2005. Assertions and Protocols for the OASIS Security Assertion Markup Language (SAML) V2.0. https://docs.oasis-open.org/security/saml/v2.0/saml-core-2.0-os.pdf (accessed Nov. 6, 2020).

[21] EUDAT Ltd. 2020. B2HANDLE – EUDAT. www.eudat.eu/services/b2handle (accessed Apr. 13, 2021).

[22] Depuydt, J. 2015. Jensd's I/O buffer – Setup a redundant PostgreSQL database with repmgr and pgpool. http://jensd.be/591/linux/setup-a-redundant-postgresql-database-with-repmgr-and-pgpool (accessed Oct. 1, 2019).

[23] SRA OSS, Inc. 2020. pgpool Wiki. www.pgpool.net/mediawiki/index.php/Main_Page (accessed Apr. 13, 2020).

[24] 2ndQuadrant Ltd. 2020. repmgr – Replication Manager for PostgreSQL clusters. https://repmgr.org (accessed Apr. 13, 2020).

[25] Russell T. 2017. SC17 Demo: Storage Tiering. https://irods.org/2017/12/sc17-demo-storage-tiering (accessed Dec. 1, 2019).

[26] Schwan P. 2003. Lustre: Building a File System for 1,000-node Clusters. In *Proceedings of the Linux Symposium (OLS 2003)*, Ottawa. www.kernel.org/doc/ols/2003/ols2003-pages-380-386.pdf.

[27] Schmuck, F. and R. Haskin. 2002. GPFS:A Shared-Disk File System for Large Computing Clusters. In *Proceedings of the Conference on File and Storage Technologies (FAST '02) – USENIX Association*, Monterey. www.usenix.org/legacy/events/fast02/full_papers/schmuck/schmuck.pdf.

[28] Allcock, W., J. Bresnahan, R. Kettimuthu and M. Link. 2005. The Globus Striped GridFTP Framework and Server. In *SC '05: Proceedings of the 2005 ACM/IEEE Conference on Supercomputing*. Seattle. https://doi.org/10.1109/SC.2005.72.

[29] Weil S. A., S. A. Brandt, E. L. Miller, D. D. E. Long, and C. Maltzahn. 2006. Ceph: A Scalable, High-Performance Distributed File System. In *Proceedings of the 7th USENIX Symposium on Operating Systems Design and Implementation (OSDI '06)*. Seattle. ISBN: 1-931971-47-1, www.usenix.org/legacy/events/osdi06/tech/full_papers/weil/weil.pdf.

[30] NVM Express, Inc. 2016. NVM Express over Fabrics 1.0. https://nvmexpress.org/wp-content/uploads/NVMe_over_Fabrics_1_0_Gold_20160605-1.pdf (accessed Nov. 6, 2020).

[31] Richards R. 2006. Representational State Transfer (REST). In *Pro PHP XML and Web Services*, ed. R. Richards, 633–672. Berkeley: Apress. https://doi.org/10.1007/978-1-4302-0139-7_17.

[32] Jones, M., J. Bradley and N. Sakimura. 2015. JSON Web Token (JWT) – Internet Engineering Task Force (IETF). https://tools.ietf.org/html/rfc7519 (accessed Apr. 6, 2021).

[33] García-Hernández, R. J., M. Golasowski. 2020. Supporting Keycloak in iRODS systems with OpenID authentication. Presentation at *CS3 2020 – Workshop on Cloud Storage Synchronization and Sharing Services (27–29 January 2020)*, Copenhagen. https://indico.cern.ch/event/854707/contributions/3681126/.

[34] LEXIS consortium. 2021. GitHub – LEXIS: Large Scale Execution for Industry & Society. https://github.com/lexis-project (accessed Apr. 6, 2021).

[35] LEXIS consortium. 2020. Zenodo Community – LEXIS project. https://zenodo.org/communities/lexis (accessed Apr. 6, 2021).

[36] Transloadit-II GmbH. 2018. tus – Open Protocol for Resumable File Uploads. https://tus.io/ (accessed Apr. 6, 2021).
[37] Brase, J. 2009. DataCite – A Global Registration Agency for Research Data. *In Fourth International Conference on Cooperation and Promotion of Information Resources in Science and Technology*, Beijing. https://doi.org/10.1109/COINFO.2009.66.
[38] DataCite – International Data Citation Initiative e.V. 2021. DataCite Metadata Schema 4.4. https://schema.datacite.org (accessed Apr. 5, 2021).
[39] Django Software Foundation. 2020. Django. www.djangoproject.com (accessed Apr. 1, 2020).
[40] Celery Project. 2020. Celery: Distributed Task Queue. www.celeryproject.org (accessed Apr. 1, 2020).
[41] Pivotal Software. 2020. Messaging that just works – RabbitMQ. www.rabbitmq.com (accessed Apr. 1, 2020).
[42] EUDAT Ltd. 2020. B2STAGE – EUDAT. www.eudat.eu/services/b2stage (accessed Apr. 13, 2020).
[43] EUDAT Ltd. 2020. B2FIND – EUDAT. www.eudat.eu/services/b2find (accessed Apr. 13, 2020).
[44] Boesch, B., S. X. Sun and L. Lannom. 2003. RFC 3650 – Handle System – Internet Engineering Task Force (IETF). https://tools.ietf.org/html/rfc3650 (accessed Apr. 27, 2020).
[45] EUDAT. 2021. B2HANDLE-HandleReverseLookupServlet. https://github.com/EUDAT-B2SAFE/B2HANDLE-HRLS (accessed Feb. 23, 2021).

5

Distributed HPC Resources Orchestration for Supporting Large-Scale Workflow Execution

Alberto Scionti, Paolo Viviani, Giacomo Vitali, Chiara Vercellino, Olivier Terzo, Stephan Hachinger, Lukáš Vojáček, Václav Svatoň, Jan Křenek, Frédéric Donnat, and François Exertier

CONTENTS

5.1 Introduction

Modern scientific and industrial applications involve the execution of complex computations, often belonging to different domains (e.g. high-performance computing (HPC) simulations, ingestion of large (unstructured) datasets, performing machine learning tasks). To support the execution of all these different types of computations, over the past decades computing infrastructures have become larger and started to integrate ever more specialized resources. As such, hardware heterogeneity in the form of multiple types of specialized accelerators have

DOI: 10.1201/9781003176664-5

become common both in HPC and in cloud facilities. While traditional complex simulations (generally, they encompass the execution of very high numbers of processes with a high level of communication between them) find a better execution environment within HPC facilities, where supercomputing resources are tightly coupled by high-speed interconnect, other (in particular, serial or embarrassingly parallel) tasks with small exchange of information among them are better tailored for running on cloud facilities. For instance, input data ingestion and preprocessing, as well as visualization tasks are representative of the kind of operations that can be executed on a modest number of CPU cores, which can be easily provisioned via a dedicated virtual machine (VM) (i.e. the cloud approach). Cloud facilities also offer an easier integration of novel hardware accelerators (e.g. FPGAs, AI accelerators, dedicated ASICs) which are today less common in the HPC world. Therefore, the execution of modern scientific and industrial applications greatly benefits from access to the full spectrum of resource types. The momentum toward hybrid execution platforms combining HPC and cloud resources is thus increased. Besides processing tasks, applications face the demands of ingesting and moving large amounts of data within the execution platform. To better support the management of large datasets, infrastructures integrate advanced storage solutions, whose purpose is to avoid application performance being I/O bound. This makes it possible to leverage the processing speed of the computing nodes and the I/O bandwidth of the storage back-end in an equilibrated manner. Burst buffer (BB) nodes are one important example for advanced data system components providing access to a faster intermediate storage layer for computations. The large availability of heterogeneous and (often geographically) distributed resources requires a simple mechanism allowing end users to split their applications into execution blocks that can be easily mapped onto these resources.

Workflows are a convenient way of expressing an application in terms of steps that must be executed on a given set of resources to get the final result, and their dependencies from each other. Moreover, the execution of these steps may be restricted to specific types of resources, including dedicated accelerators. Often, workflows are expressed as directed acyclic graphs (DAGs), where nodes represent the tasks to perform and the edges represent the (data) dependencies among them. To ease the burden on scientists and designers, the workflow management system (WMS) was devised. WMS generally provide the capability to speed up the process of definition of a workflow and its deployment on a given execution platform, by automating and hiding most of the (technical) steps needed. The user can focus on the description of the application and all the requirements in terms of execution resources (e.g. some of the workflow steps may require access to a GPU or an AI-inference chip), leaving the WMS to actually select suitable resources from those available. When a workflow mixes operations that can be clearly associated with different domains (e.g. HPC and AI) and that may require both HPC and cloud resources, it can be referred to as a multi-domain workflow, or a *hybrid* workflow.

The growing adoption of heterogeneous hardware for specialized tasks (e.g. deep learning training) and the increasing size of datasets to be ingested and processed are pushing the demand for more computational resources. Moreover, some applications may require access to computing resources in response to trigger events and to provide a result in a limited amount of time (as it is the case for urgent computing (UC) applications). In this context, while data center federation has been proposed as a way of enlarging the pool of available resources to better support such use cases, the role of WMSs and orchestration tools is particularly complex. Indeed, a WMS has not only to provide a mechanism to let the user define the application workflow, but it has to properly manage the resource allocation process. Depending on different objectives that can be set (e.g. aiming at reducing the overall energy footprint, ensuring job replication to minimize SLA violations), different approaches have been proposed and integrated in available WMSs (see Section 5.6), ranging from simple greedy allocation strategies to more complex heuristics for solving an associated combinatorial optimization problem. On the other hand, HPC resources are historically managed through batch-scheduling systems, which are in charge of assigning a set of resources to each job following cluster-wide criteria of allocation priority. In this context the role of a WMS is to correctly define its requests to the batch scheduler to satisfy objectives defined at a higher level. On the other hand, other tools can be exploited to optimize the job execution at a lower level once the resources are allocated by the batch scheduler. CompSS [1][2] and the HyperLoom [3] are two examples of tools supporting such "low-level" optimization. They provide a programming model through which relationships among different computational tasks are expressed, while incorporating mechanisms for scheduling these tasks among the available resources.

As the role of WMSs becomes more critical due to heterogeneity and scale, more sophisticated architectures and inner workings are required compared to the past: resource allocation strategies are a well-studied research topic, and many of them have been proposed in the context of cloud resource management. More specifically, a cloud-like management approach is seen as helpful in those cases where the resources are mixed (HPC and cloud) and distributed geographically (data center federation). Such is the case of execution platform and e-infrastructures developed in the context of European-funded projects, such as the LEXIS platform.[1] This chapter describes the *orchestration service* designed and implemented in the context of the LEXIS project, in order to provide the capability of distributing computing tasks among the computing resources available in a "light federation" of data centers. In the LEXIS case, HPC resources provisioned through a batch scheduler (PBS and SLURM) are available on the one hand, and cloud resources (managed by OpenStack) on the other hand. The orchestration service provides all the features needed by the users to easily define their (hybrid) workflows (see Chapters 6–8 for detailed description of the LEXIS use cases), and to define and express the required execution resources for running them. In this context, the security aspect needs to be taken into account, through the interaction with a proper security layer (i.e. the LEXIS authentication and authorization infrastructure (AAI)); similarly, the management of data represents a critical part in the overall workflow management, especially in the case where datasets may reside in a different location with regard to the location where it is foreseen to perform the computations. To this end, the LEXIS orchestration service directly interfaces with a data management layer, that is, the LEXIS distributed data infrastructure (DDI). Beyond basic storage and data management functionality, the DDI offers, for example, replication of datasets among different locations, and encryption or compression of datasets when required (see Chapter 4 for more detail).

In order to improve the performance of the user's applications, dynamicity emerged as a primary concern in the design of the LEXIS orchestration service. Through the capability of the orchestration service of dynamically meeting the application demands, that is, automatically allocating resources required by a certain task in the executed workflow, the service is able to optimize many objectives, such as resource utilization in each federated data center. Thus, it minimizes the overall costs of using resources (i.e. number of assigned credits for the cloud computing (CC) resources and core hours for the HPC ones). The time passing before a certain task is started can be minimized as well (e.g. in the case of UC workflows). Although not discussed in this chapter, many more objectives can be taken into account and optimized, such as the overall energy efficiency of the execution, by optimally exploiting the availability of heterogeneous hardware. This dynamic resource allocation works in a closed feedback loop, where execution infrastructures are continuously monitored, collecting data along with planned maintenance periods, and then combining them to generate a ranking of the possible locations for the tasks' execution. Every time the WMS execution engine finds a task that requires such dynamic resource allocation, the monitoring data is used to select the best location based on the ranking.

The remainder of the chapter is organized as follows. Section 5.2 provides a general overview of federated execution platforms, focusing on the LEXIS architecture. It also describes the AAI layer, which is of primary importance. Section 5.3 provides a detailed description of the LEXIS orchestration service architecture. It illustrates all the key components forming the service and introduces the dynamic resource allocation and the data management strategies implemented. Section 5.4 summarizes the mechanisms implemented for managing data flows within the application workflows, and clearly links to Chapter 4. Section 5.5 briefly illustrates how the presented orchestration service applies to three use cases (see Chapters 6–8 for detailed description). Section 5.6 concludes the chapter by summarizing the most relevant works and solutions.

5.2 Federated Execution Platforms

Matching the demands of hybrid workflows requires managing the distribution of the workload among different pools of resources, often taking into account the specificity of particular hardware. To better fulfill these requirements of modern scientific and industrial applications, resource pools are "virtually" aggregated from those made available by different data centers. In this regard, data center federation provides a convenient way of scaling the application needs, without impacting on the way an application works. One of the key elements of this kind of federated platform is the orchestration system. It has to manage the entire process of deploying and executing the application on the selected pool of resources. As such, it must deal with the selection of a proper set of resources (computing, memory, networking, and storage) capable of running all the application components. The execution must be monitored and all events of interest (errors, application component outputs, etc.) must be logged. Since modern applications often require dedicated hardware accelerators (e.g. GPU, FPGA), the orchestration system has to manage also the distribution of processing tasks on this type of resources, which are generally paired to dedicated frameworks and tools. To this end, generally it needs to interface with lower-level management systems, thus resulting in a flexible and modular solution. The LEXIS solution implementing the concepts laid out above is graphically depicted in Figure 5.1; it is the foundation of the LEXIS execution platform for federating IT4Innovations National Supercomputing Center (IT4I), Leibniz Supercomputing Centre (LRZ), and the Irish Centre for High-End Computing (ICHEC) resources, although it has been designed to easily include additional computing/data centers.

In this context, a successful federation of HPC centers requires to federate not only the orchestration system, but also other building blocks of the LEXIS platform – in particular the LEXIS AAI and the data management layer (i.e. the LEXIS data distributed infrastructure (DDI)). This implies the presence of a federated identity and access management (IAM) system, where the solution chosen for LEXIS provides a user-friendly single sign-on (SSO), enabling convenient usage of all platform components. Instead of implementing an entire solution from scratch, in LEXIS, the AAI is based on the Keycloak[2] IAM open-source solution, which also provides features for being deployed in a cross-data center replication mode (see Figure 5.2 for an illustration of the replication schema). Then, all the other core components of the LEXIS platform, including the orchestrator and the platform portal, interact and rely on this AAI for authenticating and authorizing users, processes, and infrastructural devices that need to execute tasks or jobs or interact with the data management layer (LEXIS DDI).

Secure communication between data centers is achieved through a dedicated site-to-site virtual private network (VPN) in LEXIS, which allows

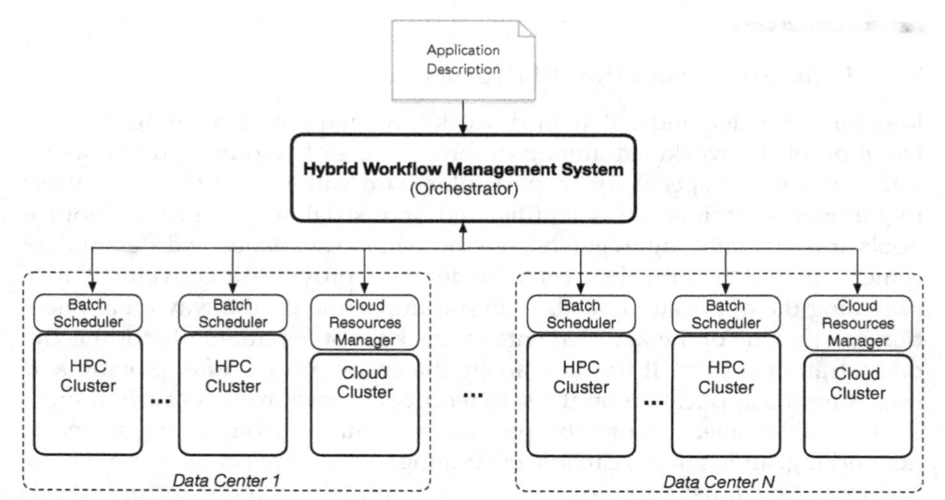

FIGURE 5.1
The relation between the orchestration system of hybrid application workflows and the underlying cloud and HPC heterogeneous resources, along with their specific management systems.

the IAM cluster (based on Keycloak) with necessary back-end components (in particular, a Galera/MariaDB[3] database and an Infinispan[4] cache cluster) to synchronize. The AAI layer described here allows, among other things, to put in place the "no-trust" security policy, whereby none of the components of the federated platform blindly rely on authentication and authorization information passed by other components, but instead validates the information against the AAI system. Finally, access to the computing and storage resources is granted through a simplified role-based access control (RBAC) model, which allows users to manage concepts such as users, organizations, and computational project contexts. This model is reflected in the Keycloak configuration. This model allows users to manage multiple security protocols such as OpenID Connect and SAML. Thus, each component of the LEXIS platform can work with suitable tokens. Keycloak OpenID Connect tokens follow a JWT standard, which represents an open method based on RFC-7519 for representing and exchanging security claims among different parties.

As described in Section 5.3, the High-End Application Execution Middleware (HEAppE [4]) provides secure remote access to HPC infrastructure while encapsulating any HPC-related functionality for the LEXIS platform's users and still following HPC center's internal security policies and ISO processes. For this purpose, independent HEAppE instances are deployed at each center.

As a consequence of the application of the "no-trust" principle, the LEXIS infrastructure monitoring system, a crucial element for dynamic orchestration policies, also relies on the LEXIS AAI layer.

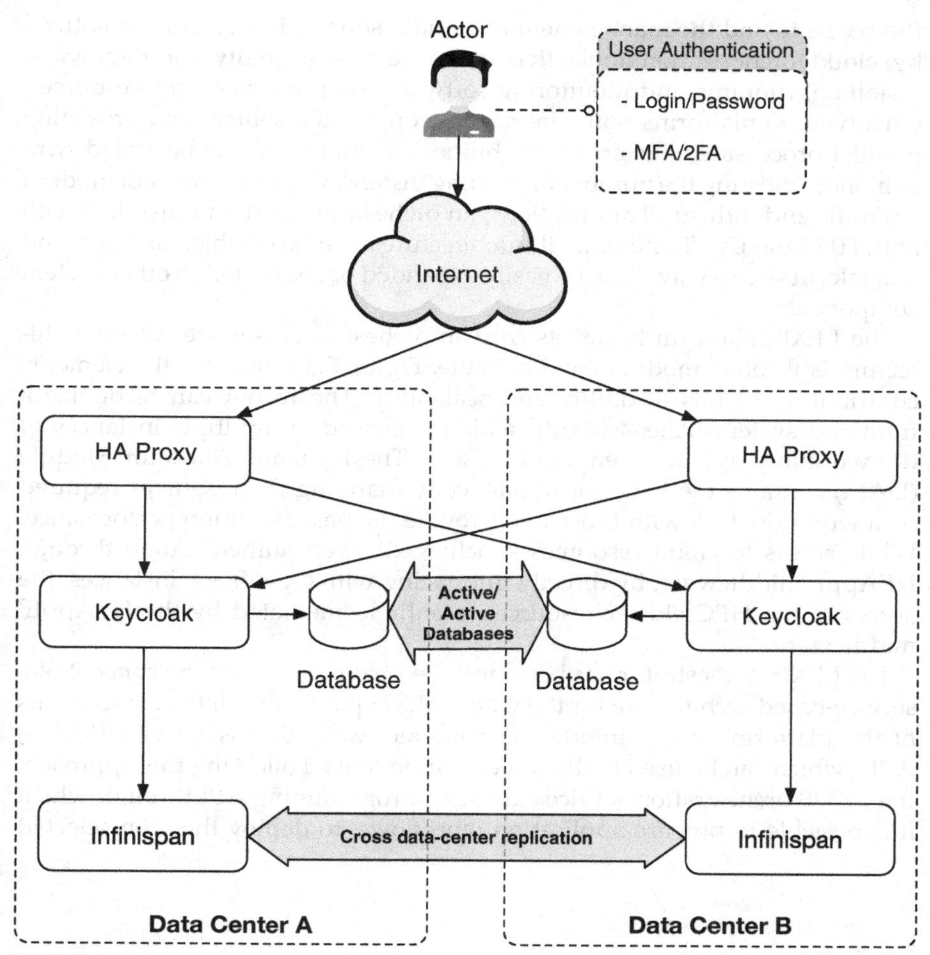

FIGURE 5.2
The high-availability configuration of the IAM system.

5.3 WMSs and Implementation in LEXIS

A WMS allows users to describe their applications in terms of DAGs to execute on a given set of resources to achieve results. When the computing resources are those of an HPC cluster, the management of the workflow life cycle is more complicated, since the WMS limits its responsibility to the correct allocation of the required resources (in a shared-usage context, often with high average cluster load). In this case, the user must take care of the necessary steps to compile the software components, to load them on the machine and to describe how to launch their execution. On the other hand, the access to and the management of cloud resources is largely automatized by cloud orchestration tools that can take responsibility for deploying, installing, running, and monitoring software components on the resources. On advanced platforms, software running on cloud resources and providing parallel processing/clustering capabilities can automatically be scaled with demand, adjusting the number of running instances. In the context of modern scientific and industrial applications, an orchestrator system must deal with both HPC and CC. To this end, its architecture must be flexible, modular, and scalable, in such a way it can be easily expanded or connected to other system components.

The LEXIS platform has at its core an orchestration service whose architecture is flexible, modular, and scalable. Figure 5.3 illustrates the elements contributing to this flexibility and scalability. The reader can recognize a front-end system (Alien4Cloud)[5] which is backed by multiple instances of the workflow execution engine (YORC).[6] The Dynamic Allocator Module (DAM) contains the logic for dynamically mapping the resources required by a workflow task with those that provide the best execution performance. While access to cloud resources is achieved, upon authentication through HEAppE middleware, by directly interfacing with OpenStack instances, the access to the HPC cluster resources is entirely mediated by the HEAppE[7] middleware.

The LEXIS orchestrator has been designed having in mind the concept of a service-based architecture for the whole LEXIS platform. All the components of the platform expose interfaces to interact with, that is, service (REST-) APIs, which can be used by the other components. Following this approach, the LEXIS orchestration service exposes a programming API through which it is possible to prepare application workflows, to deploy them on selected

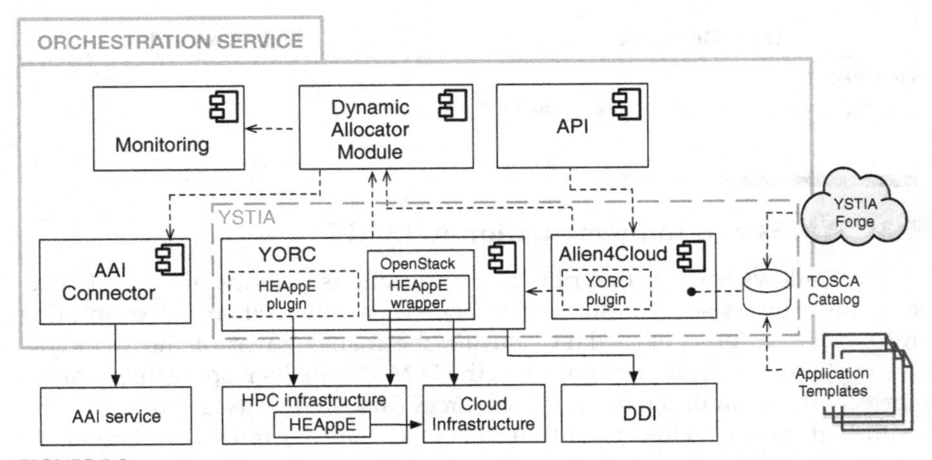

FIGURE 5.3
The architecture of the LEXIS Orchestration Service.

resources and to monitor their execution state. The core block of the LEXIS orchestration service is represented by the Alien4Cloud front-end (A4C) and the workflow execution engine (YORC). Both are available as main components of the YSTIA orchestration suite.

A4C is an open-source tool that allows users to pick up specific workflow components from a catalog and easily compose them into their application workflows. Within A4C it is possible to represents both the workflow itself with its jobs and their software components (e.g. frameworks to be installed, libraries, VMs, containers), as well as the infrastructural elements required by these jobs (e.g. a computing node with a given amount of memory and number of CPU cores). The components can also expose some capability as well as minimum requirements that need to be fulfilled in order to be used. Furthermore, the catalog can be expanded with new components and also created ad hoc to interact with specific infrastructural elements. The catalog per se can be forged from the one publicly available (YSTIA Forge)[8] and it is also used to store predefined *application templates*. The latter are parametrized applications defined in terms of catalog components, which then can be instantiated on a given set of resources. A4C offers both a graphical interface through which the user can create the application template, and a programmatic interface where the user directly defines a TOSCA blueprint (i.e. a representation of the template using an XML or YAML-based language). A4C automatically translates the graphical representation into an equivalent TOSCA blueprint, which can then be deployed on selected resources. To this end, the front-end uses the concept of locations, that is, the pool of resources belonging to a cloud cluster or an HPC cluster. During the deployment phase the front-end module matches the requested resources with those available in the selected location(s). It also generates the execution plan which contains a set of step sequences related to the operations required to install, execute, cancel, and uninstall all the software components on the location resources. Such step sequences resemble DAGs where nodes are the basic operations to be performed and edges connect consecutive operations. A4C provides all the necessary mechanisms to define the step sequences with the user interaction as required, and to associate them to an application template. The selection of the location(s) can be done statically (i.e. preselection by the user) or dynamically.

In the latter case, the DAM module of the LEXIS orchestration service automatically infers which resources best satisfy a set of evaluation criteria (see Section 5.3.1). These criteria are expressed as objectives to be minimized (costs) or maximized (rewards); a set of monitored metrics are periodically collected and combined as an input to support the cost/reward evaluation.

The orchestration engine (YORC) is responsible for carrying out the execution plans, by interacting with the lower levels of the infrastructure (on the selected location(s)) and proxies to the infrastructure (e.g. the HEAppE middleware). For this reason, it has a modular architecture: different plugins can be created and attached to it. By default, YORC comes with the capability of targeting cloud solutions such as Kubernetes and OpenStack, as well as public cloud infrastructures (e.g. Amazon AWS). Within LEXIS, its capability of interacting with computing infrastructures has been further extended by an appropriate YORC-HEAppE plugin. HEAppE provides an additional secure layer on top of the HPC clusters, so this modular architecture has been exploited to create the HEAppE extension. In LEXIS, YORC moreover relies on the DAM module to dynamically select the best set of execution locations; this provides resilience in situations where a defined location for execution becomes unavailable (e.g. owing to unforeseen outage of the computing resources). Both A4C and YORC support a deployment in a high-availability mode, to further minimize the situations where the orchestration service becomes unavailable or unresponsive.

The HEAppE middleware is IT4Innovations's in-house implementation of the HPC-as-a-service concept. HPC as a service can be understood as a popular paradigm for providing simplified access to HPC infrastructures without the need to buy and manage physical servers. HEAppE is able to submit required computational jobs on HPC infrastructure(s), transfer data, and monitor progress. It provides necessary functions for job management, monitoring and reporting, user authentication and authorization, file transfer, encryption, and various notification mechanisms.

HEAppE performs the mapping of LEXIS users to functional (non-privileged) accounts for the HPC and OpenStack infrastructures in each center. Due to this security-critical functionality, it is always deployed in an HPC center's private network. The HEAppE middleware only enables users to run a prepared set of so-called *command templates*. Each command template practically includes an arbitrary script or executable file that will be executed on the cluster, dependencies, or third-party software it might require, and the job processing queue/partition that should be used (determining, e.g. which nodes of an HPC cluster are used).

The latest version of HEAppE middleware extends HPC job support in a way that enables running more than one task within a computational job on a given cluster, with a possibility to define dependencies for an individual task. This update also contains a support for so-called long-term running jobs. This means that if the requested time allocation for a computational job exceeds the maximum wall-clock time of a selected queue, then the job is divided into a number of smaller ones with an automatically added dependency between these smaller jobs. Additionally, the definition of a computational job allows users to specify job arrays (as they are defined by batch schedulers, allowing the user to submit a bunch of jobs whose execution is parametrized) or parallelization parameters for MPI and OpenMP.

Concerning the management of datasets, YORC relies on the DDI layer, which provides mechanisms to properly handle data storage and movement, as well as other operations such as encryption and compression on the storage layer. Similar to the LEXIS orchestration service and the LEXIS AAI, the LEXIS DDI also exposes simple REST-APIs which are addressed by

YORC via CopyFromJob and CopyToJob TOSCA components, introduced in the scope of LEXIS (see Chapter 4 for more detail on the LEXIS approach to data management; Section 5.4 briefly summarizes the main policies that are taken into account by the LEXIS orchestration service to fulfill application requirements).

5.3.1 Dynamic Workflow Orchestration

Workflow orchestration can be leveraged to optimize different goals; an interesting one is the efficient use of resources, which translates into energy efficiency and shorter time to get results (from job submission to completion).

The LEXIS orchestration service targets this problem through a specific module – the DAM – to which the orchestrator delegates the decision on job placement. YSTIA, in fact, was originally designed to allow such delegation to a third-party module, however once the placement returned, it only implemented a static allocation strategy. A dynamic orchestration plugin has been developed for YORC, in order to define the job placement during the application workflow execution. Therefore, dynamic resource allocation is introduced through both the YORC dynamic orchestration plugin and the delegation of specific features to external modules/services, which in the case of the DAM as an extension module.

The DAM was developed from scratch during the project duration and represents one of the innovations of the LEXIS platform. While static allocation fulfills most of the requirements for typical applications running on public clouds (e.g. Amazon AWS, Microsoft Azure, Google Cloud), in the case of demanding scientific and industrial hybrid workflows this is not enough. For instance, UC tasks including HPC simulations may need to execute as fast as possible in response to a trigger event. Then, the system needs to select an HPC cluster with enough resources available at that specific point in time.

The DAM has been built to expose a REST-API which serves as a communication point with the orchestrator front-end/back-end for asynchronous job placement requests and for querying the result. After receiving a (YSTIA) job placement request, the module gathers monitoring information relevant to the job's resource type (for every center which is part of the federated platform), using the HEAppE API (when HPC resources are concerned), and the OpenStack API (in case of cloud resources). This information is then used to obtain a selected set of metrics (see Section 5.3.2) from which an overall score for each possible job placement location is computed and stored in memory. At this stage, only the location(s) matching the job's qualitative and quantitative requirements are considered further, discarding "incompatible" locations (e.g. systems with no GPU node available when the job requires it). Finally, the result is returned to YSTIA, following a *greedy* logic, that is, the location(s) with the highest score(s). Also, the module stores the decision in an internal database to keep track of historical information. In some use cases, replicas of the same job must be sent to more than one location, and, if so, this must be defined in the initial request. One of these special cases appears in UC in LEXIS, where replicas of the same job must be sent to several compatible locations in order to offset the risk of job failure and minimize the job's time to completion, by requiring the completion of just one of those replica jobs for the workflow's continuation. Therefore, for such cases, the DAM returns all the locations that passed the compatibility check. In this case, the application workflow also follows a well-known design pattern where a controlling task is created with the purpose of monitoring the completion status of all the replica jobs. It is responsible for triggering the next task(s) in the workflow once the first replica job completes, and for passing all the required input values to the next task(s).

Taking into consideration all of the module's necessary features, the best-suited technology for the DAM was a Python-based solution built on top of Flask libraries, connected to an InfluxDB[9] instance used as the internal database. This is a database solution tailored for storing time series, for example, complying with the fact that our logged information on the score of all computing locations is organized as a sequence of time samples. The InfluxDB solution is also used to conveniently store infrastructural information typically changing over time (e.g. the connection speed and latency among different locations, which matters if datasets need to be moved/copied).

The first step for the development of such a dynamic orchestration solution is the definition of score metrics, which is the subject of the next section.

5.3.2 Resource Management Metrics

In the LEXIS platform context, the relevant metrics for job allocation can be divided into two categories: technological metrics and cost-related metrics. The former is related to the job's requirements, the destination hardware and software capabilities, while the latter is highly dependent on the specific accounting and billing logic of the platform. In a prospective LEXIS business model, each user may have access to HPC and cloud resources according to credits, which translate to different granted usage times or intensities of the various computing and data systems.

LEXIS workflow blocks can be divided into HPC and cloud jobs. Both jobs have infrastructural constraints in the number of required CPU cores, RAM available per CPU core, or access to hardware accelerators (e.g. GPUs and/or FPGAs for accelerating specific computations). In order to select the most suitable set of resources to run the jobs – that is, the location(s) – the DAM has to: (1) combine different metrics collected at the infrastructural level; and (2) discard those locations that do not fulfill the job requirements. As such, metrics of interest are collected through internal platform monitoring tools, or also exposed by HEAppE middleware and the OpenStack management system. In this regard, the DAM does not have visibility on the resource usage at the CPU-core granularity; instead, it has the information on the node level usage, that is, the percentage of occupied nodes of a specific HPC queue.

Then, the requirement in terms of number of CPU cores is easily converted into the number of needed nodes, since all the nodes of the same HPC queue have a homogeneous composition. Based on that, the DAM evaluates:

- *location occupancy*: the ratio between the number of nodes already in use (by other jobs) and the total number of nodes on the given cluster (i.e. the location); and
- *job weight*: the ratio between the number of nodes that the job requires and maximum number of nodes available on the given cluster (i.e. the location).

Moreover, resource allocation is strongly influenced by the location and availability of input data for a given task, taking into account the possible need for moving/replicating this data among sites to identify an optimal allocation strategy. Specifically, it is defined as the *data transferring weight* as the fraction of a (generously) estimated time needed for computation (maximum wall-clock time set up for the job) over the sum of this time and the estimated time required for transferring data from the source location. By considering this weight, the DAM can choose to allocate resources on locations where the transferring time has a minimal impact with respect to the compute time. Dealing with cloud and HPC resources provided by federated HPC data centers can (depending on policies and computing-time grants of the user) require balancing their usage. Also, not all the HPC clusters expose equally up-to-date hardware resources. To cope with this heterogeneity (which appears in similar ways for systems with cloud operation), the orchestration system can, for example, take care of balancing core hours[10] used on clusters of different HPC centers. To take into account differences in performance from one system to another, core hours are normalized by using a correction factor for LEXIS accounting (for instance, the core hours on a newly acquired HPC cluster correspond to K times those of an older one).

All technological and cost-related metrics are scaled in a range between 0 and 1, and then linearly combined together in order to generate the final rank for a given location. Finally, a greedy algorithm just selects the location(s) with the highest rank as the destination for a particular job.

5.4 Workflow Data Management

As mentioned in Section 5.1, unlike traditional HPC applications, modern ones combine large-scale simulations with generating, assembling, moving, and analyzing large amounts of data, or other types of domain-specific tasks (e.g. machine learning tasks). In this context, the underlying orchestrator must be able to apply a data-oriented workflow management strategy. This kind of strategy, as implemented in the LEXIS orchestration service, automatically handles and moves data within hybrid workflows, as typical for LEXIS use cases (see Section 5.5 and Chapters 6–8 for a deeper analysis). Typical big-data characteristics, referred to as the five *V*s [5][6], are closely connected to the requirements also in LEXIS. Hybrid workflows may need to manipulate large *volumes* of data (in the range of terabytes (TBs) to petabytes (PBs)), which are generated or need to be moved and processed at a fast pace (*velocity*). Often such data is presented to the processing platform in an unstructured format (*variety*) and represents an important added *value*. Clearly, within and around LEXIS workflows, correctness and accuracy of data (*veracity*) is a central aspect. In this context, the LEXIS orchestration service provides mechanisms and policies to ensure efficient access to the data. The infrastructure layer exposes the core components that allow the orchestration service to "efficiently" manage data movement and manipulation (e.g. compression, encryption): (1) a unified, federated DDI with a common view on LEXIS data, based on iRODS[11] and EUDAT's European research data service;[12] (2) specialized and local storage libraries; and (3) staging and buffering resources, with a focus on BB I/O nodes, which allow for I/O prefetching and buffering operations with improved performance. In terms of policies, a few simple ones are fundamental to maintain the efficiency in data access. First, automation and restricted direct access to the data resources allows the orchestration service to keep the full control of the data transfers and avoid the need for manual user actions within workflows. The latter are restricted to providing input and reading output data via well-defined endpoints (mostly the LEXIS portal, and endpoints for uploading/downloading large datasets). The DDI and all related APIs ensure that data are made available to those locations where the processing will take place. Every time the DAM has to select the execution location(s), data availability is considered as one of the selection metrics, as discussed in Section 5.3.2; on the other hand, whenever data are not available in the considered location(s), then the estimated time to transfer them is considered. Of course, the orchestrator policy privileges those location(s) that provide faster access to the data. BB usage is considered in those cases where data staging, prefetching, and buffering are high priority, in order to not make computing processes stall. Last but not least, the orchestration service applies a resilience strategy in which failing data transfers from/to specific location(s) are worked around by selecting alternative ones.

5.5 LEXIS Pilot Use Cases and Orchestration

The LEXIS project comes with "LEXIS Pilot" application use cases spanning different scientific and industrial fields (weather and climate forecasting, earthquake and tsunami simulations, aeronautics). Since they involve many stringent requirements (both in terms of data management and processing), they are representative examples of modern hybrid workflows. Thus, they are at the basis of the design and implementation of all the LEXIS orchestration service features.

The aeronautics use case comprises the simulation of complex aeronautic components (turbo-machinery systems and gearboxes), requiring access to computing resources for a long time. To deal with this, the LEXIS orchestration service allows to manage long-running jobs (whose running time far exceeds the maximum wall-clock time for batch scheduler queues) via chained restart. It also has the capability to periodically check-point the application status. GPU-enhanced resources (on HPC and cloud) can be used both to boost some specific operations, in this case some part of simulations and remote visualization of the progressive results.

In the LEXIS pilot associated with earthquakes and tsunamis, simulations of tsunami on a given geographical area must be performed in response to a trigger event. The critical point here is that the LEXIS platform has to provide resources to complete UC tasks as fast as possible. As such, the application workflow should be structured in such a way that replicas of the UC task are launched on different clusters and possibly on different computing centers, thus enabling the workflow to proceed in the execution once the first replica completes. The earthquake and tsunami use case also leverages on accessing OpenBuildingMap and exposure databases to generate damage maps and shake-maps, areas of interest, loss assessments (both coarse and detailed), and a tsunami inundation map. Moreover, this use case presents tasks that are more suitable for running on HPC resources (i.e. tsunami simulations) and others that can be conveniently executed on cloud (i.e. shake-map generation and loss assessment).

Finally, the weather and climate use case provides particular requirements associated with management of external datasets (through a dedicated REST-API), and requires a reinforced strategy for distributing tasks among HPC and cloud resources. Complex simulations are performed running the Weather Research and Forecasting model [7], which can scale up to a few thousand cores. Data transfers are optimized by the orchestrator by performing multisite allocation of the different cloud and HPC tasks when possible. Here, BBs can be leveraged to prefetch input data or caching them.

5.6 Related Works

Tools for managing workflows' life cycle have been well established for a long time, although they have been tailored for specific domains (HPC, cloud) or for managing resources across distributed sites. On one hand, in the HPC domain, workflows are used to abstract all the steps associated with complex analysis and to support their execution on resources generally backed by batch schedulers. In this case, user expectations are mostly restricted to the performance aspect, rather on availability of the service and costs. On the other hand, the cloud domain makes broad use of WMSs to handle users' applications by automating operations on the underlying infrastructure as much as possible. In this case, service costs and availability become the driving factors. Although these domains remained quite separate for a long time, the recent need for accommodating rich applications involving, for example, HPC and high-performance (big-)data analytics (HPDA) components has brought about changes. It makes the world of WMS open up to HPC systems, and address the demands of all convergent domains mentioned. Also, the capability of scheduling the workload on, possibly geographically, distributed pools of resources is becoming an essential part of modern WMSs.

WMSs provide functionalities at two main levels: (1) the application/user level; (2) the infrastructural/platform level. In the former, the system has to provide mechanisms for allowing the user to express and formalize computational tasks composing the application, along with execution and data dependencies. In the latter, the system must apply specific policies to locate units of computation on the most suitable resources to fulfill all infrastructural and business requirements (e.g. for HPC applications, a minimization of overall execution time; in case of cloud applications a cost reduction and avoidance of resources being underused/overused).

The grid computing (GC) paradigm has been the basis of many large-scale computing projects, where execution resources are distributed over a large geographic area (in some cases, they were collected at world scale). In such a paradigm, the WMS dynamically creates a pool of execution resources through a *pull* mechanism, by which the single resources communicate to the management system their availability on performing computations. Then, the management system assigns workflow units of computation to them, based on cost–performance metrics (e.g. the relative performance of the resource, the cost to transfer input data). This pull mechanism has been used to implement several GC WMSs; the BOINC worldwide platform and its related projects (e.g. SETI@home, Rosetta@home) [8][9] are famous examples of such a design choice. Whenever such resources are made available on a volunteer basis, one important feature of GC-WMS is the implementation of fault-tolerant mechanisms [10] to ensure that the entire workflow execution can be completed even when some of the resources becomes unavailable. Another well-established example of GC-WMS is represented by the DIRAC project [11][12][13]. It provides the workflow management facility in the context of the CERN LHCb experiments. Here, dedicated management agents running at each resource site dynamically communicate their availability to perform specific computations to a central management system. The WMS then builds upon a central core service that provides a mechanism for submitting and storing new job requests at the user level (jobs are organized in different queues). It also implements a way of retrieving new jobs by distributed agents. Such agents run closely to the execution resources and are responsible to constantly check when they become free, and thus to retrieve new jobs from the central service. Like the DIRAC project, the CERN ATLAS experiment implemented the ATLAS PanDA WMS [14][15] to allow scientists to perform data analysis and simulations using distributed resources, also including supercomputing facilities. Although these systems have been designed to address the challenge of executing complex workflows on a various array of resources including also HPC and cloud [42], their flexibility is far from what a modern platform needs to provide. For instance, the workflow description is done via text-based files and does not conform to any specific standard [16]. Furthermore, the chosen model often does not support a flexible composition of workflows (e.g. making a new workflow from other workflows or parts thereof executed after one another), thus further limiting the usability of the solution.

The CC paradigm provides an execution environment where the resources (computing, memory, storage, networking) are dynamically acquired for the time needed to complete the assigned workflow's unit of computation, and then released. The resource provisioning is automated, so that the user can acquire and release the resources without the intervention of external actors. Such flexibility is made possible by large use of abstractions and virtualization technologies. Indeed, in a CC environment the user generally gains control over a virtual resource (VM or a container) rather than a physical one. This model is quite flexible to accommodate several application patterns, and with the growing adoption of specialized systems (e.g. GPUs, FPGAs), is becoming attractive also for HPC users. Unlike the GC paradigm, in the CC world resources tend to be aggregated in large-scale data centers, while the software management stack ensures fault tolerance and data replication among different data centers. Among the driving factors in CC, usage costs, overall energy consumption, and maximization of the usage of resources are the most relevant. WMSs generally consider these factors during the resource allocation phase. To this end, several works in the literature or on the market propose various approaches for deciding on where to run VMs or containers, based on different "optimality" criteria. In [17], the authors surveyed the main approaches to resource scheduling in the cloud domain. To mention just a few, Prodan et al. [18] proposed a bargaining-based approach in which market-based negotiation takes place between resource manager and the scheduler using self-limitation and aggressiveness; Iyer et al. [19] proposed a method based on suggested pricing resource scheduling algorithms; in [20], the authors used a cost-based approach, where a simple first-come-first-served scheduling technique was applied. Other attempts used other types of heuristics, such as genetic algorithms and other nature inspired algorithms [21][22][23] to schedule jobs based on a cost evaluation function, or transformed the scheduling/allocation problem into a combinatorial (NP-hard) optimization problem [24][25].

Although VMs offer a flexible and secure environment for the execution of computations, they introduce a large overhead due to the full hardware–software stack abstraction. To reduce performance losses, accelerators in some cases can bypass the virtualization stack and offer visibility of the hardware to the virtualized applications. Lighter virtualization approaches have existed for a long time (Solaris Zones [26], FreeBSD Jails [27] to mention but a few), and recently gained extremely strong popularity thanks to more flexible management frameworks. In this regard, Linux containers started to become the main alternative to traditional VMs. Docker[13] is a popular Linux container system; thanks to its flexibility, ease of use, and performance it quickly entered cloud-based production systems. To better support HPC-specific requirements and restrictions imposed by accessing supercomputers, the Singularity[14] project was created (which shares larger parts of the technology with Docker). Whatever container technology is used, it can help to close the gap between virtualization (as main feature of CC solutions) and performance requirements (main expectation in the HPC domain). When a large number of containers have to be deployed, different management systems can be used, for example, Kubernetes [28], Docker Compose and Docker Swarm [29], and Singularity Compose [30] to mention a few. They can also ease the distribution and control of computational tasks implemented within a large number of containers. In [31], authors implemented a management system for distributing computations across federated sites through the deployment of Docker containers. The system was implemented by directly interfacing with the Docker engines (i.e. the agents responsible for controlling containers on each specific computing node) at different sites. Although users were provided with a graphical interface and a simple CLI to control the execution of their software, the exposed features were mainly restricted to launching or cancelling single containers. When moving to workflows mixing HPC processing with artificial intelligence (AI) and data analytics jobs (hybrid workflows), containers provide large flexibility for ad hoc virtualization (even accessing hardware accelerators), while accessing to supercomputing resources can be done through traditional batch schedulers (e.g. PBS, Slurm, HTCondor). To this end, WMSs that are able to manage such mixed back-ends are of interest. Examples of such systems are the ATOS Croupier solution [32] [33][34], and StreamFlow [35]. Croupier has been developed following the concept of "meta-orchestration," that is, an orchestrator system that deals with lower-level orchestration and WMS tools. In this regard, it has the capability of managing a hybrid workflow by driving both cloud-like management systems (e.g. Kubernetes, Mesos [36]) and batch schedulers (SLURM, Torque). Like the solution proposed by the LEXIS project, it naturally connects to OpenStack for direct cloud resource provisioning (VMs), and works with containers (Singularity). Croupier is based on Cloudify,[15] a framework for operation automation in cloud environments (private, hybrid, or public); as such it exposes a TOSCA-based domain-specific language (DSL) which allows describing the workflow and controlling its execution. StreamFlow has been proposed as a system supporting the description and execution of hybrid workflows: besides the formalization of an application as a workflow using a DSL (in this case the common workflow language (CWL)), the specific sets of resources used by different pieces of the workflow are described as skeletons. A skeleton also describes how resources should interact, so that it can be easily deployed on selected (portions of) clusters, be they virtualized or available as bare metal resources. On the other hand, a solution like Croupier lacks the facilities to handle secure federation of resources and the complex authentication and authorization scenarios that are handled by the LEXIS solution. Specifically, Croupier also lacks a user-friendly interface that is, instead, provided by the LEXIS solution in the form of a GUI connected to the user portal.

For a comprehensive perspective on scientific workflow definition and the related management systems, readers can refer to a recent survey on this topic [38]. Besides WMSs, several tools have been created to support users in the phase of workflow definition. Some examples are Apache Taverna along with the SCUFL2 language [39] (designed to support the composition of remote services), Apache Airavata [40], and Triquetrum [41], to mention a few.

5.7 Conclusion

This chapter presented the LEXIS approach to workflow management and orchestration, advocating a holistic approach to scientific and industrial workflow execution, which should take into account: (1) the peculiarity of resources that are heterogeneous and geographically distributed resources; (2) the significant amount of machinery required to achieve a functional federation of those resources (e.g. authentication, authorization); and (3) the need to support data sharing among different locations, along with security and privacy constraints. The presented architecture successfully met these requirements without sacrificing the performance aspect that is important for scientific/engineering computations, as well as for data-intensive tasks; this was also due to the effectiveness of the orchestration and resource allocation components (discussed above).

Finally, this approach provides a comprehensive solution that significantly lowers the barrier for SMEs to exploit HPC and cloud resources, mainly thanks to the high-level abstractions provided by the workflow paradigm and the LEXIS stack.

Acknowledgment

This work and all contributing authors are funded/co-funded by the EU's Horizon 2020 Research and Innovation Programme (2014–2020) under grant agreement No. 825532 (Project LEXIS – "Large-scale EXecution for Industry and Society").

Notes

1 https://lexis-project.eu/web/
2 www.keycloak.org
3 https://mariadb.com/kb/en/what-is-mariadb-galera-cluster/
4 https://infinispan.org
5 https://alien4cloud.github.io/index.html
6 https://github.com/ystia/yorc
7 https://heappe.eu
8 https://github.com/ystia/forge
9 www.influxdata.com
10 One core hour corresponds to the usage of a CPU core for one hour, and is a common way of evaluating the resource usage or their assignment in the HPC domain.
11 https://irods.org
12 https://eudat.eu
13 www.docker.com
14 https://sylabs.io/singularity/
15 https://cloudify.co

Bibliography

[1] ServiceSs: an interoperable programming framework for the Cloud, *Journal of Grid Computing*, March 2014, Volume 12, Issue 1, pp 67–91, Lordan, F., E. Tejedor, J. Ejarque, R. Rafanell, J. Álvarez, F. Marozzo, D. Lezzi, R. Sirvent, D. Talia, and R. M. Badia, DOI: 10.1007/s10723-013-9272-5

[2] COMP Superscalar, an interoperable programming framework, *SoftwareX*, Volumes 3–4, December 2015, pp. 32–36, Badia, R. M., J. Conejero, C. Diaz, J. Ejarque, D. Lezzi, F. Lordan, C. Ramon-Cortes, and R. Sirvent, DOI: 10.1016/j.softx.2015.10.004

[3] Cima, Vojtěch, et al. "Hyperloom: a platform for defining and executing scientific pipelines in distributed environments." *Proceedings of the 9th Workshop and 7th Workshop on Parallel Programming and RunTime Management Techniques for*

Manycore Architectures and Design Tools and Architectures for Multicore Embedded Computing Platforms. 2018.
[4] Svaton, Vaclav, et al. "HPC-as-a-Service via HEAppE Platform." Conference on Complex, Intelligent, and Software Intensive Systems. Springer, Cham, 2019.
[5] J. Anuradha, "A brief introduction on big data 5Vs characteristics and Hadoop technology," *Procedia Computer Science*, 48: 319–324, 2015.
[6] N. Kaur and S. K. Sood, "Dynamic resource allocation for big data streams based on data characteristics (5 V s)." *International Journal of Network Management*, 27(4) (2017).
[7] Skamarock, William C., Joseph B. Klemp, and Jimy Dudhia. "Prototypes for the WRF (Weather Research and Forecasting) model." Preprints, Ninth Conf. Mesoscale Processes, J11–J15, Amer. Meteorol. Soc., Fort Lauderdale, FL. 2001.
[8] E. J. Korpela, "SETI@ home, BOINC, and volunteer distributed computing," *Annual Review of Earth and Planetary Sciences*, vol. 40, pp. 69–87, 2012.
[9] Anderson, David P. "BOINC: a platform for volunteer computing." *Journal of Grid Computing* (2019): 1–24.
[10] Kurochkin, Ilya, and Anatoliy Saevskiy. "BOINC forks, issues and directions of development." *Procedia Computer Science* 101 (2016): 369–378.
[11] Stagni, F., Ph Charpentier, and LHCb Collaboration. "The LHCb DIRAC-based production and data management operations systems." *Journal of Physics: Conference Series*. Vol. 368. No. 1. IOP Publishing, 2012.
[12] A. Tsaregorodtsev, "DIRAC distributed computing services," *Journal of Physics: Conference Series*, vol. 513, no. 3, p. 032096, 2014.
[13] Stagni, Federico, et al. "DIRAC in large particle physics experiments." *Journal of Physics: Conference Series*. Vol. 898. No. 9. IOP Publishing, 2017.
[14] Maeno, T., et al. "Evolution of the ATLAS PanDA workload management system for exascale computational science." *Journal of Physics: Conference Series*. Vol. 513. No. 3. IOP Publishing, 2014.
[15] Megino, FH Barreiro, et al. "PanDA for ATLAS distributed computing in the next decade." *Journal of Physics: Conference Series*. Vol. 898. No. 5. IOP Publishing, 2017.
[16] Carnero, Javier, and Francisco Javier Nieto. "Running simulations in HPC and cloud resources by implementing enhanced TOSCA workflows." 2018 International Conference on High Performance Computing & Simulation (HPCS). IEEE, 2018.
[17] S. Singh and I. Chana, "A survey on resource scheduling in cloud computing: Issues and challenges," *Journal of Grid Computing*, vol. 14, no. 2, pp. 217–264, 2016.
[18] R. Prodan, M. Wieczorek, and H. M. Fard, "Double auction-based scheduling of scientific applications in distributed grid and cloud environments." Journal of Grid Computing, 9(4): 531–548 (2011).
[19] G. Iyer and B. Veeravalli, "On the resource allocation and pricing strategies in Compute Clouds using bargaining approaches," in 17th IEEE International Conference on Networks (ICON), 2011.
[20] A. Oprescu and T. Kielmann, "Bag-of-tasks scheduling under budget constraints," in IEEE Second International Conference on Cloud Computing Technology and Science (CloudCom), 2010.

[21] Z. Liu, S. Wang, Q. Sun, H. Zou, and F. Yang, "Cost-Aware Cloud Service Request Scheduling for SaaS Providers," *Computer Journal*, vol. 57, no. 2, pp. 291–301, 2013.
[22] G. Xu, Y. Ding, J. Zhao, L. Hu, and X. Fu, "A novel artificial bee colony approach of live virtual machine migration policy using Bayes theorem (Article ID 369209)," *Science World Journal*, vol. 13, 2013.
[23] X. Song, L. Gao, and J. Wang, "Job scheduling based on ant colony optimization in cloud computing," in International Conference on Computer Science and Service System (CSSS), 2011.
[24] M. Somnath and M. Pranzo, "Power efficient server consolidation for cloud data center," *Future Generation Computer Systems*, vol. 70, pp. 4–16, 2017.
[25] M. Somnath, A. Scionti, and A. S. Kumar, "Adaptive resource allocation for load balancing in cloud," in *Cloud Computing*, Springer, Cham, 2017, pp. 301–327.
[26] Tucker, Andrew, and David Comay. "Solaris Zones: Operating System Support for Server Consolidation." *Virtual Machine Research and Technology Symposium*. 2004.
[27] McKusick, Kirk. "The jail facility in FreeBSD 5.2." *The USENIX Association newsletter* 29.4 (2004).
[28] Bernstein, David. "Containers and cloud: From LXC to Docker to Kubernetes." *IEEE Cloud Computing* 1.3 (2014): 81–84.
[29] Smith, Randall. *Docker Orchestration*. Packt, 2017.
[30] Sochat, Vanessa. "Singularity compose: orchestration for singularity instances." *Journal of Open Source Software* 4.40 (2019): 1578.
[31] Scionti, Alberto, et al. "Demogrape: managing scientific applications in a cloud-federated environment." 2016 10th International Conference on Complex, Intelligent, and Software Intensive Systems (CISIS). IEEE, 2016.
[32] Carnero, Javier, and Francisco Javier Nieto. "Running simulations in HPC and cloud resources by implementing enhanced TOSCA workflows." 2018 International Conference on High Performance Computing & Simulation (HPCS). IEEE, 2018.
[33] H2020 SODALITE project: https://sodalite.eu/content/orchestration-cloud-and-hpc
[34] Croupier docs: https://croupier.readthedocs.io/en/latest/
[35] Colonnelli, Iacopo, et al. "StreamFlow: cross-breeding cloud with HPC." *IEEE Transactions on Emerging Topics in Computing* (2020).
[36] Kakadia, Dharmesh. *Apache Mesos Essentials*. Packt, 2015.
[37] P. Amstutz, M. R. Crusoe, N. Tijanić, et al. Common Workflow Language, v1.0, 2016.
[38] M. Kowalik, H.-F. Chiang, G. Daues, and R. Kooper. 2016. DMTN-025: A survey of workflow management systems. https://dmtn-025.lsst.io/
[39] SCUFL2 Taverna Language (2018). https://taverna.incubator.apache.org/documentation/scufl2
[40] M. Pierce, S. Marru, L. Gunathilake, T. A. Kanewala, R. Singh, S. Wijeratne, C. Wimalasena, C. Herath, E. Chinthaka, C. Mattmann, A. Slominski, P. Tangchaisin. 2014. Apache Airavata: Design and Directions of a Science Gateway Framework, *Proceedings of the 2014 6th International Workshop on Science Gateways*, pp. 48–54.

[41] Christopher Brooks. 2015. Triquetrum: Models of Computation for Workflows. www.eclipsecon.org/na2016/session/triquetrum-models-computation-workflows

[42] Megino, Fernando Barreiro, et al. "PanDA: Exascale Federation of Resources for the ATLAS Experiment at the LHC." *EPJ Web of Conferences*. Vol. 108. EDP Sciences, 2016.

6

Advanced Engineering Platform Supporting CFD Simulations of Aeronautical Engine Critical Parts

Donato Magarielli, Ennio Spano, Tommaso Fondelli, Daniele Pampaloni, Antonio Andreini, Paolo Savio, Michele Marconcini, and Francesco Poli

CONTENTS

DOI: 10.1201/9781003176664-6

6.1 Introduction: Background and LEXIS Aeronautics Pilot

Reducing fuel consumption of aircraft engines is a key requirement for players in the aeronautic industry. Significant efforts are underway to produce reliable turbo engine performance predictions, striving to perform more realistic physics-based simulations allowing engineers/designers to help anticipate problems typically encountered in the detailed design phases. This demands the adoption of CPU-intensive and time-consuming CAE simulations based on advanced numerical solvers. The synergy among next-generation HPC/cloud/big data management technologies and their coupling with sophisticated CFD software solutions in the advanced engineering platform provided by LEXIS are opening new scenarios for the aircraft engine design and optimization in the domain of CFD engineering analyses, enabling innovative and faster investigation strategies, and providing unprecedented levels of accuracy and detail. In this context, the Aeronautics large-scale pilot led by Avio Aero[1] in LEXIS aims to significantly improve the feasibility and exploitation of advanced CFD numerical modeling capabilities able to predict the fluid-dynamic behavior of aircraft engine critical components. From both a digital technology and business perspective, a marked step change is here envisaged: faster and more accurate CAE analyses that exploit newly deployed HW/SW resources in an innovative cross-converged HPC/cloud/big data environment enabling the implementation of greatly improved or newly designed CFD-based engineering methodologies. To meet this ambitious objective, the industrial applicability of the LEXIS advanced engineering platform is under investigation through two aeronautical engineering case studies that are described here.

6.2 Engineering Case Studies in the LEXIS Aeronautics Pilot

Avio Aero is leading two different aeronautics pilots in LEXIS,[2] one regarding turbomachinery and the other one referring to rotating parts representing gearboxes, both requiring hardware-intensive and time-consuming CAE simulations but based on different application software.

6.3 The Turbomachinery Case Study

There are currently significant efforts being made to achieve increasingly reliable performance prediction, using more and more physics-based solutions and a multi-physics simulation approach able to anticipate problems, typically encountered in the aero engine detailed design phases. To cope with these strategic goals, the improvement of a CFD code named TRAF, developed by the University of Florence, is envisaged and specifically designed to help turbomachinery designers. Extensively validated against several turbomachinery configurations, a high level of parallelization is provided in the reference (starting) version of the code by means of a hybrid parallel programming model (MPI/multi-platform shared-memory parallel), but the increasingly complex and detailed analyses performed during the design process require a step change in terms of computational job duration. For this reason, development activities for porting this code from a pure CPU-based computing platform to a GPU-accelerated one were foreseen in the LEXIS project, aimed to drastically reduce the execution time and to make the code fully deployable for the current higher computationally intensive industrial needs.

6.3.1 Engineering Context

The turbomachinery pilot test case is representative of a multistage turbine, shown in Figure 6.1. Due to the high number of computational blocks, the underlying CFD model is very HPC-demanding, particularly in terms of both required memory resources and time duration needed to perform the calculations. More deeply, since the memory available in the LEXIS GPU-accelerated HPC platform is not sufficient to manage this model, a reduced version of the test case will be used.

The rolled-out annulus of the test turbine illustrated in Figure 6.1 is composed of four airfoil rows (two stators and two rotors) followed by a final row of struts. The shown representations of the test turbine are marked as "Complete" and "Reduced" depending upon whether used on a pure CPU-based system or a GPU-equipped one, respectively. To be more specific, the reduced model needs to comply with the available memory on the provided GPU-equipped systems.

To assess the performance of the TRAF code, the original test case will be deployed on a solely CPU-based cluster targeting to evaluate the scalability of the application on an increased number of cores while trying to reduce the run time according to the expected goal. After that, the opportunity to further decrease the duration of the CFD computational phase will be assessed over all the solely CPU-based HPC clusters available in LEXIS. Finally, using the GPU-enabled release of the code, the TRAF-based CFD simulations will be executed with the goal to reduce by at least five times the running time, that is one of the KPIs set in the LEXIS project scope.

6.3.2 Digital Technology Deployment

In this section, the digital technology implementation perspective will be illustrated for the turbomachinery use case providing some details about the underlying application workflow and the HW/SW requirements.

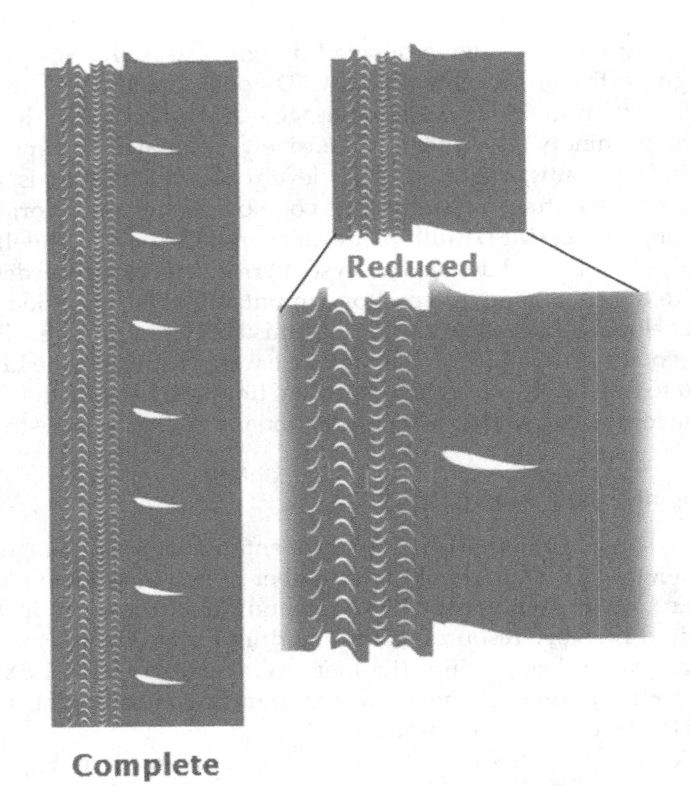

FIGURE 6.1
Turbomachinery test case LPT turbine.

6.3.2.1 Application Workflow

As in any standard CAE analysis, the CFD simulations that the aeronautics turbomachinery use case relies on include the following three stages:

- preprocessing;
- analysis solver execution; and
- post-processing of results.

Figure 6.2 provides a high-level overview of the deployed CFD workflow, illustrating the tasks of the above-mentioned steps and the software application used in the computational phase.

The preprocessing stage is beyond the scope of the LEXIS project and is performed locally on the workstation of end users, while the computational phase, which is strongly HPC-demanding and time-consuming, needs to leverage the advanced capabilities provided by LEXIS in terms of

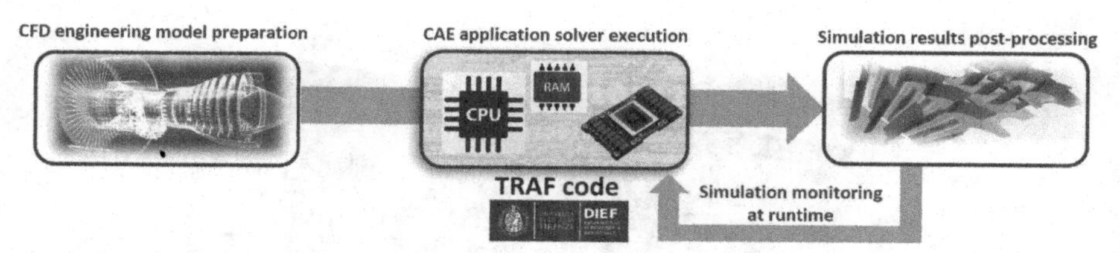

FIGURE 6.2
Three-stage application workflow for the CFD simulations in LEXIS.

state-of-the-art supercomputing centers, SW enhancements, and HW/SW systems integration. More specifically, the computational phase is based on TRAF code, a CFD application solver developed by the University of Florence to investigate fluid-dynamics phenomena with a special focus on turbomachinery simulations. The output files from the TRAF simulations are stored on an NFS file system available on all computational nodes and, in the post-processing phase, the results from such CFD simulations are visualized using large-memory GPU-equipped visualization nodes in LEXIS with the help of a proper visualizer tool.

The deployment of the CFD simulations involved in the workflow presented relies on having at least one core for each MPI process and, due to the complexity of the input model, a high number of computational nodes. To resolve the underlying Navier–Stokes equations, the computational domain was spatially discretized by means of a grid, which is divided into different blocks. Figure 6.3 reports in different colors the computational blocks adopted in the turbomachinery use case (just a limited angular sector of the turbine is shown).

In order to try to reduce as much as possible the duration of the computational phase, first of all, extensive scalability and performance tests were performed on the solely CPU release of TRAF code running on pure CPU-based resources at the two LEXIS supercomputing partners. This did not prove sufficient to meet the industrial target of reduction of TRAF running time by at least a factor of 5×, so a newly designed GPU-enabled release of the code was developed and tested on the GPU-accelerated HPC resources provided by IT4Innovations supercomputing partner.

6.3.2.2 *Main Application Software and HW Resources*

Specifically developed to assist turbomachinery designers, the main application software used here is the TRAF 3D CFD numerical solver that solves the steady/unsteady, three-dimensional, Reynolds-averaged Navier–Stokes equations in the finite volume formulation on multi-block structured grids [T1]. Convective fluxes are discretized by a second-order TVD-MUSCL strategy built on the Roe's upwind scheme [T2]. A central difference scheme

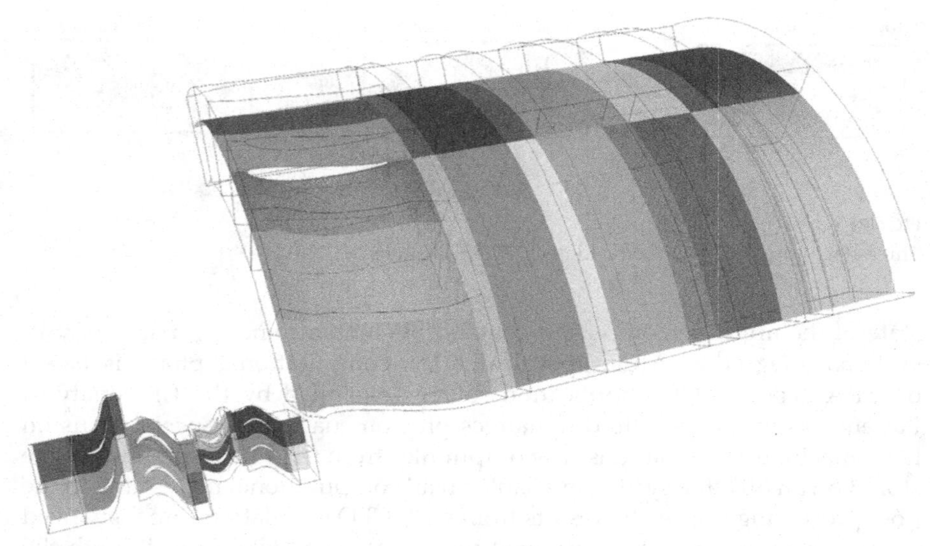

FIGURE 6.3
Representation of the computational blocks adopted in the turbomachinery use.

is used for the viscous fluxes. The turbulence closure is based on different turbulence models. Time-accurate calculations are performed by means of a dual-time stepping approach [T3]. The coupling between consecutive rows is handled by sliding interfaces that use phantom cells lying on the adjacent blade passage. Linear interpolations are used to exchange the flow variable values.

The last consolidated version of TRAF currently running in the Avio Aero digital technology production environment relies on a hybrid MPI/multi-platform shared-memory parallel computational model and runs on solely CPU-based HPC resources [T4][T5]. With the aim of identifying the best LEXIS HW platform able to optimize the application solver execution, this release was deployed throughout different HPCs in LEXIS: Salomon and Barbora supercomputer clusters at IT4Innovations (IT4I), CoolMUC-2 and SuperMUC at Leibniz-Rechenzentrum (LRZ).

More specifically, the Salomon cluster was used to assess the scalability of the code, using from 200 cores up to 1788 cores. After determining the best number of cores to be used to handle the considered computational blocks, the CPU-based TRAF release was also run on the Barbora cluster at IT4I, CoolMUC-2 and SuperMUC at LRZ with the target of evaluating the most suitable HPC platform in LEXIS able to minimize the duration of the computational phase.

Then, aiming at further accelerating the execution of the computational phase in the application workflow, a newly developed release of TRAF enabled to run on GPUs was deployed on the GPU-equipped nodes of

Barbora, leveraging up to 16 state-of-the-art GPUs and only one core per GPU to manage the HW-accelerated computations. At the present time only an early pre-release of the GPU-enabled version of TRAF solver was used, but software development activities are currently underway to further improve and consolidate it.

6.3.3 First Results

In the context of the LEXIS project, various activities were carried out to assess the performance of the TRAF solver.

One of the first objectives was the scalability evaluation of TRAF on CPU-based platforms. An initial set of tests were performed on the Salomon HPC cluster with the aim of assessing the performance of the code when running in parallel on several compute nodes. The TRAF solver was compiled and linked with MPI and multi-platform shared-memory parallel programming libraries for parallel execution. The scalability tests were run on the "complete" use case provided by Avio Aero: the domain consisted of more than 700 computational blocks, requiring an overall memory amount of about 1 TB. By scalability, we mean the capability of a system to efficiently integrate more computing resources in order to perform the same task in a reduced amount of time. This was assessed by launching the same analysis with different numbers of CPU cores. The test covered three time-steps of an unsteady calculation: a representative, although limited, effort to get reliable information on the code performance. Special attention was paid to minimizing the load unbalance among the involved CPU cores. The first configuration was based on 200 MPI processes executed by 200 cores distributed among 9 nodes ($22 \times 7 + 23 \times 2$). The second test used 447 MPI processes on 447 cores ($24 \times 10 + 23 \times 9$), a configuration of special interest, because the largest grid block was assigned to one MPI process dealing with this block only. Since this block cannot be subdivided into more MPI processes due to the coarse-grain parallelism implemented in the code, this represents the most efficient configuration for the MPI parallel computation: any further increase in the number of MPI processes would only decrease the load balance, with no benefit on the computation time. To achieve additional speedup, the third test considered the use of hybrid parallelism: 447 MPI processes, each with two multi-platform shared-memory parallel programming threads managed by two cores, resulting in a total of 447×2 cores and 38 nodes in use ($12 \times 2 \times 29 + 11 \times 2 \times 9$). Other tests used 447×3 cores over 56 nodes and 447×4 cores over 75 nodes. To evaluate the level of exploitation of computational resources, we can define the "resource use" as the ratio of the number of used cores to the number of reserved cores (which is the number of reserved nodes multiplied by 24, number of cores per node). An assessment of the efficiency of the resource exploitation can be based on the "computational efficiency," defined as the ratio of the total computational time in the baseline configuration (number of baseline cores, i.e. 200, multiplied by the baseline execution time) to the

TABLE 6.1
Scalability test results

Number of cores	Number of nodes	Execution time[s]	Resource use [%]	Computational efficiency [%]
200 (baseline)	9	6247	92.59	92.59
447	19	3781	98.03	72.47
447 × 2	38	2294	98.03	59.72
447 × 3	56	1759	99.78	52.85
447 × 4	75	1446	99.33	48.00

overall reserved computational time (number of reserved cores multiplied by the test execution time). Table 6.1 summarizes the scalability test results.

As can be seen, reducing the execution time and maximizing the computational efficiency are two competing goals: the best trade-off seems to be the 447 × 2 configuration. This same configuration was tested on other available HPC clusters: on CoolMUC-2 (where the execution time was identical to Salomon, surprisingly), on Barbora (where the time was lower: 2069 s) and on SuperMUC (where the time was comparable with Barbora: 2064 s). After that, TRAF MPI communications were profiled with a specific profiler tool, in order to measure the overhead due to inter-process communications and process workload balancing: it was concluded that the MPI communication contributes a significant amount to the overall execution time. Scalability is limited by a communication bottleneck, thus explaining the plateaus reached in the above-described tests.

After assessing TRAF performance on CPU-based architectures, the focus shifted to evaluating the benefit that can be obtained by running TRAF on GPU-accelerated platforms. The porting of TRAF to GPU-accelerated architectures is carried out by University of Florence and began at the outset of the LEXIS project. A first preliminary release of the ported code was completed by modifying data structures to accommodate GPU onboard memory counterparts (with the "device" attribute) of all the main arrays needed for the computation. Device arrays are allocated and initialized at the beginning of the execution, as much as possible. All subroutines and modules of the code were modified to implement the offloading of their loops into kernels to be executed on the GPU. The original structure of the code was preserved as much as possible, by making use of conditional compilation directives, where needed: the code can therefore be compiled for different HPC architectures from a single source base, thus easing the code maintenance and development. As a second step, the parallel MPI capability of the code was enabled for the GPU version: initially, the exchange of boundary condition data was implemented by copying array slices from device to host memory, sending/receiving data between MPI processes, and copying slices back from host to device memory on the receiving side. This manual data

TABLE 6.2

Benchmark results shortly after the beginning of the optimization effort

Number of nodes	MPI processes per node	Hardware resources	Execution time[s]	Speedup
1	13	13 CPU cores	2975	(baseline)
2	1	2 GPUs	1030	2.89
2	3	2 GPUs	918.2	3.24
2	7	2 GPUs	759.8	3.92

TABLE 6.3

Benchmark results at the end of the optimization effort

Number of nodes	MPI processes per node	Hardware resources	Execution time[s]	Speedup
1	13	13 CPU cores	2975	(baseline)
2	1	2 GPUs	298.6	9.96
2	3	2 GPUs	353.0	8.43
2	7	2 GPUs	337.7	8.81

transfer limited the code performance, as was expected. Once a functional release of the code was available, a profiling and optimization effort was started, beginning from the execution on a single GPU, then focusing on MPI communications for execution on multiple GPUs on the same node and/or on multiple nodes.

During the initial porting, the preliminary MPI enablement and the first few code optimizations, the execution correctness and performance were checked only on small tests. After the first few optimizations, a more representative benchmark was defined and run on the Barbora HPC cluster: in this benchmark, grid block dimensions are similar to those found in production calculations, but only 18 grid blocks are present, requiring an overall memory amount of about 32 GB (thus filling the onboard memory of 2 GPUs on Barbora). Table 6.2 shows the benchmark results obtained with the TRAF release including only the first few code optimizations.

The speedup, selected as the main KPI, is here defined as the ratio of the CPU (baseline) execution time to the GPU execution time. The GPU-accelerated computation was run on a number of GPUs compatible with the required overall memory, while the CPU computation was carried out with a number of MPI processes each dealing with a number of cells comparable to that of the biggest grid block and assigning one CPU core for each MPI process.

At the end of the optimization effort, the benchmark was repeated in order to assess the obtained improvements. Table 6.3 shows the results.

As can be seen, the maximum speedup has increased by about 154% (from 3.92 to 9.96).

TABLE 6.4

"Reduced" case test results

Number of nodes	MPI processes x multi-platform shared-memory parallel threads	Hardware resources	Net time per step[s]	Speedup
3	108	108 CPU cores	689.4	(baseline)
6	108 × 2	216 CPU cores	399.0	1.73
4	16	16 GPUs	132.0	5.22

After assessing the progress on the TRAF GPU performance with the above-described benchmark, an even more realistic test was performed, based on the "complete" case provided by Avio Aero (the one previously used for the scalability tests). Unfortunately, the overall required memory amount of the "complete" case was excessive: a "reduced" case was thus created, corresponding to a quarter of the original domain, compatible with the total onboard memory of 16 GPUs on the Barbora cluster. The test covered a few time-steps of the unsteady calculation and the average net execution time per time-step was measured. The baseline configuration corresponded to 108 MPI processes executed by 108 CPU cores on three nodes. A second configuration also used two multi-platform shared-memory parallel programming threads per MPI process, using 108 × 2 CPU cores on six nodes. The GPU-accelerated configuration employed 16 GPUs on four nodes. Table 6.4 shows the performance figures.

As can be seen, the obtained speedup (5.22) is greater than 5 (which was the minimum goal decided at the beginning of the porting effort), although less than what was achieved in the smaller benchmark. The greater amount of MPI communications involved in the test might explain the performance reduction. Further optimizations may improve these results even more.

6.3.4 Benefit–Cost Analysis of HW Acceleration

Starting from the very promising speedup value obtained through the GPU-enabled release of TRAF code, a benefit–cost analysis was carried out to measure the benefits of HW acceleration in terms of computational time reduction and in relation to the increased costs of a GPU-equipped HPC system compared to a solely CPU-based one.

To perform this analysis, a benefit–cost index was defined as the ratio between the benefits from the computational speedup and the impacts in terms of cost increase, according to the following formula:

$$\text{Benefit–cost index} = \text{computational speedup} \, / \, \text{cost increase factor} \quad (6.1)$$

where the computational speedup is the ratio of the execution time on a pure CPU-based HPC system (baseline) to the one on GPU-equipped resources, while the cost increase factor is the ratio of core-hours cost on GPU-equipped resources to the one on a solely CPU-based computational system.

Depending on the value of this index, the two HPC solutions to be assessed can be easily compared as follows:

- index = 1: the two solutions are equivalent from a mixed technical/cost perspective;
- index > 1: the accelerated solution is better than the pure CPU-based one; and
- index < 1: the pure CPU-based one is better than the accelerated one.

In order to calculate the cost increase factor, the increment of costs from a GPU-equipped system compared to a pure CPU-based one was weighed by the amount of cores actually allocated in the two compared systems according to the following formula:

$$\text{Cost increase factor} = \text{normalization factors ratio} * \text{CPU cores usage weight} \quad (6.2)$$

where the normalization factors ratio is calculated by dividing the normalization factor of the GPU-equipped system used by the pure CPU-based system one, while the weight is calculated by dividing the amount of CPU cores used in the GPU-equipped system by the one used in the pure CPU-based system.

Introduced by IT4I to treat systems of different age on equal footing, the normalization factors that underlie this cost model are dependent on the cluster adopted and allow the accounting of computer utilization time depending on the HPC system used.

Considering now the 5.22 computational speedup obtained from the GPU-enabled release on Barbora and based on the information (summarized in Table 6.5) about the amount of allocated cores and normalization factors of the HPC systems used at IT4I:

the benefit–cost index was calculated at 1.83, so revealing that the GPU-accelerated solution deployed for the engineering computational phase here illustrated is also beneficial from a cost perspective.

TABLE 6.5

Cores allocated and normalization factors at the two assessed HPC solutions

HPC cluster	HPC subsystem	CPU cores allocated	Normalization factor
Barbora	Pure CPU-based	108	1.40
Barbora	GPU-equipped	96	4.50

6.3.5 Next Steps

Although the minimum goal of a five times speedup on GPU-accelerated platforms was reached, more optimizations are underway to further improve the obtained results. The main efforts are focusing on the GPU-enabled release of the CFD solver used to take advantage of latest compiler enhancements, to minimize remaining performance bottlenecks, and to exploit the best GPU-to-GPU communication channels. Moreover, the opportunity to further speed up the execution of the most HW-intensive subroutine of the adopted CFD solver will be investigated as well through FPGA-based HPC platforms. The long-term goal is to develop and advance the unsteady Reynolds-averaged Navier–Stokes method implementation, in order to enable aeronautical turbomachinery design for unsteady analyses in industrial practice.

6.4 The Rotating Parts Case Study

Beyond standard CFD simulated products afforded in the turbomachinery use case, today challenges are arising when studying complex flow fields in mechanical parts that rotate at high speed in the presence of air and lubricating oil. Nowadays, this kind of engineering analysis is at the leading edge of numerical technology and perfectly fits the needs of designing gearboxes capable of withstanding high transmission efficiency. The challenge here is to predict, with increasingly accuracy, the flow field operating inside the gearbox where the combination of jet lubrication and high tangential speeds precludes the possibility of neglecting the interactions between liquid and gaseous phases.

The simulation of these phenomena typically requires a large amount of computational resources and may take considerable running time. In this context, the commercial code Altair nanoFluidX™ will be used, since it was widely proven in different conditions, in order to provide a high-quality analysis and minimize the computational time as much as possible. nanoFluidX® is a smoothed-particle hydrodynamics (SPH) simulation tool able to predict the flow fields in complex geometries with moving or rotating parts.

Avio Aero, inside LEXIS workscope, has undertaken the challenge to validate this SW and to push it to its best possible performance from both numerical modeling and hardware integration perspectives. The objective is to make consistent progress in predicting the flow field inside gearboxes, together with a drastic reduction of computing time.

6.4.1 Engineering Context

The main engineering goal here is replacing, or at least supporting, the traditional gearbox design approach based on correlations, expertise, and legacy data with a completely new CFD numerical method capable of validating the industrial solutions that Avio Aero is going to insert in the next-generation gearboxes, quickly and reliably. Checking the general behavior of the air–oil mixture, its distribution, the scavenging capabilities as well as keeping the resistant torque levels under control are the most important issues to be addressed during the design and analysis phases. For this reason, securing the accuracy and reliability of the numerical solver is the top priority.

In the global vision of improvement, simulation speed is the second objective that can be identified in the optimization of the fluid structure of the gearbox's components, as the key factor enabling the provision of state-of-art products on the international market. For the players in this field, the optimal design of the next-generation engines has become a priority and, to reach this goal faster, calls for state-of-the-art computational activity: a step change in the numerical investigations of aeronautical engines' critical parts is envisaged as part of the vision of getting real-time simulation available.

Having these two objectives in mind, the present case study will be devoted to innovating the framework of gearbox engineering and will be based on a set of different test cases aimed to assess the predictions of the simulated CFD models in comparison to the measured data from experiments. The identified steps of simulation are reported hereafter. First, simulations will be run considering oil only and air only fluids, in a logic to split the problems into separate contributions. Finally, real conditions will be simulated merging oil and air media flowing in a real gearbox environment. The selected test cases are listed below.

- **Single high-speed wheel**: SPH oil phase simulation and SPH air flow simulation will be run separately, with the target of determining the proper discretization level for reproducing liquid and gas motion as separate effects. Figure 6.4 shows the experimental setup for single wheel.
- **Gear pair with oil-jet lubrication,** with the target of assessing nanoFluidX on the simplest gearing geometry and applying multiphase approach in simplest geometries. The experimental setup for gear pair characterization is shown in Figure 6.5.
- **Simulation of an engine-like gearbox,** with the target of validating the nanoFluidX predictions on a more complex environment (air–oil mixture) and comparing them with experimental and finite volume CFD results. The planetary gearbox setup is shown in Figure 6.6.

6.4.2 Digital Technology Deployment

In this section, a digital technology implementation perspective will be illustrated for the rotating parts case study providing some details about the underlying application workflow and the HW/SW requirements.

FIGURE 6.4
Single wheel experiment carried out at University of Florence.

FIGURE 6.5
Double wheels experiment carried out at the University of Florence.

FIGURE 6.6
Planetary gearbox (EU project).

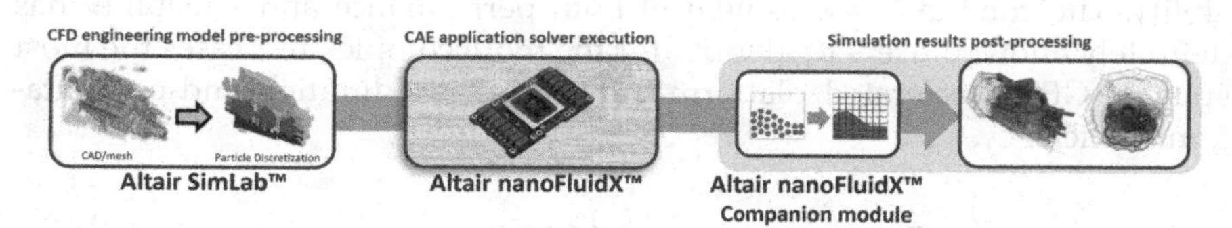

FIGURE 6.7
Three-stages application workflow for the CFD simulations in LEXIS.

6.4.2.1 Application Workflow

The CFD simulations that the Aeronautics rotating parts case study relies on includes the three phases of any standard CAE analysis: preprocessing, analysis solver execution, and post-processing of results. All these phases are very computationally intensive and need to leverage the advanced capabilities provided by LEXIS in terms of state-of-the-art HW resources available and HW/SW systems integration skills.

A high-level overview of the deployed CFD workflow is depicted in Figure 6.7, which illustrates the tasks and the main software applications involved in the above-mentioned phases.

The pre-processing phase is executed on large-memory GPU-equipped visualization nodes and leverages Altair SimLab™ software to generate high-resolution input models. The computational phase required from the analysis solver execution is run using GPU-accelerated HPC systems and nanoFluidX software application. The output files from nanoFluidX simulations are stored on an NFS file system mounted on all computational nodes and, in the post-processing phase, the results from such CFD simulations are visualized using large-memory GPU-equipped visualization nodes in LEXIS and leveraging a proper visualized tool after a data interpolation process performed by nanoFluidX companion module on solely CPU-based HPCs.

With the aim of tuning a new CFD-based engineering practice to simulate the multiphase flow inside gearboxes, the application workflow is being deployed over the three engineering case studies mentioned above. The deployment of the CFD simulations involved in the workflow requires the availability of several GPUs due to the complexity of the input models. Different GPU-accelerated HPC resources in LEXIS were assessed and used to figure out where to execute nanoFluidX, first of all depending on the specific HW prerequisites needed to run the considered test cases and, then, in a reasonable time. More specifically, to assess the performance of the nanoFluidX execution, the same computational job, related to a well-identified rotating parts test case, was deployed and run on different GPU-equipped HPC clusters. Once identified, the best GPU-accelerated HPC cluster, the selected Rotating parts test case was executed again on the same cluster increasing the number of GPUs used, with the aim of evaluating the application scalability. The combined assessment of both performance and scalability has definitely allowed users to identify for the more complex test cases the most suitable GPU-accelerated platform in terms of job duration and computational efficiency.

6.4.2.2 Main Application Software and HW Resources

Enabled and optimized for use on clusters of GPUs, the main application software is nanoFluidX. This is an extremely fast SPH simulation tool conceived in the last years and for typical gear-train applications, without the need to simplify geometries, the code can run an order of magnitude faster than an equivalent finite-volume code. It is designed to leverage the computing power of GPUs, therefore achieving high levels of parallelism and excellent scalability on both single and multi-node GPU servers. GPU adoption in scientific and engineering computing is rapidly progressing and nanoFluidX is one of the first commercial software packages to use this technology, bringing a significant speed-up in the overall product development. Thanks to this approach, the software is ready for industrial applications, with no need for simplification in the computational domain. The discretization process in the computational domain is based only on a set of independent points, named particles, and different numbers of particles were used depending on the considered phase simulation and test case, requiring different running times. Referring to the simulation of the most complex test case included in the pilot, the needed number of particles and running time are expected to be up to 290 M and 1000 hours, respectively, requiring to leverage in this case up to 16 state-of-the-art GPUs.

With the aim of identifying the proper LEXIS HPC system able to support the application solver execution depending on the specific test case considered, nanoFluidX was deployed throughout different GPU-accelerated systems in LEXIS: the Anselm, Barbora, and DGX-2 systems at IT4I, and the DGX-1 one at LRZ.

6.4.3 First Results

This section reports the main results achieved with nanoFluidX applied to the single high-speed wheel test case, which was characterized experimentally at the Technology for High Temperature Laboratory (University of Florence) [R1][R2][R3]. Such test case was exploited to assess the discretization level suited for reproducing successfully both the liquid and gas phase motion separately, by means of two different single-phase simulations:

- SPH liquid-phase simulation
- SPH gas-phase simulation

The high-speed gearing systems designed for aircraft applications are generally cooled and lubricated by oil-jet systems, where the jets diameter is about 1 mm while the gear casing dimensions are two to three orders of magnitude higher. This leads to a significant numerical effort since nanoFluidX needs a single particle size for both fluid and solid bodies discretization, therefore the jet-diameter fixes the maximum spatial discretization of the model. Consequently, the definition of the maximum particle diameter needed to reproduce both the air and liquid phase motion is a paramount point of this investigation.

As shown in Figure 6.8, by the side and frontal pictures of the experimental test chamber, the oil is injected toward the wheel through a spray bar having a single orifice with d as diameter, the wheel is enclosed inside a wide rectangular box 550·d wide (W), 380·d high (H) and deep 212·d (L). The wheel pitch diameter (Dp) is 126.6·d, the face width is 45.8·d, the module is 3.33·d, while 38 is the teeth number. Power losses due to the air phase were measured separately by carrying out an experimental test without oil injection. On the contrary, the power losses due to the oil-jet lubrication cannot be measured individually, because, as the wheel runs, the power losses related to the air motion always occur; however, the latter can be significantly reduced by decreasing the air density, and this was experimentally achieved by decreasing the test chamber pressure level through a vacuum

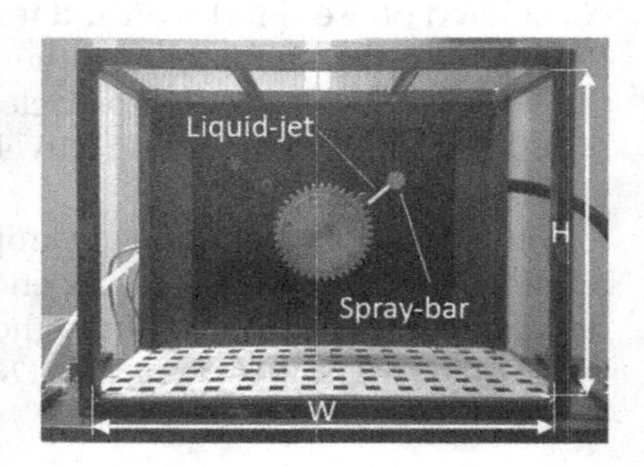

FIGURE 6.8
Single wheel test case geometry [R1].

pump, allowing power losses related to the oil-jet lubrication only to be extrapolated.

The average resistance torque due to the oil-jet lubrication (T_0) can be estimated through a zero-dimensional model (Equation 6.1) presented by Fondelli et al. [R4], which computes the oil momentum variation during the interaction with the tooth; ρj, d, and Uj are the density, diameter, and the jet injection velocity, whereas Dp and Up are the pitch diameter and the pitch line velocity of the wheel. All torque values reported in Section 6.4.3 were normalized by T_0.

$$T_0 = (\rho j \cdot Uj \cdot \pi d^2/4) \cdot Up \cdot Dp/2 \tag{6.3}$$

Experiments performed by Massini et al. [R1] have shown how such formulation generally overestimates the resistant torque since it does not take into account the reduction in the oil amount reaching the gear surface due to break-up phenomena.

6.4.3.1 SPH Liquid-Phase Simulation

Since the air phase was not simulated, the external casing was not modeled as well as the wheel side faces and the rotating shaft; moreover, to further minimize the computational effort, the wheel axial dimension was reduced; in fact, the gear face width in the model is 22 times the jet diameter. After those simplifications, the computational domain was as shown in Figure 6.9. The boundaries consist of a bounding box where a "simple outlet" boundary condition was imposed to each face, that deletes all the particles crossing such boundary. The wheel surface is modeled as moving wall, considered smooth with a no-slip condition, while the liquid injection is achieved through a circular area (the yellow one in Figure 6.9) located 16 jet diameters above the wheel, assigning the velocity components and the liquid density.

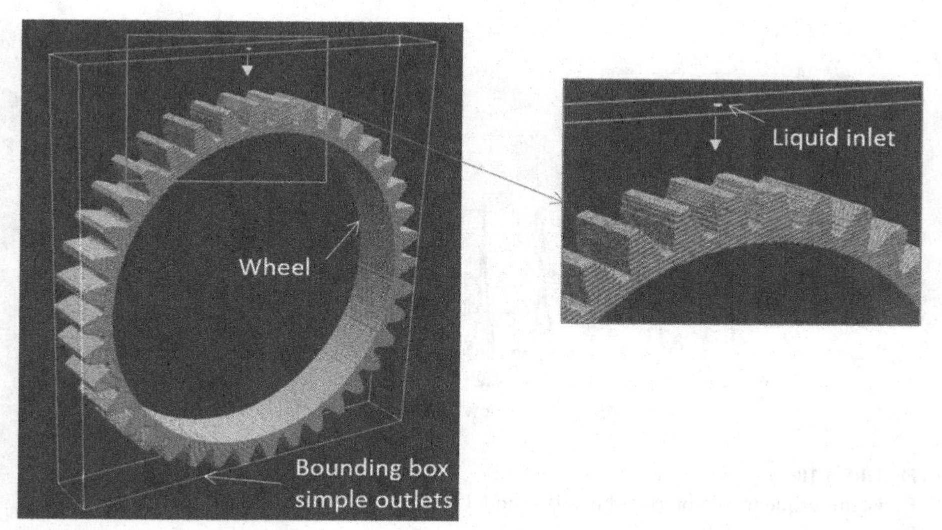

FIGURE 6.9
SPH liquid-phase simulation computational domain.

TABLE 6.6
Liquid-phase simulation models

Model	Oil-jet diameter fraction	Initial particles number
Mesh0	1/5	$3.8 \cdot 10^6$
Mesh1	1/10	$16 \cdot 10^6$
Mesh2	1/20	$65.8 \cdot 10^6$
Mesh3	1/30	$149 \cdot 10^6$

In order to carry out a result sensitivity to the particle size, four models were generated whose characteristics are summarized in Table 6.6; the particle size was chosen on the basis of the oil-jet diameter. The simulated operating condition has 918.5 kg/m3 liquid density, 0.0035 Pa·s liquid dynamic viscosity, 100 m/s as wheel pitch velocity and an oil-jet velocity by 25 m/s. At the initial time the computational domain consists only of solid particles, while the liquid ones will enter the domain during the simulation. The ramp-up time for both wheel rotating speed and oil-jet injection velocity was set to 1 ms, since both are zero at initial time, while the simulation total time was fixed to 8 ms so that 55 oil-jet impacts against the wheel will be simulated.

The resistant torque due to the oil-jet impact against the gear teeth, computed by the SPH model with Mesh2, is shown in Figure 6.10 in the time range 1.6–3 m·s, while the liquid motion as a function of time is shown through the snapshots in Figure 6.11, where the fluid particles are colored

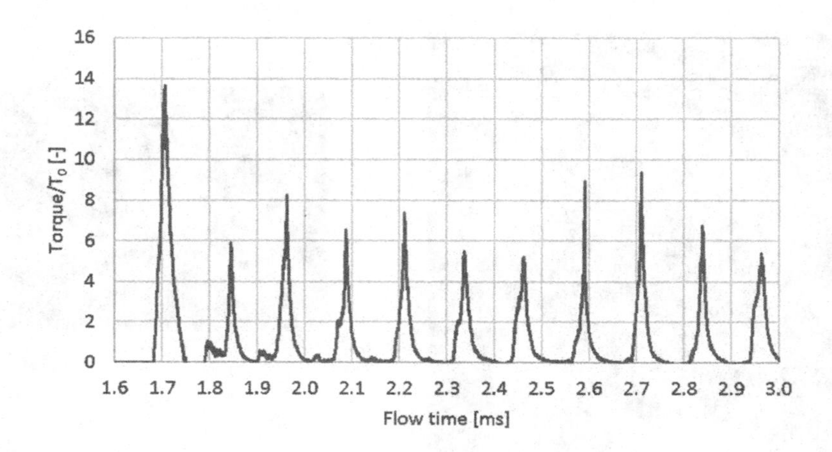

FIGURE 6.10
Resistant torque trend computed by SPH model Mesh2.

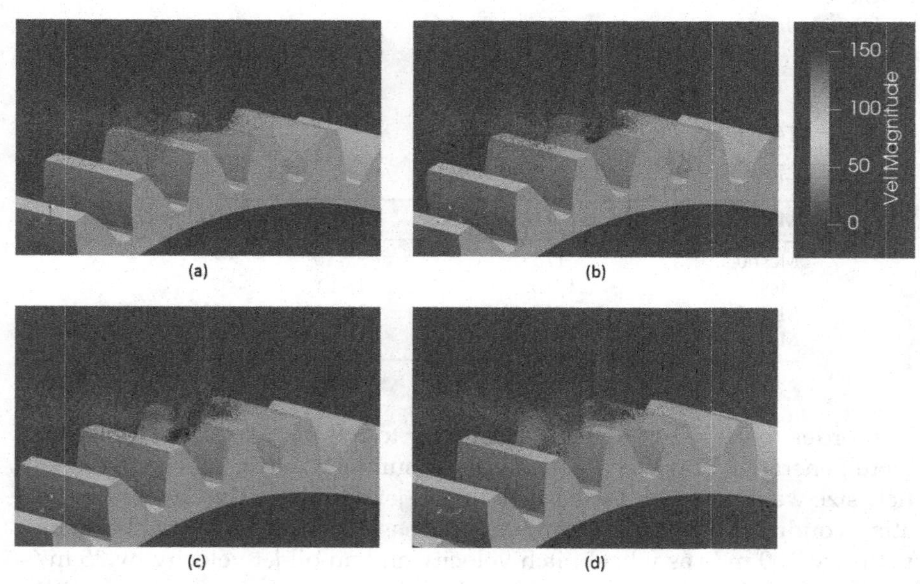

FIGURE 6.11
Liquid motion as a function of time: Mesh2.

by the velocity magnitude. The torque trend successfully reproduces the periodic nature of the jet lubrication, in agreement with previous numerical results by Fondelli et al. [R4] achieved with a finite volume CFD solver using the volume-of-fluid (VOF) approach.

The oil jet approaches the tooth face (a), impacts against the tooth (b), then impinges on the gear top land (c) before impacting with the next tooth; the

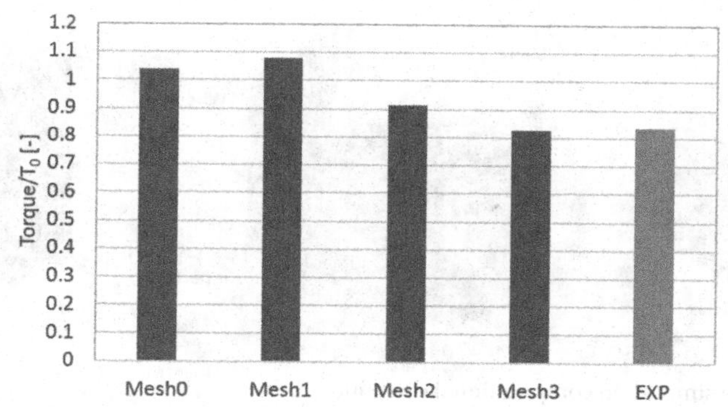

FIGURE 6.12
SPH liquid-phase model: average resistant torque as a function of particle size.

oil on the tooth face is thrown away in tangential direction breaking up into ligaments and droplets (c–d). During the oil-jet impact with the tooth flank, a really fast momentum transfer occurs between the wheel tooth and the liquid, therefore the oil is accelerated up to the tooth tangential speed causing the torque peaks shown in Figure 6.10.

Concerning the average resistant torque predicted by SPH models, it is compared with experimental data in Figure 6.12. The SPH model with the lower resolution overestimates the experimental value by 24%, the Mesh2 model over-predicts about 9%, while the finest resolution compares very well with the experiment, in fact the difference is about 1%. As far as the computational time is concerned, the finest model took 145 hours on DGX-2, whereas the Mesh2 model about 28 hours; the latter model, having a particle dimension $^1/_{20}$ of the oil-jet diameter, is a good trade-off between numerical effort and accuracy.

6.4.3.2 SPH Gas-Phase Simulation

In this section, the SPH model developed to simulate the air motion inside the single wheel test case is discussed. The test chamber of the experimental rig was numerically reproduced, therefore the computational domain, shown in Figure 6.13, consists of a single wheel, clamped on a shaft, which is enclosed within a test box, much bigger than the wheel.

The domain consists of solid particles representing the moving parts, namely the wheel and the shaft, solid particles representing the stationary walls of the containment box and fluid particles discretizing the air that fills the domain. To perform the mesh sensitivity evaluation, the particle dimension was referenced to the gear tooth module, considered the characteristic length of the system. Three models were generated with a particle diameter of ½, $^3/_8$, and $^3/_{10}$ of tooth module, as shown in Table 6.7. The simulated operating condition has 1.068 kg/m^3 air density, 1.98 E-05 Pa·s air dynamic viscosity, and 100 m/s as wheel pitch velocity.

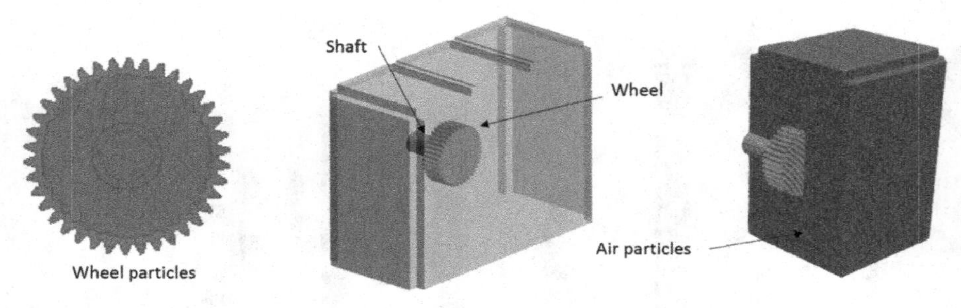

FIGURE 6.13
SPH gas-phase simulation computational domain.

TABLE 6.7
Gas-phase simulation models

Model	Oil-jet diameter fraction	Initial particles number
Mesh0	½	$12.8 \cdot 10^6$
Mesh1	3/8	$30.5 \cdot 10^6$
Mesh2	3/10	$59.3 \cdot 10^6$

The ramp-up time for wheel rotation was set to 1 ms and the simulations were stopped after 2 seconds of flow time, when the resistant torque had reached a steady-state condition, namely after about 400 wheel revolutions.

As described by Fondelli et al. [R5], when the wheel starts spinning, the air is drawn into the tooth passage because of the low pressure that arises in that region, and consequently the flow impinges on the incoming tooth surface that transfers momentum to the air. In the configuration without containment walls, such a process always acts on the new fluid that is drawn from the environment and then is expelled radially from the tooth passage at high speed. When the wheel is surrounded by a casing, the boundary wall keeps the fluid near the tooth surfaces, confining the angular momentum impressed by the gear tooth, therefore the fluid progressively gains angular momentum, increasing its tangential velocity until a stationary condition is achieved. This mechanism is reproduced well by the SPH model, and it takes a lot of computational time, as proved by the high number of wheel turns needed to reach a steady condition.

As far as the resistant torque is concerned, the results are summarized in Figure 6.14, where the average values computed by the different models are compared with experimental data. To sum up, the SPH models overestimate the experiment: in detail, the coarse model prediction is about 30% higher than the measure, the difference decreases to 22% with the mid model, whereas a further refinement improves the torque prediction, in fact the difference is

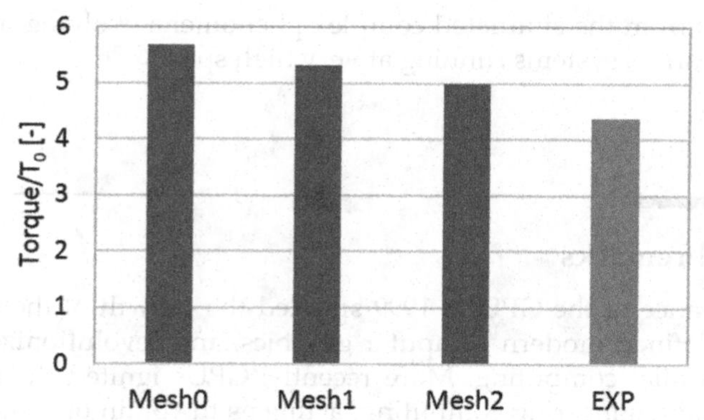

FIGURE 6.14
SPH gas-phase model – average resistant torque as a function of particle size.

14%. In any case, the monotonic trend of the numerical result suggests that further refinements can fill this gap. Regarding the computational time, each model takes a significant time before resistant torque achieve steady-state condition, mainly because of the great test box dimensions; the finest model took about 300 hours on DGX-2.

6.4.4 Next Steps

As reported in the previous paragraphs, the first phase of investigation can be considered accomplished. Main outcomes are the following: SPH methodology is really promising especially for what concerns the preprocessing phase, simple and very short. Alternative CFD methods, like VOF, require longer simulations (about 10 times more time-consuming). Some best practices were defined to push air–oil investigation at the best and, in fact, good agreement with experimental data in terms of torque and flow physics were obtained.

The next steps will focus on industrial-like simulations by identifying the specifically needed HW and SW requirements. Points of attention will be:

- implementing advanced multiphase approach in the planetary gearbox study (see Figure 6.6);
- investigating correct granularity of SPH particles involved in this simulation aiming at correctly predicting physical phenomena while balancing/limiting it with required computational time; and
- identifying the HW specifications of the GPU-accelerated HPC system required to delimit close-to-reality simulations inside an industrial timeframe. This means that the number of GPUs is expected to significantly rise from the value currently used (=16) to capture the needed

details from the simulated complex phenomena evolving in aeronautical gearbox systems running at very high speed.

6.5 Final remarks

The appearance of the GPU in 1999 sparked the growth of the PC gaming market, redefined modern computer graphics, and revolutionized general-purpose parallel computing. More recently, GPUs ignited the next era of scientific and engineering computing, acting as the brain of computers, and providing in all engineering fields a powerful boost to achieve both deep insight into the physics of the phenomena and to enlarge the design space of the most innovative industrial applications. GPU computing for modern engineering is mentioned here but, obviously, the discussion must be considered from a wider perspective including the world of supercomputers, as first-hand experienced in the Lexis program where we met experts and truly innovative HPC products and services for advanced computer-aided engineering. Without them, the development of faster and more realistic physical modeling would not be able to achieve the levels of excellence required and expected of the industry.

In this chapter, the Aeronautics pilot led by Avio Aero inside the Lexis project was illustrated and its two main project streams were described: the porting of a CFD code and the development of a newly designed engineering methodology, always CFD-based, dedicated to the analysis of gearboxes, a realm until recently based only on past decades' expertise.

The first results and progress achieved so far are very promising and applicable in the short term, but still need to be further improved in order to finally integrate them inside the Avio Aero's design systems.

However, the road to achieving this has been identified.

Acknowledgment

This work was supported by the LEXIS project funded by the EU's Horizon 2020 Research and Innovation Programme (2014–2020) under grant agreement No. 825532 and is making use of various computing and data facilities at LRZ and IT4I. More specifically, the core hours and the work at IT4I are supported by the Ministry of Education, Youth and Sports of the Czech Republic from the Large Infrastructures for Research, Experimental Development, and Innovations project "e-INFRA CZ – LM2018140."

Notes

1 Avio Aero is a GE Aviation global business which operates in the design, manufacture, and maintenance of civil and military aeronautics subsystems and systems.
2 The related data presented here are not subject to export control constraints.

References

[T1] A. Arnone, "Viscous analysis of three-dimensional rotor flow using a multigrid method," J. Turbomach. 116(3) (1994) 435–445.
[T2] R. Pacciani, M. Marconcini, A. Arnone, "Comparison of the AUSM+-up and other advection schemes for turbomachinery applications," Shock Waves 29(5) (2019).
[T3] Arnone, M. Liou, L. Povinelli, "Integration of Navier–Stokes equations using dual time stepping and a multigrid method," AIAA J. 33(6) (1995) 985–990.
[T4] M. Giovannini, M. Marconcini, A. Arnone, A. Dominguez, "Hybrid parallelization strategy of a CFD code for turbomachinery applications," in: 11th European Turbomachinery Conference, Madrid, Spain, 2015.paper ETC2015-188
[T5] Burberi C, Michelassi V, Scotti Del Greco A, Lorusso S, Tapinassi L, Marconcini M, Pacciani R. 2020. "Validation of Steady and Unsteady CFD Strategies in the Design of Axial Compressors for Gas Turbine Engines." Aerospace Science and Technology. 107:106307.
[R1] D. Massini, T. Fondelli, B. Facchini, L. Tarchi, and F. Leonardi, "Experimental investigation on power losses due to oil jet lubrication in high speed gearing systems," Proc. ASME Turbo Expo, Vol. 5B-2017, 2017, GT2017–64703.
[R2] D. Massini, T. Fondelli, A. Andreini, B. Facchini, L. Tarchi and F. Leonardi, "Experimental and numerical investigation on windage power losses in high speed gears," J. Eng. Gas Turbines Power, vol. 140, no. 8, 2018.
[R3] Massini, D, Fondelli, T, Facchini, B, Tarchi, L, and Leonardi, F. "Windage Losses of a Meshing Gear Pair Measured at Different Working Conditions." Proc. ASME Turbo Expo, Vol. 5B, 2018, GT2018–76823.
[R4] T. Fondelli, A. Andreini, R. Da Soghe, B. Facchini, and L. Cipolla, "Numerical simulation of oil jet lubrication for high speed gears," Int. J. Aerosp. Eng., vol. 2015, 2015.
[R5] T. Fondelli, A. Andreini, and B. Facchini, "Numerical investigation on windage losses of high-speed gears in enclosed configuration," AIAA J, vol. 56, no. 5, 2018.

7

Event-Driven, Time-Constrained Workflows: An Earthquake and Tsunami Pilot

Rubén J. García-Hernández, Thierry Goubier, Danijel Schorlemmer, Natalja Rakowsky, Sven Harig, Andrea Ajmar, Lorenza Bovio, Stephane Louise, and Tomáš Martinovič

CONTENTS

7.1 Introduction

A strong concern about disasters and the emergency response is the ability to have early situation assessments and short-term forecasts, since obtaining a precise and detailed view from the ground is difficult, particularly if one considers that communication breakdowns are a possibility for disasters. Therefore a remote understanding of the situation and short-term forecasts can be key tools, and this is what our current systems are able to provide, by

DOI: 10.1201/9781003176664-7

combining multiple technologies and resources, such as detailed simulations and large datasets describing the affected areas, allowing us to project the effect of an event on what is a kind of digital twin of the ground truth.

To try to answer this initial intuition, an earthquake and tsunami large-scale pilot is being implemented, combining simulation, large datasets, and a combined workflow to provide up-to-date results and estimates very early on after an event in the projected impacted area, and where the emergency teams should concentrate their efforts on.

7.2 Event-Driven, Time-Constrained Workflows

Between the moment an earthquake is detected, and the moment a potential tsunami reaches the coastal areas, is often a matter of mere minutes, or hours in the best case. Especially for tsunamis with the largest economic and human impacts the tendency would be on the lowest values of this interval. This is a very short time frame within which any forecast as to when and where the largest impact could happen can make a difference on the number of victims or on the scale of the economic impact the tsunami resulting from the earthquake would have.

The key elements that are exercised here is that we need, at an unknown starting point in time (i.e. an event), forecasts and estimates that feeds into the emergency response decision process within a precise, and short, time frame. This calls for an automated workflow, that is event-triggered (here the occurrence of an earthquake) and time-constrained for the forecasting and estimates results to be used in the emergency decision process.

7.2.1 Requirements

The starting point of what would constitute a tool for emergency prediction and management should be the occurrence of an earthquake. From the first second of such an event to the date an emergency response would be useful to mitigate losses in the most effective way, is a matter of minutes. The primary requirement of such a tool is therefore event-driven with strong latency constraints to the point a first conclusion can be given to emergency teams and policymakers, and additional time constraints to ensure that further decisions from the first responders are made with the best available estimates and forecasts.

Nonetheless, as the precision for a rapid damage assessment would be rough and coarse-grained compared to what can be done on precise simulations, but also as a more precise simulation would be required as soon as available, we can track two parallel simulation paths from the time of detection of an earthquake: a first workflow would be constituted by a (set of) quick and latency constrained simulations, coarse-grained to give emergency responders an early and timely warning about where the most important damages, risks, and fatalities could be situated, and a second workflow offering a fine-grained simulation with refined assessments but that can be delayed by several minutes, up until the fine-grained simulations are fully run and their results provided to the emergency responders.

7.2.2 Background

Numerous approaches to workflows do exist, from business process management to embedded systems and high-performance computing. They allow the expression of processing steps and the flow of data through those steps, allowing efficient allocation of the steps onto computing resources since it becomes possible to predict when each step will happen, and how much it will require from previous steps. Such models allow the computation of properties on the workflow, ensuring, for example, that no resource limitation could hamper execution of the workflow in any situation.

In the context of a workflow linked to natural disasters and emergencies, three properties of those models (that we will name "model of computation," or MoC for short) stand out. First, the fact that we can chain execution of all elements while verifying that we do have the minimum set of resources (compute and storage) that are needed for the workflow to complete in all cases. Second, that we can define and propagate time constraints over this flow, and so defining when processes or tasks have to start and complete, when the data items must be available, and so on. Third, that we can predict in advance what are the steps that will take place, and so improve performance and efficiency by pre-allocating resources and setting up steps, as well as using intelligent allocation policies to select the most adequate resources.

7.2.3 Overall View of the Workflow

As described in Figure 7.1, the workflow is organized around three main elements: an exposure dataset update part, a tsunami and earthquake event response part, and a satellite-based emergency mapping (SEM) part.

First, an exposure dataset is maintained up to date with OpenStreetMap (OSM). OSM emits updates every minute; each update is processed, filtered, and enriched before being added to the OpenBuildingMap database. Then, at regular intervals, an exposure dataset is released out of the OpenBuildingMap database. This part of the workflow runs on a permanent basis, 24/7, so as to keep an always up-to-date view of the global dataset.

Second, when an event occurs, the tsunami and earthquake event response part is activated. It is composed of two branches: the first is activated when the early information about the earthquake event is known (the earthquake

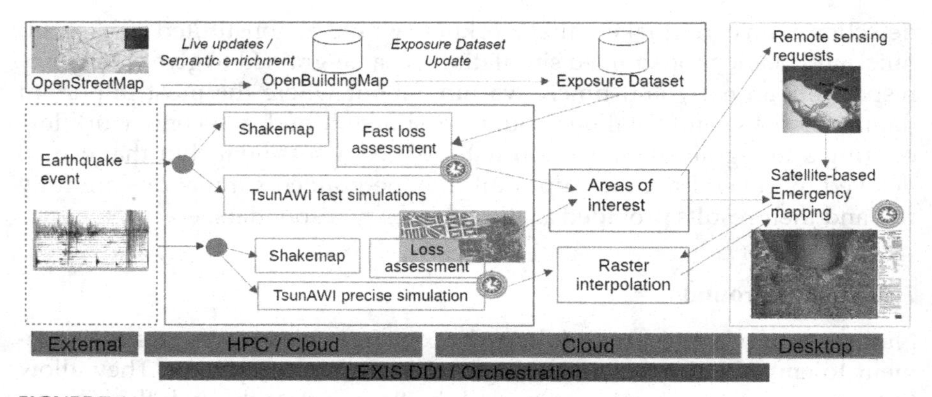

FIGURE 7.1
A schematic description of all parts of the earthquake and tsunami workflow. With map and map data from OSM and OpenStreetMap Foundation, image from Copernicus Emergency Management Service (© 2021 European Union), EMSR317.

hypocenter and magnitude) and it triggers both fast tsunami inundation simulations with TsunAWI, and the generation of a shakemap which is fed to a fast loss assessment computation. The latter requires access to an aggregate level of the exposure dataset. This fast branch is tied to a first deadline, suitable for triggering an emergency warning.[1] Upon reception of additional information about the earthquake event, namely the moment tensor, the second branch is activated. Similarly to the fast branch, the precise branch will compute a shakemap, a loss assessment, and tsunami simulations, but all of those at a higher precision level given the availability of a more accurate source characterization of the event. The tsunami simulation will be done with a higher resolution mesh and a finer time step; the loss assessment will be a detailed one, at the building level.

The SEM part will also be triggered by the earthquake event, but won't proceed until the first results from the tsunami and earthquake part are available. This flow relies on combining multiple sources to produce emergency mapping products; in particular it requires post-event remote-sensing products of the affected areas, and the determination of the affected areas is based on the results of the simulations and loss assessments: a procedure for determining the areas of interest is used there to extract priority areas from both the fast and the precise simulation paths.

Not shown here, but implicit in that workflow is a feedback process which incorporates input from the emergency response teams through updates to OSM, so an extension of that workflow could be used to keep a situation assessment up to date during the emergency response phase. Such an update can already be considered possible today, by reactivating by hand the precise simulation path and providing as input an event file, which may contain updated information about the source of the earthquake event: in that case simulation and assessment may benefit from an updated source and an updated exposure dataset.

7.3 Workflow Components

7.3.1 Shakemap and Exposure Dataset

The computation of the ground-motion field (shakemap) of an earthquake is performed using the *shakyground2* library developed at GFZ Potsdam. This library is encapsulated into a web application for easy exploration of the effects of different earthquake and ground-motion parameters on the resulting shakemap and in a docker container for integration in the workflow. The shakemap codes take earthquake parameters in QuakeML format as input and use preset ground-motion models to compute the expected levels of shaking in the target area. The same earthquake-parameter set is used by the tsunami-modeling software as described in the next section.

A rapid damage-and-loss assessment is calculated using the *losscalculator* library developed at GFZ. The new aspect of this library is that it extends classic damage assessments of aggregated exposure models to computing damage probabilities for each single building if such data is available.

The exposure model used here extends the classic exposure models from aggregated exposure data with open building information. The aggregated exposure is distributed to a regular grid based on globally available settlement information. The model then pulls data from OSM and assigns them to the grid cells depending on their footprint centroid. In the next step, the number of buildings assigned to the grid cell are reduced by the number of buildings from OSM to keep the overall building count in balance. In areas where a manual completeness assessment of OSM buildings was conducted, this procedure can change for cells considered complete. In this case, the OSM buildings are completely replacing the building information assigned to the grid cell. Thus, over time and with increasing contributions to OSM, more and more cells will have complete building coverage, gradually replacing the estimated building counts based on global settlement data with on-the-ground voluntarily collected data in OSM. Each building is assigned at first the probability distribution of building types (e.g. wood-frame house, reinforced concrete with four stories) with their relative frequency as defined in the aggregated exposure model. For each building for which a building dataset exists in OSM, this probability distribution is narrowed down by the building properties given in OSM (e.g. number of stories, roof shape). The resulting probability distributions therefore provide the best possible knowledge of each building while simultaneously covering the uncertainties around the type determination.

The new damage-and-loss calculator computes damage for each building separately, thereby taking the probability distribution of different possible building types into consideration. This results in a probabilistic estimate of different damage levels possible to be observed for each building. In the case of cells not completely covered with building footprints, aggregated building information has been assigned to the cell for which a probabilistic assessment of possible damage states is also computed. These damage assessments can now be either aggregated over the cells or shown separately for each building (with a residual damage per cell in the case of incomplete cells).

7.3.2 Tsunami Simulations

The tsunami simulation is performed with the open-source code TsunAWI (https://tsunami.awi.de) which is in operational use in the Indonesia Tsunami Early Warning Center (Harig et al. 2020). TsunAWI discretizes the nonlinear shallow water equations with wetting and drying on an unstructured triangular finite element mesh with varying resolution. Usually, the deep ocean can be resolved with 10km–20km edge length, while 500m allow for a rough estimate of the wave height at the coast, and about 20m are required for a realistic assessment of inundation (Griffin et al. 2015). The code is written in Fortran90, initial conditions are usually given as the bottom displacement due to an earthquake (Okada parameters, idealized source, or prescribed at the mesh vertices), and optional output formats are netcdf on mesh vertices, ascii on tide gauge locations, and an interpolation to raster (Golden Software Surfer grid). Until recently, the focus was on pre-computed tsunami scenarios and TsunAWI was optimized with regard to throughput. In particular, OpenMP parallelization was sufficient. With modern hardware, real-time computation of high-resolution simulations became feasible, and adding optional MPI support further decreased the time to solution for a single scenario, with hybrid runs with four to six OpenMP threads and hyper-threading giving shortest run times.

For the LEXIS workflow, regional setups with both a coarse mesh for a very fast first assessment of the inundation and a fine mesh with the required 20m resolution for a more realistic simulation are provided. This also takes into account that in the case of a real event, first measurements of the earthquake magnitude and location come with a high uncertainty. One should wait for the more accurate simulation until the momentum tensor – information on the mechanism and the geometry of the fault – is available.

Table 7.1 summarizes the characteristics of the fine and the coarse mesh for the Padang area, with the computational mesh and an example of a simulated inundation shown in Figure 7.2. Similar pairs of meshes were set up for Coquimbo, Chile, and the Mediterranean. Extensive test runs are essential to determine the optimal time step and the characteristics of tsunamis in the respective region. For example, while in Padang, two hours of

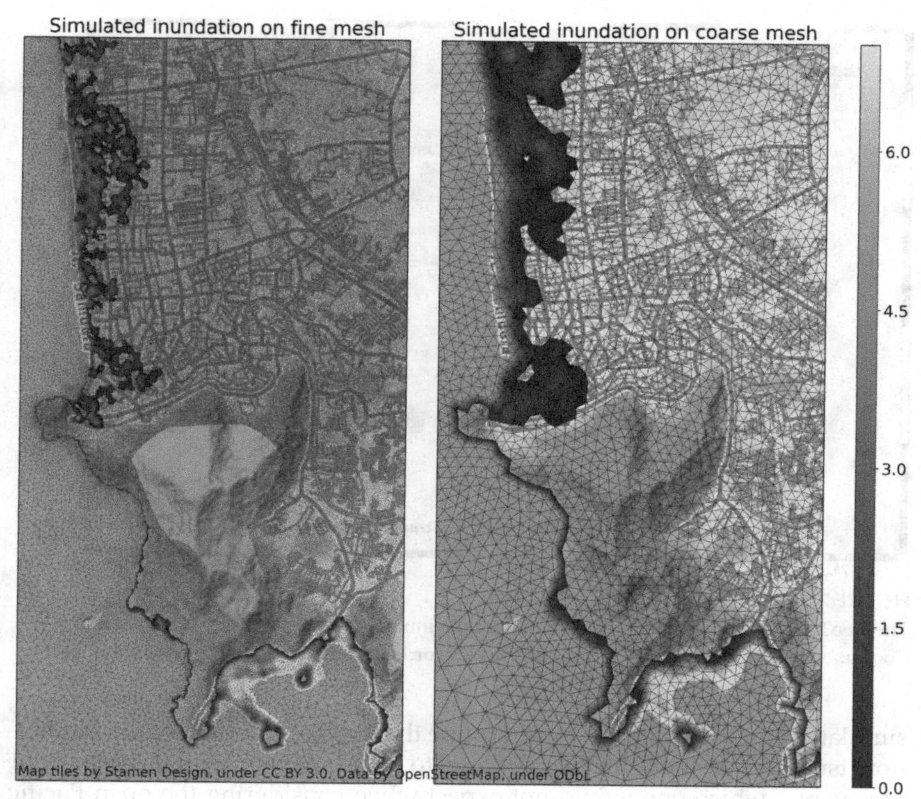

FIGURE 7.2
Inundation and mesh over the city of Padang for the test scenarios, left with the fine mesh, right with the coarse mesh.

TABLE 7.1
Comparison of fine and coarse mesh for Padang

		Fine mesh	Coarse mesh
Number of mesh vertices		1,242,653	231,586
Resolution	In Padang	20m	200m
	In the ocean	5,000m	15,000m
Timestep		0.15s	1.5s
Compute time for a 2h simulation	48 MPI tasks, 4 OpenMP threads, 2x Intel Xeon Platinum 9242	2:57 min	4s
	192 MPI tasks, 4 OpenMP threads, 4 nodes with 2 Intel Xeon Platinum 9242	1:06 min	—

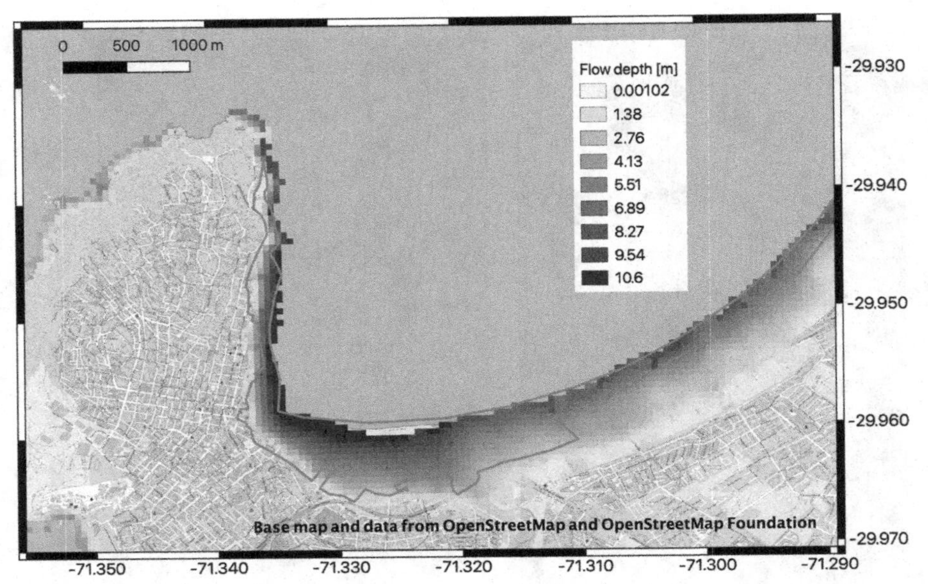

FIGURE 7.3
The GeoTIFF produced by the interpolation of the inundation mesh as computed for the Coquimbo tsunami overlaid with the historical recorded inundation extent (Coquimbo, 2015)

simulation time is sufficient to determine the maximum extent of the inundation, tsunamis on the Chilean coast tend to have large wave crests hours after the event – which one would not expect when considering the open Pacific Coast compared to the more complex coast with the Mentawai islands in front of Padang.

For the workflow within LEXIS, which aims at the inundation, the result is a netcdf file with the triangulation and the inundation height at the vertices on land. A post-processing step (Python, GDAL) interpolates from the unstructured data to a GeoTIFF raster of desired resolution as presented in Figure 7.3.

7.3.3 SEM

SEM mechanism aims to improve disaster relief effectiveness through the provision of information derived by the analysis of satellite imagery acquired in the aftermath of an event.

The Copernicus Emergency Management Service (CEMS), the European Commission SEM service, is composed of an on-demand mapping component providing rapid maps for emergency response and risk–recovery maps for prevention and planning in support of disaster management activities not related to immediate response.

The CEMS Rapid Mapping service provides geoinformation within hours or days, immediately after a disaster event. The service with its 24/7 availability covers all the phases of the emergency management cycle, from the satellite tasking and image acquisition until the delivery of vector data and ready-to-print maps required by the user.

All maps, in particular post-event ones, are provided within 12 hours after image reception and quality acceptance.

Since time is the major constraint in a response phase immediately after a disaster, this specific phase can have great advantages in using automatic models and procedures.

Two are the specific CEMS phases that can be significantly improved by the exploitation of models and automatic procedures integrated into the LEXIS platform:

1. early-tasking of satellite acquisitions, exploiting the output of LEXIS model;
2. automatically generate a first estimate product (FEP), an early information product providing an extremely fast assessment of most affected locations, combining model outputs with exposure data.

Satellite images are the main source of information and it is clear that early tasking of satellite acquisitions can improve the time delivery of the SEM products. In case of satellite platforms acquiring data on demand, image request cut-off time is specific for each satellite data provider and missing these cut-off times can mean missing the opportunity for image data acquisition, leading to a consequent delay in the delivery of the products.

In the specific case, the output of LEXIS models can be used to trigger satellite tasking over a possible affected area, after an alert and before the activation of the service by an authorized user, moving up the delivery of the products.

For this purpose, a core component of this part in the workflow is a procedure for quickly identifying potential areas of interest to be submitted to the satellite data provider, by crossing event-related model outputs (shakemaps and inundation maps) with exposed assets and prioritizing them based on the potential impact.

The fast identification of those areas based on objective criteria, provides a more aware and solid satellite images acquisition process limiting false positive and false negative in the identification of affected areas.

Additionally, the procedure is exploited to generate a quick damage assessment estimation for an FEP, based uniquely on modeled data and to support the following damage assessment based on the interpretation of post-event satellite images.

7.4 Technological Layers

The implementation of the workflow is done along three layers, each layer contributing to the overall result. The first layer is the orchestration, enabling the scheduling of the various components of the workflow so as to ensure all dependencies and deadlines are fulfilled. The second layer is the heterogeneous distributed compute, which both enables performance and reliability of the workflow execution. The third layer is the data layer, where the geographically distributed availability of the necessary datasets and output storage, as well as ensuring the performance needed are met.

7.4.1 Technology Layer 1: Orchestration

This layer relies on two technologies: one is an MoC named PolyGraph (Dubrulle et al. 2019) which enables the specification and analysis of the workflow specifics and requirements; the second is YSTIA, an orchestration layer that allows the integration of all computing components and data movements, as well as their dependencies, and the effective execution of the related components according to the schedule.

The MoC specifies a workflow as a set of actors, dependencies between those actors, and time constraints applied to some of those actors. Time constraints allow the description of elements such as: should have completed before a deadline, should not start before a specific time. Time is expressed as a clock (a sequence of ticks) and a phase (an offset in time to a tick); the phase allows the expression of a latency such as the fact a certain number of steps must complete before a delay to a clock tick. Those concepts allow the expression of periodic workflows, such as the exposure dataset update part of the workflow.

When dealing with event-driven parts, the clock ticks are merged with events in the following way. An event is considered as a clock tick and all derived clocks and phases are understood as relative to the event. To match a periodic schedule and to be able to discuss resources needed for such a permanently ready workflow, we artificially map the event-driven, sporadic nature of the disaster considered in a pseudo-period defined as the shortest period for which we expect at most one event to happen. In the context of earthquakes with tsunamis anywhere on earth, such a period cannot be formally defined; it cannot even be defined if we consider a single earthquake event and aftershocks. However, from an operational point of view, one can define such a period, and the MoC can use it to simulate the occurrence of an event at every period, and so compute the necessary resources for ensuring that the workflow is able to fire once every period.

7.4.2 Technology Layer 2: Heterogeneous Compute

In the various computational elements of the earthquake and tsunami pilot, a significant concern is performance. Not in the sense of benchmark-beating throughput, but in ensuring the deadlines are met, and that, if possible, the computed results are of the best quality considering that a compromise between the resolution of the situation and the time needed to simulate, for example, a tsunami inundation, can be made. As seen above, coarse-resolution simulations are extremely fast and so amenable to almost any deadline constraint.

Knowing that the performance problem of the minimum result has been attained, the issue is now to extract higher performance from the code and the infrastructure, by using more resources and in particular heterogeneous resources.

For this, the main computationally heavy parts of the workflow have been studied and modified to make use of accelerators, that is specific hardware components such as GPU and FPGA that are able to increase all or parts of the computation. The main driver here has been the tsunami simulation code, TsunAWI, for which two things have been done.

First, a post-processing step of TsunAWI has been extracted as an independent routine, and this routine, originally CPU only, is being ported to GPU and FPGA: the raster interpolation step. That job can then be allocated freely on one of the available accelerated nodes, the workflow being able to provide all three different implementations.

Second, TsunAWI itself has been considered for acceleration. Early results indicate that an FPGA accelerator is a possible target for the unstructured mesh computations of TsunAWI, and work is underway to explore that path. Once completed, it would allow for two different targets for high-performance runs of TsunAWI: either a group of CPU nodes or an FPGA-accelerated node.

Using additional resources is also considered, because of two requirements that are inherent in this workflow. The original source, on the fast simulation path, is imprecise, and its deformation mechanism unknown at that step in time. To compensate for that, multiple runs of TsunAWI are used, each one on a source generated in the uncertainty radius of the event source itself. The approach here is one suitable to urgent computing: the more simulations complete, the better is the picture of the possible tsunamis around the source. And, the strict requirement is to have at least one result by the deadline. That requirement translates into the following orchestration choice: multiple jobs are created and submitted at that stage, using for that a script that generates artificial, additional sources around the original one in the uncertainty radius. All these jobs are submitted on multiple centers and technologies (e.g. cloud resources, HPC centers). In the HPC centers, a specific approach is used which is to over-submit over the minimum resources needed, with a timeout (i.e. many jobs, each of them on a different source).

We expect each HPC center submission to happen over a queue, where a minimum set of nodes will be guaranteed, hence ensuring we have a minimum set of results by our deadline (in the future, this could be a dedicated urgent computing queue provided by HPC centers). If the center manages to allocate additional resources during our run to the deadline, then some (maybe all) of our over-submitted jobs will also be submitted and hopefully completed. After that step, a post-processing job is allocated on a burst buffer, starting as soon as a first result is made available and aggregating all additional results that are made available during the run to the deadline, to complete an aggregate map of the inundation risk associated with all those sources in the uncertainty circles. Once the deadline is reached, all currently running jobs and all not yet submitted jobs are killed: their eventual result will never be used.

The same approach is used with the cloud job submissions and also submitting on an additional computing center, to ensure redundancy: there, the result aggregation rule at the deadline is: use the best result available by then, and kill everything else still being processed or submitted. The resulting workflow is presented in Figure 7.4.

The target for the first, fast simulation path in the workflow is therefore expressed by the following formula:

$$\left|k \times [n - m] + c\right| \text{ number of results by } T_0 + \Delta T \text{ is greater than 1}$$

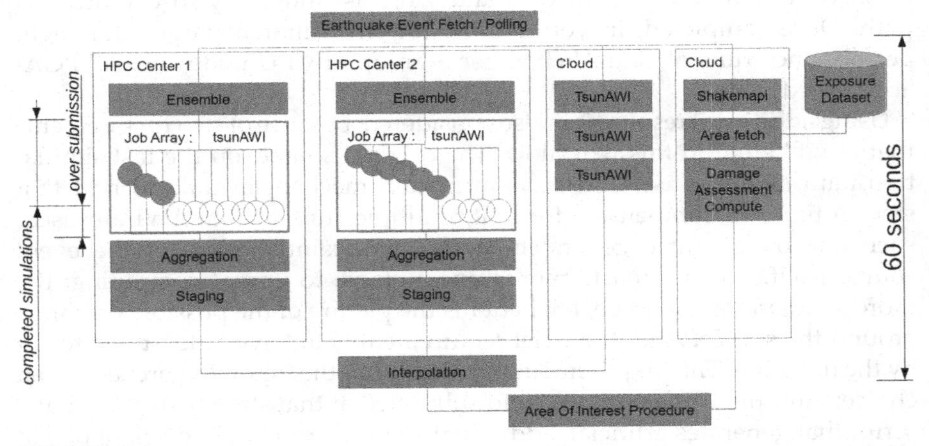

FIGURE 7.4
The fast simulation path of the workflow, showing how ensemble tsunami simulations are dispatched on two HPC centers, completed with cloud tsunami simulations and the shakemap and aggregate loss assessment computations.

where k is the number of HPC centers, n is the guaranteed number of nodes in a center, m is the maximum number of nodes the HPC center can allocate from T_0 to $T_0 + \Delta T$, and c is the number of cloud jobs that can be submitted and run during ΔT. The $\lfloor n - m \rfloor$ results from the HPC centers are aggregated by center, as explained above; so each HPC center run will provide a single result. T_0 is the earthquake initial time and ΔT is the delay before the deadline.

7.4.3 Technology Layer 3: Data

The distributed data infrastructure (DDI) provides APIs to organize datasets and their associated metadata in an efficient manner, and provide methods to transfer the datasets needed so that they are available at the compute infrastructure needed. Datasets may also be replicated within the federation (using EUDAT/B2STAGE); this is especially important in urgent computing scenarios since having the data already at the compute site saves precious time. We plan to use the burst buffers to accelerate optional encryption and compression, since datasets with many small files result in inefficient transfers across zones.

The pilot workflow understands the DDI structure to ensure proper runs, and is codesigned as well to maximize its use of the storage layer: relying on the DDI capabilities to enable in particular transparent, location-independent processing (a key element in reaching guaranteed availability of the workflow) and at the same time expect the layer to efficiently exploit local resources to ensure I/O heavy, massively parallel access to datasets and post-processing of simulation results do not overload the DDI with requests.

First, the large exposure model dataset that is used by the workflow is organized so that the workflow can fetch, out of the dataset, the relevant subset, through the quadtree decomposition of the global dataset, as a single request. That decomposition is also optimized so that the smallest unit of data is sized to match the DDI preferred size to reach maximum transfer performance. In the case of an earthquake (and tsunami), it is possible to decide which subset of that exposure dataset has to be fetched very early on, as soon as the hypocenter of the earthquake is known.

Second, the workflow is organized to benefit from the burst buffers, by orchestrating jobs so that they read their datasets needed from the burst buffer, and store and post-process their results on the burst buffer. The burst buffer processing capabilities also allow additional optimizations to be done, such as compression either at the DDI request level (transparent compression/decompression to reduce request size and reduce also the number of transferred objects by gathering them into a compressed archive), or at the dataset level (compress in advance the dataset while maintaining its capability to directly access subsets, and use the burst buffer decompression capability upon workflow execution), but this is undertaken in a transparent way for the workflow.

Third, the workflows rely on the ability of the DDI to indicate when a specific file or dataset has been uploaded or made available. This allows us to maintain the event-triggered nature of the workflow: steps in a pipeline have dependencies to the previous steps expressed as data items needed, that are stored by the DDI.

7.5 Conclusion

In disaster response workflows, a combination of elements are necessary to ensure the effective use of the results for disaster early warnings and emergency response. A fusion of multiple sources and simulations are needed to properly identify the affected areas, so combining datasets and computing resources for simulation. In the case of an earthquake, the workflow has to be event-triggered and time-constrained, with deadlines for the availability of results. A combination of technologies are then involved to ensure the implementation of such a workflow: the orchestration, heterogeneous, multi-site execution of components, and the data infrastructure. We have shown how the technologies in orchestration, the compute federation, and the DDI of the LEXIS project enables such a workflow.

Acknowledgment

This chapter was supported by the LEXIS project, funded by the EU's Horizon 2020 Research and Innovation Programme (2014–2020) under grant agreement no. 825532.

Note

1 This deadline provides information to experts so that they can decide on an evacuation warning for people close to the epicenter.

References

Dubrulle, P., Gaston, C., Kosmatov, N., Lapitre, A., and Louise, S. (2019). A Data Flow Model with Frequency Arithmetic. *FASE*: 369–385.

Griffin, J., Latief, H., Kongko, W., Harig, S., Horspool, N., Hanung, R., Rojali, A., Maher, N., Fuchs A., Hossen, J., Upi, S., Edi Dewanto, S., Rakowsky, N. and Cummins, P. (2015). An evaluation of onshore digital elevation models for modeling tsunami inundation zones. *Frontiers Earth Science* 3:32. doi: 10.3389/feart.2015.00032

Harig, S., Immerz, A., Weniza, G. J., Weber, B., Babeyko, A., Rakowsky, N., Hartanto, D., Nurokhim, A., Handayani, T., and Weber, R. (2020). The Tsunami Scenario Database of the Indonesia Tsunami Early Warning System (InaTEWS): Evolution of the Coverage and the Involved Modeling Approaches. *Pure and Applied Geophysics* 177, 1379–1401. https://doi.org/10.1007/s00024-019-02305-1

8

Exploitation of Multiple Model Layers within LEXIS Weather and Climate Pilot: An HPC-Based Approach

Paola Mazzoglio, Emanuele Danovaro, Laurent Ganne, Andrea Parodi, Stephan Hachinger, Antonella Galizia, Antonio Parodi, and Jan Martinovič

CONTENTS

8.1 Introduction: Background and Driving Forces

The H2020 LEXIS project aims to design and develop an advanced engineering platform at the confluence of high-performance computing (HPC), big data (BD), and cloud solutions which leverages large-scale geographically distributed resources from existing HPC infrastructure, employing BD analytics solutions and augmenting them with cloud services (Parodi et al. 2021). The emphasis of LEXIS is on the interaction between HPC and cloud systems, built on top of data sharing and methods to compose workflows of tasks running on both cloud and HPC systems, understood as LEXIS distributed computed infrastructure (DCI). A platform has been developed to enable these workflows and demonstrate its abilities through three large-scale socioeconomic pilots, targeting aeronautics, weather, and earthquakes

DOI: 10.1201/9781003176664-8

and tsunamis (Parodi et al. 2021). This chapter reports the key challenges and results concerning the Weather and Climate pilot which fully benefits from the project results in terms of orchestration system, distributed data infrastructure (DDI), DCI, and workflows management.

8.2 The Weather and Climate Pilot

The Weather and Climate pilot focuses on a complex system to provide a set of forecasts concerning weather, flood, forest fire, air pollution, and agriculture by means of several complex workflows each consisting of various meteorological, hydrological, and air quality components (Parodi et al. 2021). These workflows include ingestion of conventional and unconventional observations, global weather models, regional weather models, application models, and socioeconomic impact models (Figure 8.1). The workflows are run across disjoint computing resources (LEXIS DCI): global weather models are executed on ECMWF's HPC in the UK while regional weather models run on HPCs in Italy (CIMA), Germany (LRZ), or Czechia (IT4I). Application models and socioeconomic impact models are instead executed on cloud-based resources in Germany or Czechia.

LEXIS is managing several weather and climate modeling tasks, namely WRF Model, RISICO, Continuum, ADMS, and ERDS, which are described below.

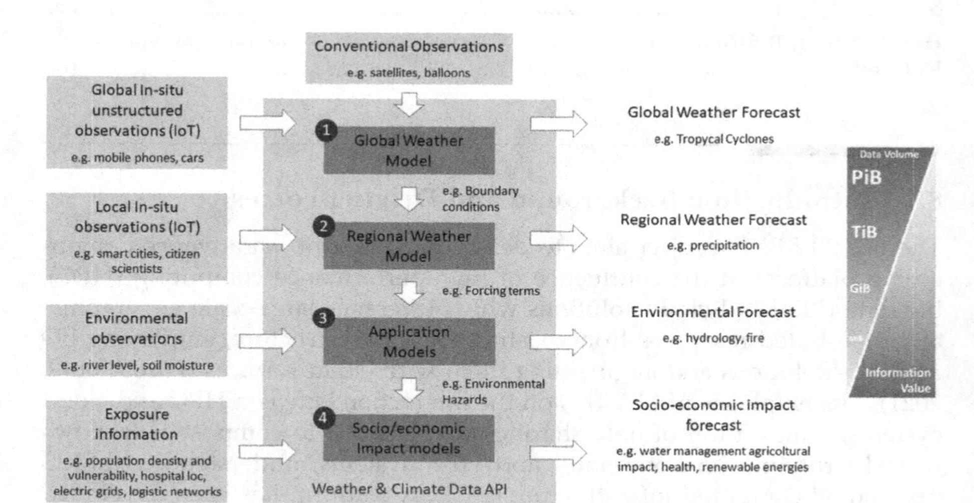

FIGURE 8.1
LEXIS Weather and Climate complex workflows.

The Weather Research and Forecasting (WRF) model is a proven mesoscale numerical weather prediction (NWP) system, designed to serve both operational forecasting and atmospheric research needs (Powers et al. 2017). It features multiple dynamical cores, a 3D variational (3DVAR) data assimilation system, and a software architecture allowing for computational parallelism and system extensibility (Lagasio et al. 2019a, 2019b). WRF is suitable for a broad spectrum of applications across different scales, ranging from thousands of kilometers to meters.

RISICO (RISchio Incendi e COordinamento/Fire Risk and Coordination) is a mathematical model developed by the CIMA Research Foundation to support operators in forest fire prevention activities (Fiorucci et al. 2008). RISICO processes a continuous data flow consisting of meteorological information as weather forecast and satellite records. Parameters such as the moisture content of the vegetation, the wind and the orography of the territory allow users to quantitatively assess the danger resulting from the eventual triggering of a forest fire both in terms of propagation speed and linear intensity of the flame front.

Continuum is a hydrological model developed by the CIMA Research Foundation to reproduce the flow of water within a basin (Silvestro et al. 2015). The model has a reduced number of parameters and is able to work both in the pre-event analysis and forecast phase and in the monitoring stage for the active control of hydrological events taking advantage of all the information available via in situ weather stations and satellite data.

ADMS (atmospheric dispersion modeling system) models are managed by NUMTECH to perform atmospheric dispersion calculation depending on meteorological conditions and emission release (Brocheton et al. 2008; Carruthers et al. 2011). Two applications are performed. The first one (industrial case) aims to forecast SO_2 impact at ground from industrial release to prevent pollution peak, while the second one (urban case) has been developed to forecast at high-scale NO_2 and PM concentration over a full city (Paris).

The Extreme Rainfall Detection System (ERDS – http://erds.ithacaweb.org), developed and implemented by ITHACA, is an early warning system for the monitoring and forecasting of rainfall events, with near-global spatial coverage (Mazzoglio et al. 2019a, 2019b, 2019c). The system is able to provide alerts about heavy rainfall events using both near real-time measurements and rainfall forecasts. The information is accessible through a WebGIS application, developed in an open-source environment.

8.3 Observational Data

The Weather and Climate pilot considers different observations such as authoritative and personal weather stations (PWS) as well as meteorological radar data to be assimilated in the WRF model and used for validation purposes.

Concerning authoritative weather stations, namely the Italian Civil Protection Department (ICPD) dataset, ICPD is designing and managing in real-time risk reduction actions over the national territory, determined by high-impact adverse weather, through its Centro Funzionale Centrale (Central Functional Center), and coordinating a federated national early warning system, in collaboration with regional authorities. In this framework, CIMA archives and curates, on behalf of ICPD, a large number of in situ authoritative weather stations: 6059 rain gauges, 2299 hydrometers, 4373 thermometers, 1270 barometers, about 2500 anemometers, and 2683 hygrometers (Parodi et al. 2021).

CIMA has also developed a partnership with IBM to obtain real-time measurements (hourly temperature, wind, rainfall, relative humidity, and pressure) acquired by about 150,000 PWS over the globe (32,000 from Europe, Figure 8.2). In addition, historical data acquired by about 13,000 PWS over Europe are available for the time period June 1–November 30, 2018.

Two different radar meteorological datasets are available. Reflectivity radar CAPPI (500 m vertical resolution over 500–12,000 m, 2.5 x 2.5 km grid spacing) are available over France for the period June 1–November 30, 2018. Furthermore, for the years 2018, 2019, and 2020 reflectivity radar CAPPI (2000–3000–5000 m, 1 km x 1 km grid spacing) is available over Italy (Figure 8.3).

8.4 LEXIS DDI and Weather and Climate Data API

The LEXIS DDI is based on an iRODS (https://irods.org) data management system with fault-tolerant setup. One of the key strengths of iRODS is represented by its seamless integration of the services B2SAFE, B2STAGE, and B2HANDLE of the European Collaborative Data Infrastructure EUDAT (Parodi et al. 2021). The LEXIS iRODS backend is structured in federated zones (one per supercomputing center, namely LRZ and IT4I) in order to ensure for the users the data accessibility independently of its actual location as well as automatic replication and migration of data on an as-needed basis.

LEXIS has designed and implemented a rich set of modern JSON-based REST APIs as a front-end to the iRODS backend to facilitate upload, download, and deletion of datasets, managing access rights and staging of datasets into LEXIS DCI. Datasets (e.g. weather forecast outputs) include relevant metadata, so that data can be findable according to FAIR principles (Wilkinson et al. 2016). The data volumes involved in LEXIS are managed by means of asynchronous transfers over the LEXIS DDI: an appropriate queuing system

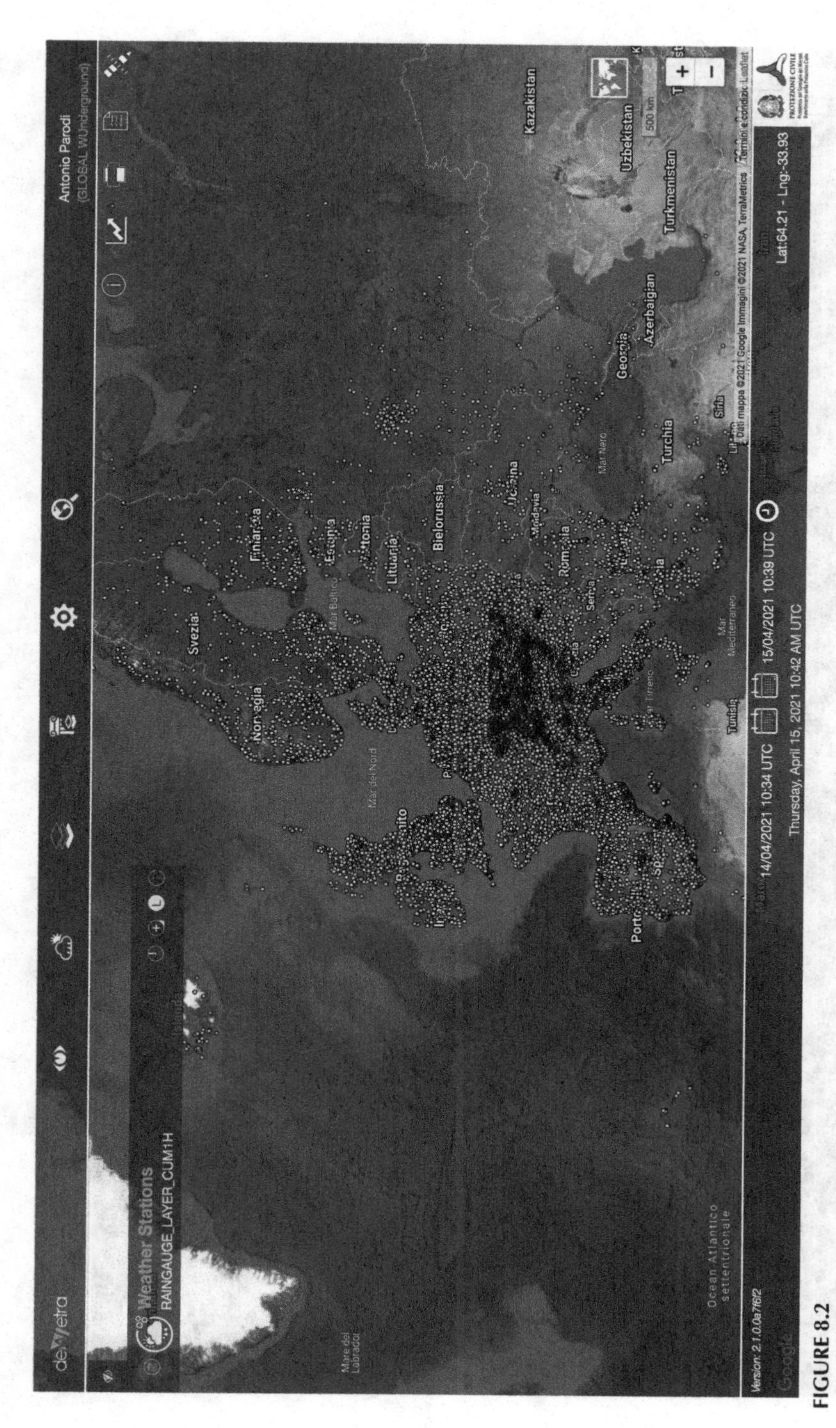

FIGURE 8.2
Weather Underground PWS on MyDewetra platform (courtesy of ICPD).

FIGURE 8.3
Radar data mosaic over Italy published on MyDewetra platform (courtesy of ICPD).

allows the user to follow the progress of their requests. The DDI security is ensured by means of the global LEXIS authorization and authentication infrastructure (AAI), which is based on Keycloak4 instances, and it interfaces to via OpenID3. The setup uses token hashing, as a security extra layer, to allow iRODS and Keycloak to interact (Garcia-Hernandez and Golasowski 2020).

However, the LEXIS DDI does not exist in a vacuum and independently from other data sources requested by the different workflows: some datasets required for the Weather and Climate large-scale pilot have been stored (and continue to be stored) in previously developed domain-specific storage libraries, as for example datasets at ECMWF and at CIMA Research Foundation. These systems can often provide a more feature-rich view of the datasets, due to their domain knowledge.

The LEXIS project is implementing the "Weather and Climate Data API" (WCDA) in order to deliver a state-of-the-art domain-specific storage library for curated weather and climate data (Parodi et al. 2021). WCDA stores and organizes weather observations from a variety of sources (including in situ unstructured observations at the European level), as well as NWP outputs and intermediate weather data. Data are indexed according to domain-specific metadata, to efficiently support metadata-based queries. Additionally, WCDA has been designed to provide efficient distributed access to ECMWF's MARS, which, to our knowledge, can be considered the largest European meteorological archive. MARS stores more than 300 PB of meteorological and climatological data, from observations to global model outputs. WCDA, in addition to MARS, utilizes a FDB (Fields DataBase) for storing, indexing, and retrieving GRIB data. The FDB is an internally provided service used as part of ECMWF's weather forecasting software stack: it operates as a domain-specific object store, designed to store, index, and serve meteorological fields produced by ECMWF's forecast model and able to support different storage systems (parallel FS, Ceph cluster). The FDB serves as a "hot-object" cache inside ECMWF's HPC facility (HPCF) for accessing MARS Archive. Each instance of WCDA can provide seamless access to local and remote data by contacting other WCDA instances available on the LEXIS platform (Parodi et al. 2021). The WCDA interface is RESTful and the backend is based on a fully scalable architecture with containerized components. Moreover, it can be deployed with Docker Compose or in a Kubernetes cluster. Figure 8.4 represents the WCDA instance deployed at ECMWF.

Around 80% of MARS requests are served directly from the FDB, typically for very recently produced data (Gogolenko et al. 2020). A subset of this data are later reaggregated and archived into the permanent archive for long-term availability. Usage of the FDB allows the WCDA to meet the requirements of data sizes.

The current release of WCDA, based on FDB approach, is designed to efficiently handle global NWP model outputs, encoded in GRIB file format. GRIB (General Regularly distributed Information in Binary form) is a concise

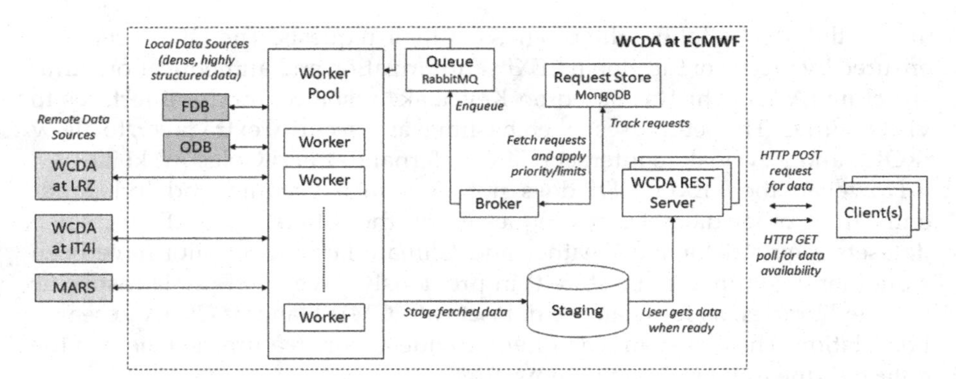

FIGURE 8.4
WCDA at ECMWF.

data format widely used in meteorology to store weather data standardized by the World Meteorological Organization's Commission for Basic Systems. GRIB files are a collection of self-contained records of 2D data (GRIB messages). The individual records stand alone as meaningful data, with no references to other records or to an overall schema. Each GRIB record has two components: the first one is the part that describes the record (i.e. the header) while the second part is the actual binary data itself. The data in GRIB-1 are typically converted to integers using scale and offset, and then bit-packed while with GRIB-2 the possibility of compression is available.

Recent versions of FDB, and thus of the WCDA, have been extended to handle observational data encoded in Observation DataBase (ODB) format, a World Meteorological Organization's standard for meteorological observations. ODB is a file-based database-like system developed at ECMWF to store and retrieve efficiently large volumes of meteorological observational and feedback data.

To accommodate intermediate output files encoded in NetCDF format, the WCDA FDB instance has been further extended to store a selected set of fields generated by the WRF mesoscale NWP model and required by the downstream Risico and Continuum applications.

For intellectual properties rights motivations, the ICPD observational data will be served, solely for LEXIS project research purposes, via a set of dedicated and secured APIs directly accessing CIMA Foundation databases. These data will be transformed into input data necessary for the WRF data assimilation module. The Weather and Climate large-scale pilot has thus two data infrastructures it can rely on – the specialized WCDA for efficient handling of meteorological data, and the DDI which facilitates data exchange as well as general-purpose sharing and publications of results. The usage of both systems will be combined for maximum efficiency.

8.5 LEXIS Orchestration System

The LEXIS platform tightly couples and federates multiple heterogeneous resources to facilitate workflows mixing HPC, IaaS-Cloud, and BD requirements. The workflow orchestration is built on a flexible orchestration solution, Ystia (https://ystia.github.io/), developed by Atos, which combines a front-end system, Alien4Cloud (https://alien4cloud.github.io), and an orchestration engine, Yorc (Ystia orchestrator, https://github.com/ystia/yorc).

Applications and workflows are modeled using the TOSCA (Topology and Orchestration Specification for Cloud Applications, www.oasis-open.org/committees/tc_home.php?wg_abbrev=tosca), an OASIS consortium standard language to describe an application made of components, with their relationships, requirements, capabilities, and operations (Atrey et al. 2015). The TOSCA description of an application includes its life cycle support, as deployment workflow, execution workflow, etc.

The front-end Alien4Cloud, provides a studio allowing users to create applications from an extensible catalog of TOSCA components, deploy these applications, run, and monitor workflows through its backend Yorc. Yorc provides the ability to allocate computing resources on Iaas-Cloud Infrastructures, to manage applications life cycle on these compute instances, and to submit/run jobs on HPC infrastructures. A Yorc plugin was developed in LEXIS so that the orchestrator can access LEXIS HPC infrastructures through IT4Innovation's HPC-as-a-service framework HEAppE (high-end application execution middleware; https://heappe.eu; Svaton et al. 2019). HEAppE middleware allows users to run complex calculations on HPC infrastructure via the user interface of a client application, without the necessity to connect directly to the HPC cluster. HEAppE is able to provide information about the status of submitted jobs and to ensure data transfer between HPC infrastructure and a client-side app. HEAppE also provides a mapping of LEXIS user identities to the user accounts on HPC systems. The HPC-center account to which the LEXIS user is mapped is a particular account (functional account usually not personalized) associated with an HPC project ID as per approval of the project's principal investigator (PI). For this mapping and the HEAppE mechanism to work, the project PI does not have any affiliation with LEXIS whatsoever.

The LEXIS orchestration system can dynamically select available and appropriate computing resources from all LEXIS sites for each modeling task of the Weather and Climate large-scale pilot workflow (Parodi et al. 2021). This approach allows each of the modeling tasks to execute as fast as possible, while the integration with different European computing centers increases the redundancy, allowing for a reliable execution. The Weather and

Climate large-scale pilot workflow implementation includes a set of tasks for cloud as well as HPC systems and is thus ideally suited for execution on the hybrid platform (Parodi et al. 2021). The proposed orchestration system solution features an open and user-friendly design that allows users to maximize the compatibility, minimizing the effort for porting workflows to the LEXIS platform and facilitating a future integration of more computing and data centers with LEXIS. Indeed, the LEXIS orchestration solution, based on TOSCA, allows users to define portable workflows (templates) that can be easily customized; any additional computing or data resources integrated in the LEXIS platform will automatically be taken into account for deploying such workflows.

8.6 Weather and Climate Pilot Workflows

As mentioned in the Introduction, the LEXIS Weather and Climate large-scale pilot includes modeling tasks from weather prediction to hydrological prediction, forest fire risk forecast, air quality, and industrial pollution forecasting, as well as extreme rainfall identification.

The hydrological prediction workflow involves the WRF model, including a WRFDA data assimilation system, and the fully distributed hydrological model Continuum (Figure 8.5), which have already demonstrated their combined potential in research activities (see, e.g. Lagasio et al. 2019b). The WRF model is executed on HPC facilities at IT4I and LRZ after the preparation of initial and boundary conditions provided by the WPS (WRF preprocessing system). The WPS task is executed, as a container, on cloud-computing facilities (at LRZ and IT4I) and it allows processing both ECMWF IFS (Integrated Forecasting System) and NCEP (National Centers for Environmental Prediction) GFS (Global Forecast System) global circulation model data to generate input fields for the WRF model (Parodi et al. 2021).

The WRFDA task is a flexible, state-of-the-art atmospheric data assimilation system that is portable and efficient on available parallel computing platforms. WRFDA is a task executed on cloud-computing facilities while the Continuum hydrological model is executed as sequential tasks (container) on cloud-computing facilities. Both the models are executed using IT4I and LRZ HPC capabilities. Hydrological prediction workflow results are published on the MyDewetra platform (www.mydewetra.org), a web-based real-time system for hydrometeorological forecasting and monitoring developed by CIMA on behalf of the Italian Civil Protection Department (ICP Department et al. 2014).

The complete forest fire risk prediction workflow involves the aforementioned meteorological task WRF, including the WRFDA data assimilation

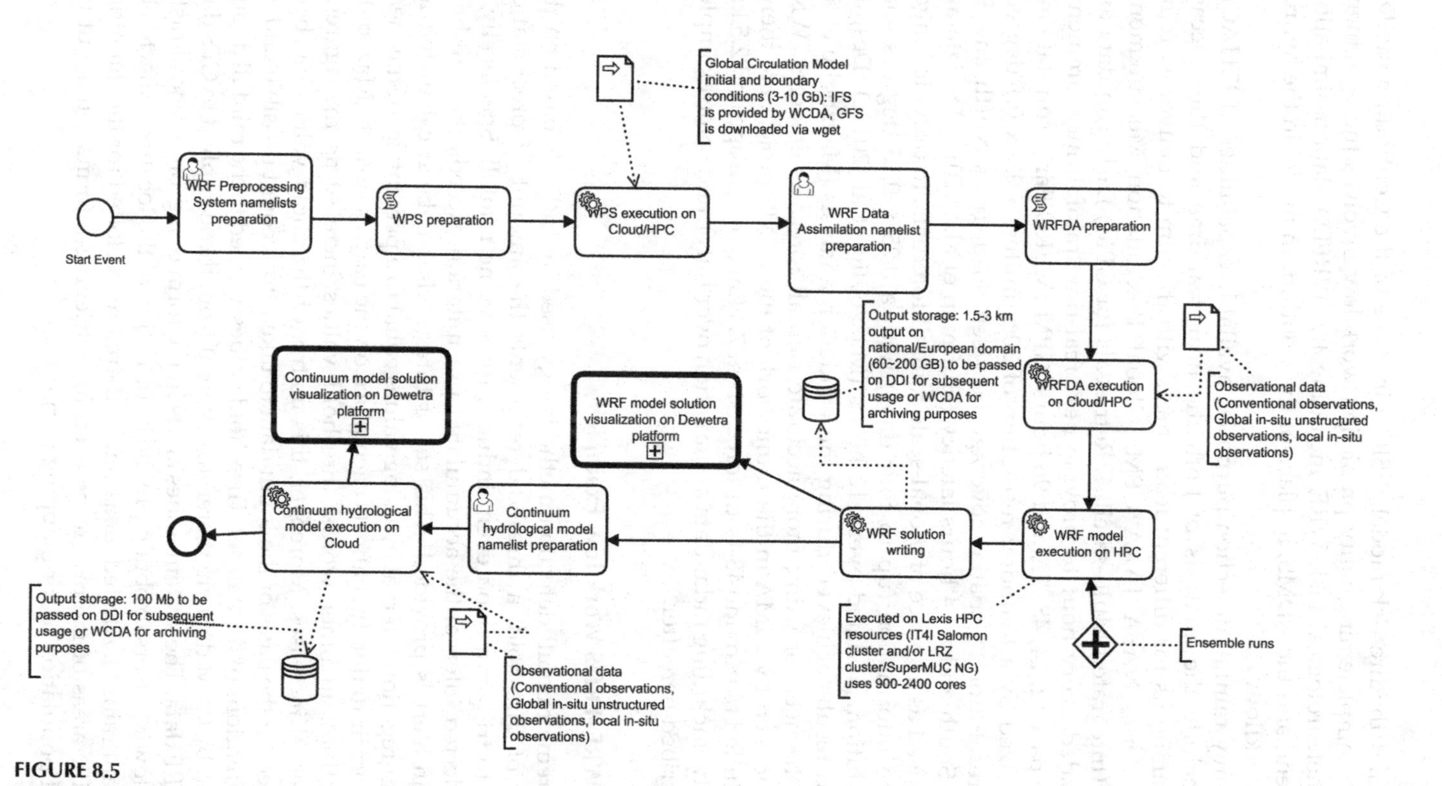

FIGURE 8.5
WRF–Continuum workflow.

system, and the fire risk model RISICO in place of the Continuum model. Also, the complete air quality prediction workflow involves the aforementioned meteorological task WRF, including the WRFDA data assimilation component, and the ADMS, in place of the Continuum model in the hydrological workflow.

For heavy rainfall detection purposes, available capabilities of ITHACA ERDS include the analysis of both the near real-time and the forecast rainfall amounts for different lead times, with the aim to deliver extreme rainfall alerts. NASA/JAXA GPM (Global Precipitation Measurement) IMERG (Integrated Multi-satellitE Retrievals for GPM) Early run data are downloaded every hour to provide near real-time rainfall measurements over the past 12, 24, 48, 72, and 96 hours. GPM IMERG Early run data are characterized by a temporal resolution of 30 minutes, a 0.1 x 0.1 degrees spatial resolution, a spatial coverage between 90 degrees North and 90 degrees South and a ~4 hours latency (Huffman et al. 2020). GFS data are instead used as a source of global-scale rainfall forecasts, to provide longer lead-time information (up to four days) with a 0.25 x 0.25 degrees spatial resolution, updated every 12 hours (Mazzoglio et al. 2019c). Despite the good results achieved during the validation analysis (Mazzoglio et al. 2019a), the need of using more accurate rainfall forecasts emerges. WRF data produced by CIMA in the framework of the LEXIS project are therefore included to provide 48-hour forecasts at 7.5 km over Europe and 2.5 km over Italy (including radar data assimilation over Italy), and some examples are described hereafter.

8.6.1 WRF–ERDS Workflow Examples

The extreme rainfall detection methodology used by ERDS is based on the concept of threshold: a threshold represents the amount of precipitation needed to trigger a flood event induced by extreme rainfall. Specifically, if for a selected interval the accumulated precipitation exceeds the threshold value, an alert is provided. This set of thresholds has been calculated for every aggregation interval by using values equal to a percentage of the mean annual precipitation that affects each place on the earth's surface (Mazzoglio et al. 2019b). In other words, threshold values increase as the aggregation interval increases. Moreover, this matrix of threshold values has been calibrated on the basis of the input data to take into account possible under/overestimations of the dataset. Threshold values applied to the rainfall depths retrieved by GPM data are therefore different from those applied to GFS data or to WRF data. The system issues alerts in the form of a georeferenced raster map, allowing users to obtain precise information of the locations affected by significant rainfall. Alerted areas can be therefore exploited for the definition of specific areas of interest, to be used for retrieving information about the affected population or for mapping purposes.

The entire WRF–ERDS workflow has been tested over different heavy rainfall events that affected Italy during 2020, both in the case of convective and stratiform rainfall events. In this chapter, the results related to two case studies are reported.

The first case study is related to a heavy rainfall event that affected Tuscany on June 4, 2020 (Figure 8.6). More than 100 mm of rainfall was recorded in the northern part of Tuscany (Figure 8.6a), corresponding to an estimated return period of about 200 years (Centro Funzionale della Regione Toscana 2020) in a very small area and > 50 years in a larger area (Figure 8.6b). Despite the slight underestimation of the rainfall depth (Figure 8.6c), WRF model was able to properly identify the most affected areas (Figure 8.6d). Thanks to the WRF-based analysis, information about the locations that would be affected by heavy rainfall was available in the early morning of June 4, several hours before the event.

An intense convective event affected the city of Palermo (Sicily, South of Italy) during the afternoon of July 15, 2020. According to SIAS (Servizio Informativo Agrometeorologico Siciliano) more than 130 mm of rainfall was recorded in about 2.5 hours, causing urban flooding phenomena and damages. ERDS was not able to detect the event using GFS data due to severe underestimation of the forecast. A WRF modeling experiment based on three nested domains (with 22.5, 7.5, and 2.5 km grid spacing), innermost over Italy, was executed by assimilating the national radar reflectivity mosaic and in situ weather stations from the Italian Civil Protection Department. Good results were achieved using WRF data at a 2.5 km resolution: a peak rainfall depth of about 35 mm in 1 hour and 55 mm in 3 hours were predicted roughly 30 km from Palermo (Figure 8.7).

8.7 Conclusion

The chapter describes the first results achieved with the framework of LEXIS Weather and Climate large-scale pilot. LEXIS DDI and Weather and Climate Data API are described in Section 8.4 while LEXIS orchestration system is described in Section 8.5. All the Weather and Climate pilot workflows are described in Section 8.6, together with a specific focus on WRF–ERDS workflow. Preliminary results obtained with the WRF–ERDS workflow highlight that improved rainfall forecasts obtained by using HPC resources significantly increase the performance of an early warning system as ERDS. Global-scale low-resolution rainfall datasets as the GFS one are often characterized by poor performances, especially in the case of very intense and localized convective rainfall that occurs in the summer season. Further experiments will be performed to assimilate atmospheric data from PWS over Europe to increase WRF accuracy.

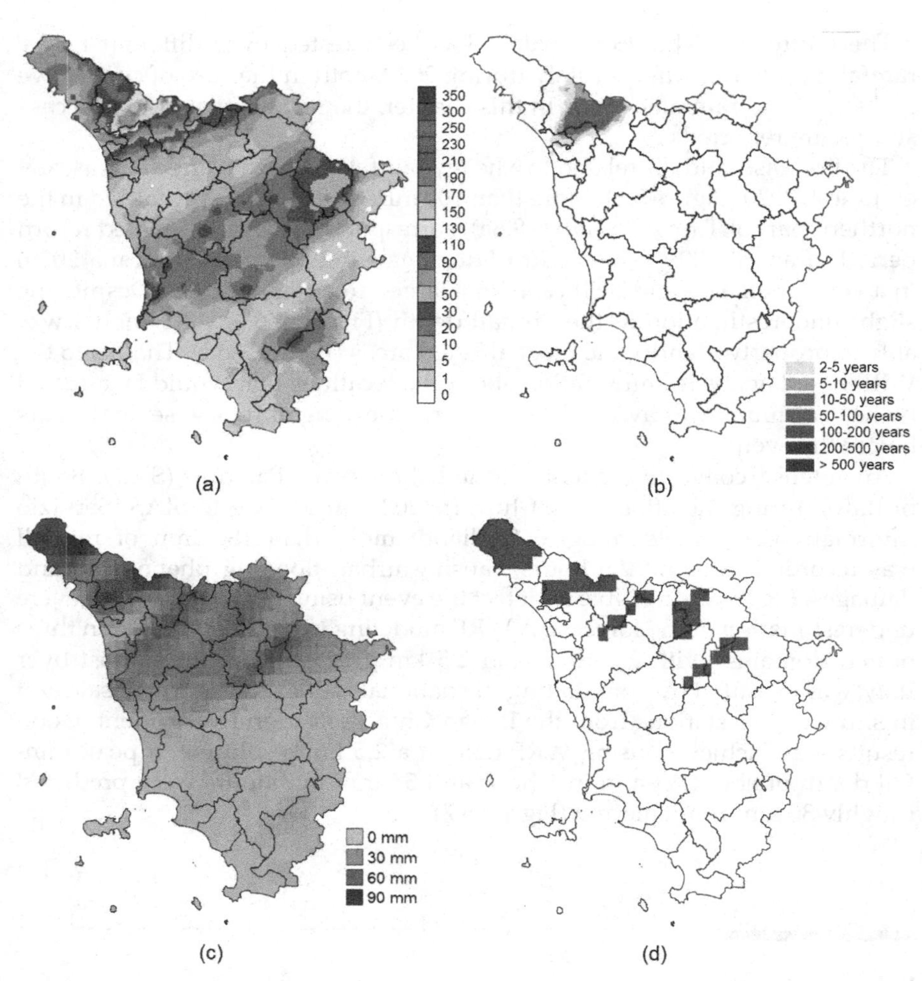

FIGURE 8.6
24-hours rainfall depth recorded by rain gauges on June 4, 2020 (figure a). 24-hours rainfall return period (figure b). 24-hours rainfall forecast provided by WRF model on June 4, 2020 (figure c). Heavy rainfall alerts provided by ERDS using WRF data as input (figure d). Contains rain gauge and return period information retrieved by Centro Funzionale della Regione Toscana (2020).

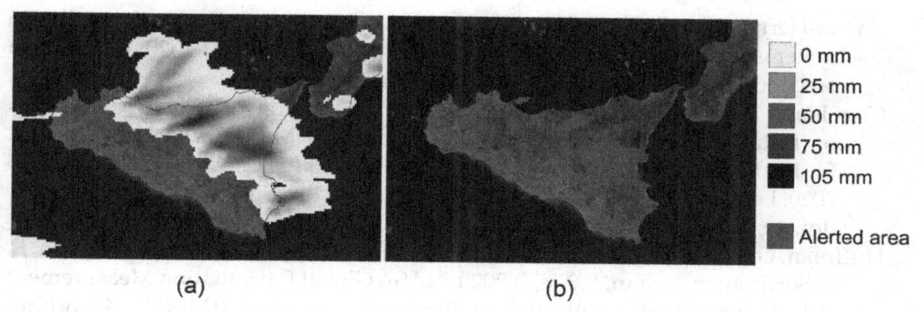

FIGURE 8.7
WRF 24-hours forecast at 2.5 km resolution (a) and heavy rainfall alerts (b) provided by ERDS using WRF data as input.

Acknowledgment

The authors acknowledge the use of imagery from the NASA Worldview application (https://worldview.earthdata.nasa.gov/), part of the NASA Earth Observing System Data and Information System (EOSDIS). This chapter was supported by the LEXIS project, funded by the EU's Horizon 2020 Research and Innovation Programme (2014–2020) under grant agreement No. 825532.

References

Atrey, A., Moens, H., Seghbroeck, G., Volckaert, B., Turck, F. 2015. An overview of the OASIS TOSCA standard: topology and orchestration specification for cloud applications. Technical Report, IBCN-iMinds, Department of Information Technology.

Brocheton, F., Armand, P., Soulhac, L., Buisson, E. 2008. A methodology to characterise the sources of uncertainties in atmospheric transport modelling. *Hrvatski meteorološki časopis* 43(43/1), 78–82.

Carruthers, D., Seaton, M., McHugh, C., Sheng, X., Solazzo, E., Vanvyve, E. 2011. Comparison of the complex terrain algorithms incorporated into two commonly used local-scale air pollution dispersion models (ADMS and AERMOD) using a hybrid model. *Journal of the Air & Waste Management Association* 61(11), 1227–1235. https://doi.org/10.1080/10473289.2011.609750

Centro Funzionale della Regione Toscana 2020. Disaster report June 4–5, 2020. Available online: www.cfr.toscana.it/supports/download/eventi/report_evento_04-05_giugno_2020.pdf (accessed March 14, 2021).

Fiorucci, P., Gaetani, F., Minciardi, R. 2008. Development and application of a system for dynamic wildfire risk assessment in Italy. *Environmental Modelling & Software* 23(6), 690–702. https://doi.org/10.1016/j.envsoft.2007.05.008

Garcia-Hernandez, R. J., Golasowski, M. 2020. Supporting Keycloak in iRODS systems with OpenID authentication. Presented at the CS3 2020 – Workshop on Cloud Storage Synchronization and Sharing Services.

Gogolenko S. et al. 2020. Towards Accurate Simulation of Global Challenges on Data Centers Infrastructures via Coupling of Models and Data Sources. In: Krzhizhanovskaya V. et al. (eds) *Computational Science* – ICCS 2020. ICCS 2020. Lecture Notes in Computer Science, vol. 12142. Springer, Cham. https://doi.org/10.1007/978-3-030-50433-5_32

Huffman, G. J., Bolvin, D. T., Braithwaite, D., Hsu, K., Joyce, R., Kidd, C., Nelkin, E. J., Sorooshian, S., Tan, J., Xie, P. 2020. NASA Global Precipitation Measurement (GPM) Integrated Multi-satellitE Retrievals for GPM (IMERG) Algorithm Theoretical Basis Document (ATBD) Version 06. https://gpm.nasa.gov/sites/default/files/2020-05/IMERG_ATBD_V06.3.pdf (accessed April 14, 2021).

ICP Department, CIMA Research Foundation 2014. The Dewetra platform: a multi-perspective architecture for risk management during emergencies. In: *Proceedings of the First International Conference Information Systems for Crisis Response and Management in Mediterranean Countries* (ISCRAM-med 2014), Toulouse, France, October 15–17, 2014, vol. 1, pp. 165–177. Springer.

Lagasio, M., Parodi, A., Pulvirenti, L. et al. 2019a. A synergistic use of a high-resolution numerical weather prediction model and high-resolution earth observation products to improve precipitation forecast. *Remote Sensing* 11(20), 2387.

Lagasio, M., Silvestro, F., Campo, L., Parodi, A. 2019b. Predictive capability of a high-resolution hydrometeorological forecasting framework coupling WRF cycling 3dvar and continuum. *Journal of Hydrometeorology* 20(7), 1307–1337.

Mazzoglio, P., Laio, F., Sandu, C., Boccardo, P. 2019a. Assessment of an Extreme Rainfall Detection System for Flood Prediction over Queensland (Australia). *Proceedings* 18(1):1. https://doi.org/10.3390/ECRS-3-06187

Mazzoglio, P., Laio, F., Balbo, S., Boccardo, P., Disabato, F. 2019b. Improving an Extreme Rain-fall Detection System with GPM IMERG data. *Remote Sensing* 11(6), 677–677. https://doi.org/10.3390/rs11060677

Mazzoglio, P., Laio, F., Balbo, S., Boccardo, P. 2019c. ERDS: an Extreme Rainfall Detection System based on both near real-time and forecast rainfall measurements. *Annual of the University of Architecture, Civil Engineering and Geodesy (Sofia)*, 52, Issue S1, 1423–1433. https://uacg.bg/UserFiles/File/UACEG_Annual/2019/%D0%91%D1%80%D0%BE%D0%B9%20S1/19-3.pdf

Parodi, A., Danovaro, E., Hawkes, J., et al. 2021. LEXIS Weather and Climate Large-Scale Pilot. In: Barolli L., Poniszewska-Maranda A., Enokido T. (eds) *Complex, Intelligent and Software Intensive Systems. CISIS 2020. Advances in Intelligent Systems and Computing*, vol. 1194. Springer, Cham. https://doi.org/10.1007/978-3-030-50454-0_25

Powers, J. G., Klemp, J. B., Skamarock, W. C., et al. 2017. The weather research and forecasting model: Overview, system efforts, and future directions. *Bulletin of the American Meteorological Society* 98(8), 1717–1737. https://doi.org/10.1175/BAMS-D-15-00308.1

Silvestro, F., Gabellani, S., Rudari, R., Delogu, F., Laiolo, P., Boni, G. 2015. Uncertainty reduction and parameter estimation of a distributed hydrological model with ground and remote-sensing data. *Hydrology and Earth System Sciences* 19(4), 1727–1751. https://doi.org/10.5194/hess-19-1727-2015

Svaton, V., Martinovic, J., Krenek, J., Esch, T., Tomancak, P. 2019. HPC-as-a-Service via HEAppE Platform. In: *Conference on Complex, Intelligent, and Software Intensive Systems*, pp. 290–293. Springer.

Wilkinson, M. D., Dumontier, M., Aalbersberg, et al. 2016. The FAIR Guiding Principles for scientific data management and stewardship. *Sci. Data 3*, 160018. https://doi.org/10.1038/sdata.2016.18

9

Data Convergence for High-Performance Cloud

Christian Pinto, Antony Chazapis, Christos Kozanitis, Yiannis Gkoufas, Panos Koutsovasilis, Srikumar Venugopal, Jean-Thomas Acquaviva, and Angelos Bilas

CONTENTS

DOI: 10.1201/9781003176664-9

9.1 Introduction

In our attempt to build a high-performance data analytics (HPDA) platform, where large scale data analytics algorithms run on high-performance computing (HPC) infrastructure, we had to cope with conflicting storage abstractions provided by the available hardware and system software, and respective assumptions used by different cloud and big data tools and frameworks.

On one hand, HPC applications typically rely on a shared filesystem to play the role of a reliable communication channel between the moving parts of highly distributed applications. This facilitates both the exchanging of information across many asynchronous processes, as well as the periodic checkpointing of the state of the application itself. Such shared, distributed, filesystems are capable of sustaining high data rates and many concurrent IO operations. Files are historically easy to manipulate via simple system calls and APIs, which are fundamental in nearly all programming languages and operating systems. Moreover, in a shared filesystem files appear in the same location regardless of the node used for accessing the data, including the login node of the cluster. This is a valuable feature, because it makes manually moving data unnecessary, thus helping with the overall system's usability and the productivity of its users, while also allowing the convenient exchange of data between the HPC cluster and the outside world.

On the other hand, data analytics pipelines are commonly expressed as complex directed acyclic graphs (DAGs) that run on cloud-based or on-premises virtualized resources. Pipeline stages may use diverse software frameworks that deal with storage in a completely different manner. The key characteristic of such deployments is that storage volumes are treated as per-stage assets with a limited lifetime; they are only available to a workflow stage for the life span of the respective VM or container on which they are mounted. Data sharing across different stages is achieved through object stores, which are provided as external, persistent storage services. Unlike the universal file abstraction of HPC storage systems, the use of external object stores introduces a heterogeneous collection of APIs and programming libraries to interact with data. Also, it is common for HPDA pipelines to explicitly fetch data when and where it is needed, and write it back if modified.

In this chapter we present a new abstraction, called *unified storage layer* (USL) [1], which tackles the heterogeneity of converged HPDA applications at the storage level. USL aims to enable applications to be designed following the cloud-ready paradigm and to transparently benefit from potentially any storage solution available, ranging from high-performance file systems to cloud-based object stores and key–value databases, without the need for adapting applications to different storage access mechanisms.

USL is implemented as a set of Kubernetes [2] microservices; our platform of choice to provide a unified execution environment for HDPA applications on HPC infrastructure. As a scalable, container-based substrate, Kubernetes hides the complexity of managing massive compute node counts, and thus it is quickly becoming the industry standard for supporting large-scale applications and complex processing pipelines dealing with enormous and diverse datasets. With the seamless data abstraction provided by the USL, we have experienced that workflows become simpler to develop and deploy, portable, and users experience the whole system as more familiar and friendly.

9.2 Motivations

EVOLVE aims at combining the best features of HPC, big data, and cloud to create an infrastructure and ecosystem, where users deploy complex applications processing large amounts of data with the simplicity and efficiency provided by cloud offerings. The majority of the EVOLVE pilot applications process large unstructured or semi-structured datasets, such as images and sensor data, and after a set of extract–transform–load (ETL) routines, they run large amounts of parallel machine learning prediction steps. Afterwards, some of those applications implement their business logic by running complex numeric algorithms, which process the output of the machine learning stages.

To make application pipelines portable, scalable, and reproducible, the EVOLVE consortium opted for containerized application components and to use a high-level workflow orchestration framework. The choice is to use Argo Workflows [3], a Kubernetes native workflow management engine. In EVOLVE, users synthesize and submit complex execution pipelines as container-based DAGs. Each node of the graph may use a completely different language and high-level programming framework to perform one step of data processing. The whole graph is submitted to Argo, which in turn uses Kubernetes for execution. It is very common that such heterogeneous steps work on the same data. Argo refers to data exchanged between workflow steps as *artifacts* and includes plugins to manage artifact movement. However, initial runs showed that copying data in and out of containers introduced a significant overhead. In addition, moving data between different storage services proved to be frustrating to users, and contradicting to the requirement of portability, as it required to design the whole pipeline for specific storage configurations. USL was designed to solve these problems, by presenting all available storage endpoints in a unified and portable manner to all workflow stages.

9.3 Design and Implementation

USL consists of a set of three main components designed to run as microservices on a Kubernetes-managed cluster:

1. Karvdash [4]: The USL front end, which provides a user interface for configuring DataShim and H3 storage attachments. Karvdash also wires up a private and shared dataset per user by default.
2. DataShim [5]: The USL core, mounting the actual datasets to containers, thus unifying access to a diverse set of actual storage protocols and technologies.
3. H3 [6]: An object store library, backed by a high-performance key–value store. H3 can either be embedded into applications or accessed through DataShim as part of the unified storage offering.

In essence, USL simplifies workflows by completely replacing multiple data movement stages with a set of dataset configuration directives that are defined as part of workflow initialization. To this end, USL provides an abstract form for defining varying types of storage endpoints, and attaches the corresponding data collections as filesystem mounts inside containers, preserving the files-based data access pattern. These configuration directives can either be constructed by users manually, or set up through Karvdash. The latter, as part of the default deployment, automatically configures one private and one shared "home" folder per user and makes them accessible through a web-based front end. DataShim performs storage attachments by utilizing low-level, Kubernetes-compatible container storage interface (CSI) plugins. H3 is meant to ease application transition to high-performance key–value stores, by offering a simple object store API that translates respective commands to key–value operations, in order to decrease the latency of small data operations and facilitate the use of node-local storage devices and memory in workflow steps. However, H3 also implements a CSI plugin for integration with DataShim, for allowing existing pipeline steps to seamlessly exploit the functionality without code changes.

Each microservice is presented in detail in the following paragraphs. Note that each USL component can also function autonomously in any Kubernetes environment, but they have all been designed from ground up with the overall integration in mind and the goal of offering the combined feature set of the USL. All USL components are open source.

9.4 Karvdash

Karvdash is a service for facilitating data science in Kubernetes-based environments, by supplying the landing page for users working on the platform, allowing them to launch services, design workflows, request resources, and specify other parameters related to execution through a user-friendly interface. Karvdash aims to make it straightforward for domain experts to interact with resources in the underlying infrastructure without having to understand lower-level tools and interfaces.

In summary, Karvdash provides a web-based, graphical interface to:

- Manage services or applications that are launched from customizable templates.
- Securely provision multiple services under one externally accessible HTTPS endpoint.
- Organize container images stored in a private Docker registry.
- Perform high-level user management and isolate respective services in per-user namespaces.
- Interact with DataShim datasets that are automatically attached to service and application containers when launched.

Karvdash serves as the USL front end via its "Datasets" and "Files" tabs (Figure 9.1), but also performs important integration tasks. From the storage management viewpoint, it allows the graphical configuration of DataShim, while it also bootstraps the process of mounting datasets to containers, by setting up the appropriate DataShim-specific Kubernetes labels in user namespaces and applying the selected configuration to their pod[1] deployments.

Furthermore, Karvdash applies the notion of a "home folder" in the containerized execution environment, by explicitly mounting preconfigured "private" and "shared" datasets to all containers. Thus, when a user runs a workflow, the same data are *always* available at a well-known path – as it would be in a bare-metal HPC setup. The "private" dataset holds files private to the user and is mounted in containers under /private, while the "shared" dataset is meant to ease data sharing among users (e.g. for storage of large reference data collections) and is mounted under /shared. These two "special" datasets can either reside in a platform-wide, local shared folder (mounted over NFS, Lustre, Gluster, etc.), or in some S3 service, so that Karvdash can also provide a web-based file browser as part of its GUI. This makes interacting with files easier, and lets application developers to quickly provide inputs and collect outputs through a familiar interface.

To wire-up datasets to containers, Karvdash implements a Kubernetes mutating admission webhook which intercepts all calls to create pods or

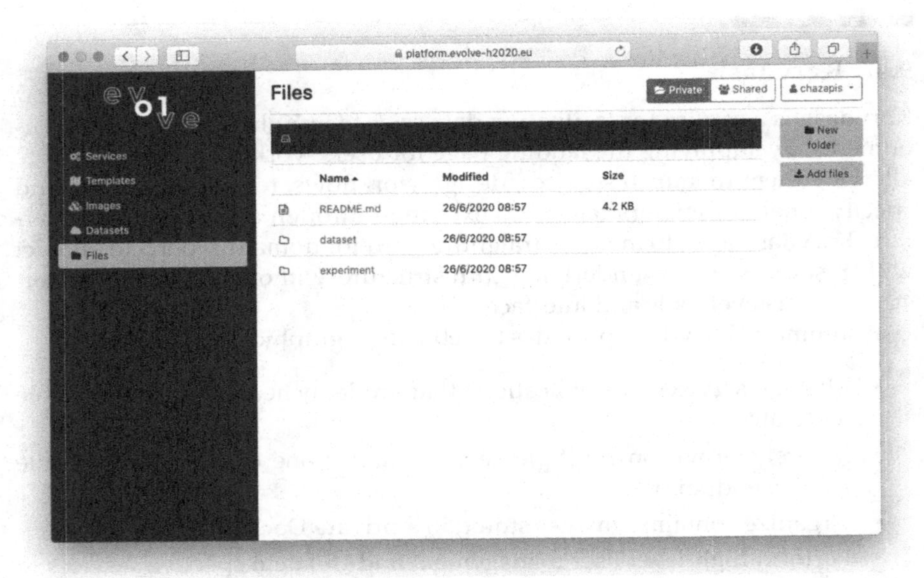

FIGURE 9.1
The files tab in Karvdash allows users to directly interact with their home datasets via a convenient interface.

deployments and injects the appropriate DataShim metadata before they are applied. As a security precaution, another validating admission webhook verifies that only allowed host paths are mounted in pods.

9.5 DataShim

The interface between Karvdash and the actual storage back ends is enabled via DataShim, which introduces in Kubernetes a new data abstraction called *Dataset*. DataShim plays the middle-layer role in the USL infrastructure and enables from one side easy interfacing of data sources to containers, and on the other side it takes care of materializing the connection with the underlying storage technology (H3 in USL).

9.5.1 Overview

DataShim is a cloud-ready software framework targeting Kubernetes (Figure 9.2), aiming to abstract low-level storage-related details from the user, allowing attachment of potentially any storage solution to a

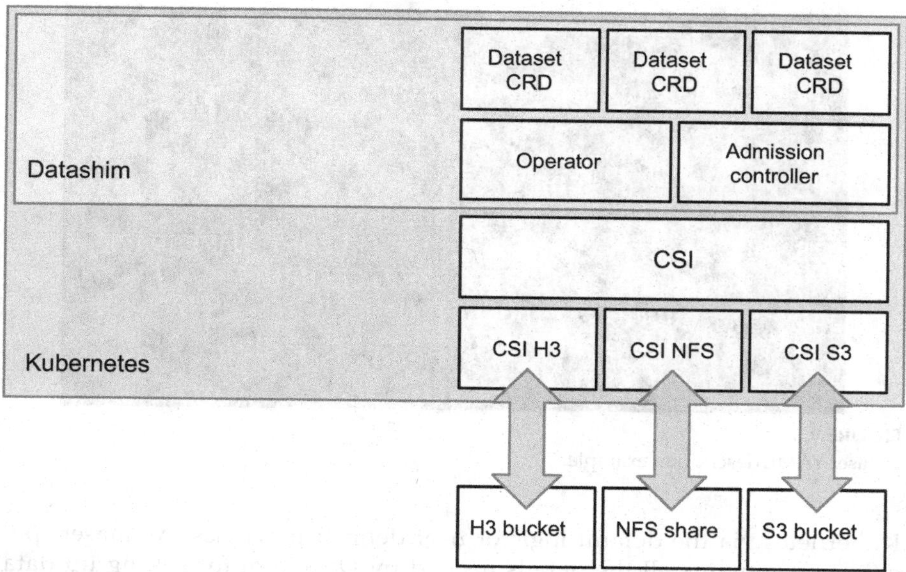

FIGURE 9.2
DataShim software stack.

containerized application via a common interface. So far, Kubernetes users have to deal with each storage solution they need in a separate way that is often customized to reflect the underlying storage infrastructure. DataShim introduces data as a first class citizen in Kubernetes and relieves users from integrating data-related information in their application code or container definitions. Basically, through a dataset cloud administrators can more easily control who accesses data and allow data sources to change without affecting the applications that use them. By using DataShim, users do not access a specific NFS share, or an S3 bucket anymore; they instead request access to a specific dataset they know (or are told by the administrator) contains their data. DataShim takes care of presenting the data the way the user wants: as a POSIX mount point, or user credentials injected as environmental variables. DataShim supports multiple storage back-ends in parallel by relying on the CSI [7], and as of today it supports H3, NFS, S3 buckets and gzip archives. The integration of new storage technologies depends only on the availability of the respective CSI plugins.

9.5.2 Dataset Custom Resource Definition

DataShim defines the *dataset* abstraction via a Kubernetes Custom Resource Definition (CRD). CRDs are the standard way resources are handled in

```
1  kind: Dataset
2  metadata:
3    name: example-dataset
4  spec:
5    local:
6      type: "COS"
7      secret-name: "{SECRET_NAME}"
8      secret-namespace: "{SECRET_NAMESPACE}"
9      endpoint: "{S3_SERVICE_URL}"
10     bucket: "{BUCKET_NAME}"
11     readonly: "true" # default is false
12     region: "" #it can be empty
```

FIGURE 9.3
Dataset YAML description example.

Kubernetes via the default logic or user defined processes. A dataset specification contains all the details needed by DataShim for setting up data access: reference to a storage source (potentially populated with preexisting data), credentials whenever needed, and some access control information (some datasets can be read only). Even though a dataset can be created by a user, the preferred embodiment envisioned is for cluster administrators to create them and give access to specific users. This is to completely remove the burden from the final user and enforce centralized access control. The access scope of a dataset is that of a Kubernetes namespace, therefore an administrator can have users access only namespaces they have access rights on.

Datasets follow a predefined YAML schema whose fields depend in part on the specific data source type. After its creation, a dataset is stored in the Kubernetes persistent storage and is ready for use from a pod. Figure 9.3 shows an example dataset definition referring to an Amazon S3-like bucket. The main fields to consider are the *endpoint* (line 9), the *bucket* (line 10), and the user credentials (lines 7–8) that for privacy reasons are kept within a Kubernetes secret. The *name* field at line 3 is the only information a user needs to embed in their pod definitions to request access to that specific dataset.

9.5.3 DatasetInternal Custom Resource Definition

When a dataset is created the dataset controller creates also PVC associated with it. However, before creating the PVC, the framework would need to check if caching is to be enabled and eventually provision the cache for that specific dataset, which could take time in the order of minutes. For this reason, DataShim uses the DatasetInternal CRD. While the Dataset CRD remains the user-facing object, the DatasetInternal is used as a "shadow" object of a dataset and is the one that will finally provide the details for the creation of a PVC. In terms of fields and specification dataset and DatasetInternal are identical.

9.5.4 DataShim Operator and Admission Controller

DataShim implements the logic for handling datasets into a so-called *operator*, a template for building software components adopted by the Kubernetes community. Attachment of datasets to pods is instead handled by a component called *admission controller* that is constantly monitoring newly created pods, searching for dataset-related information.

The operator is constantly monitoring for events on *datasets*, and creates the necessary connections whenever a matching reference is found (i.e. a pod references an existing dataset). The operator relies on CSI for implementing the actual connection with the specific storage solution backing datasets. This results in DataShim being able to handle datasets for potentially any storage solution that provides a CSI plugin implementation. A CSI plugin instructs Kubernetes on how containers can physically access the storage medium, and whether they support dynamic provisioning (i.e. automated creation of volumes dedicated to specific pods).

The admission controller monitors all newly submitted pods, looking for a specific set of labels that specify if a dataset is to be attached to the pod. After a dataset is created, pods request access by simply referencing it via its mnemonic name. If an existing dataset is referenced by a pod, the admission controller takes care of creating the necessary connection between the containers created and the storage back end. Datasets can be accessed by containers as POSIX mount points or, in case of API-based storage solutions such as Amazon S3, the access credentials are injected into the running pod as environment variables or configuration files mounted in the container's local filesystem. In this latter case it is up to the applications running inside containers to retrieve the access credentials. Figure 9.4 shows an example pod that wants to use the dataset named "example-dataset," and it wants it mounted through a POSIX mount point. This is expressed in lines 6–7 with the *id* and useas fields. Setting useas: "mount" instructs the admission controller that the pod wants the dataset with the specified id mounted via a POSIX mount point. The exact mount location is configured at line 13, where users describe a volume whose name is matching the ID of the dataset. The admission controller, once the pod is submitted, will make sure the volume is created and that the mount point is successfully linked to the data source upon execution of the containers.

```yaml
apiVersion: v1
kind: Pod
metadata:
  name: nginx
  labels:
    dataset.0.id: "example-dataset"
    dataset.0.useas: "mount"
spec:
  containers:
    - name: nginx
      image: nginx
      volumeMounts:
        - mountPath: "/mount/dataset1"
          name: "example-dataset"
```

FIGURE 9.4
Example YAML description of a pod using a dataset.

9.5.5 Caching Plugin

Within our updated architecture, the caching plugin is the component which is responsible for cache provisioning for the datasets. The Kubernetes administrator can install at any point a caching plugin and therefore provide caching capabilities for specific types of datasets. Upon the creation of a dataset, the dataset controller queries the cluster for available caching plugins for the specific type of datasets. In cases where is no available caching plugin, it proceeds to create the mapped DatasetInternal object which in this case would be identical with the original dataset. Where there is a caching plugin available then the dataset controller delegates the creation of the corresponding DatasetInternal object. Now the caching plugin handling that dataset would proceed to provision the cache and take all actions necessary (in terms of provisioning new pods, services, endpoints) so that the workloads using that dataset access the cached version of it. The only requirement for a caching plugin implementation is to return a DatasetInternal object after it has successfully provisioned the necessary infrastructure to support caching of a dataset. After that step, the DatasetInternal controller would proceed to create the native Kubernetes components like PVCs. One important advantage of this approach is that a caching plugin implementation doesn't need to include any logic regarding enabling access to S3/NFS endpoints, since this is part of the core framework, thus making it easy for any external cache provider to be transparently supported within DataShim.

9.5.6 Objects Caching on CEPH

With the goal of evaluating the caching framework in DataShim we have selected Ceph as a candidate for playing the role of the cache. Ceph [8] is a distributed storage solution providing native support for S3-based objects among others. The object interface is exposed by the Rados Gateway (RGW) [9] that is running as a separate component with respect to the core Ceph infrastructure. Moreover, RGW relies on Ceph Object Storage Daemons (OSDs), deployed through the cluster, to store and load objects.

We have extended RGW (Figure 9.5) to act as a cache between applications and a remote S3 enabled data bucket. An application would access data via the local (to the cluster) S3 endpoint exposed by RGW, instead of using the remote endpoint provided by the cloud provider (e.g. IBM's cloud object storage). Our extension, which we have submitted as a public pull request2 to the original CEPH project, introduces a "lazy" caching architecture to RGW and synchronizes with the remote cloud object bucket only when there is a need to, such as a local object missing (cache-miss). We did not choose "eager" caching as it mandates taking complex decisions such as when to eager load a dataset or not, how multiple concurrent dataset loadings will affect the performance of the cache, and whether to charge the user for the whole dataset that the workload at hand may not need. That said, as shown in Figure 9.5, our extension code footprint of RGW can be divided into three categories of operations. The first category are the operations that we want the client to always get the latest info based on the remote bucket, such as list buckets and list objects operations. The next two categories are both in the case of a cache-miss and they are divided into operations that require local changes or can be directly served based on the info of the remote bucket. As an example, in the case of an object cache-miss, our RGW extension, as shown

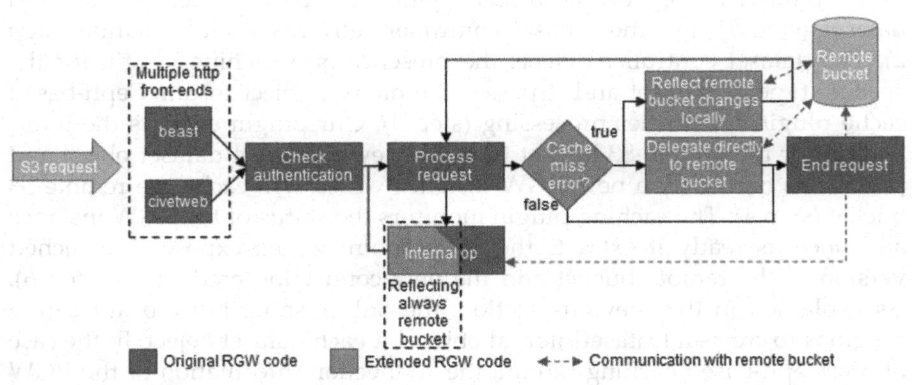

FIGURE 9.5
High-level overview of the RGW CEPH-based cache extension code.

in Figure 9.5, intercepts the corresponding error and instead of returning it to the client, checks the remote bucket, fetches the object locally and responds accordingly to the client. However, for a delete object request, in case of a cache-miss, our extension delegates directly this to the remote bucket, without downloading the object, and responds to the client based on the answer of the remote bucket. Our extensions to RGW make caching completely transparent to any application that accesses the S3 endpoint. Moreover, by serving client requests directly from the remote bucket, without requiring fetching the object into to the cache first, we minimize the performance overhead of the caching mechanism but also, we do not perform unnecessary actions that would result in unnecessary charging of the user. For example, it is common for multiple S3 accessing mechanisms, namely goofys, etc., to initially perform head requests for the objects of a bucket and, thus, extract initial information, such as the size and the etag of each object. If our caching mechanism could not serve these requests solely based on the remote bucket and, instead it had to fetch all these objects locally, this would result in a significant performance penalty and yet again unnecessary charging of the user.

9.5.7 Ceph-Based Caching Plugin Implementation

The assumption for our Ceph-based caching plugin is that the Kubernetes administrator has deployed a Ceph cluster on the same Kubernetes. To that end, we leveraged Rook [10] which enables a Kubernetes administrator to provision Ceph distributed storage on Kubernetes using CRDs. In addition, it provided us with high-level APIs to use in our Ceph-based caching plugin implementation regarding provisioning on-the-fly instances of our modified version of RGW. We have detailed instructions about the requirements and the installation process in our wiki3.

In Figure 9.6 we outline the end-to-end flow of the framework for caching user's datasets. The flow is initiated when the user creates an S3-based dataset (step 1) and the dataset controller retrieves its information (step 2). The dataset controller detects the presence of a caching plugin for the specific type of dataset and it passes the dataset object to our Ceph-based cache plugin for further processing (step 3). Our plugin extracts the information for the remote S3 bucket from the newly created dataset object and provisions on-the-fly a new RGW instance which will cache the remote S3 bucket (step 4). The caching plugin monitors the status of the RGW instance and once its ready it extracts the S3 endpoint which exposes the cached version of the remote bucket and the new connection credentials (step 6). As explained in the previous section, the only responsibility of a caching plugin is to create a DatasetInternal object for each dataset object. In the case of our Ceph-based caching plugin, the connection information of the RGW instance are used to create the corresponding DatasetInternal object (step 6). Now the DatasetInternal controller of the core framework would process the DatasetInternal object (step 7) and proceed in creating the PVC (step 8). The

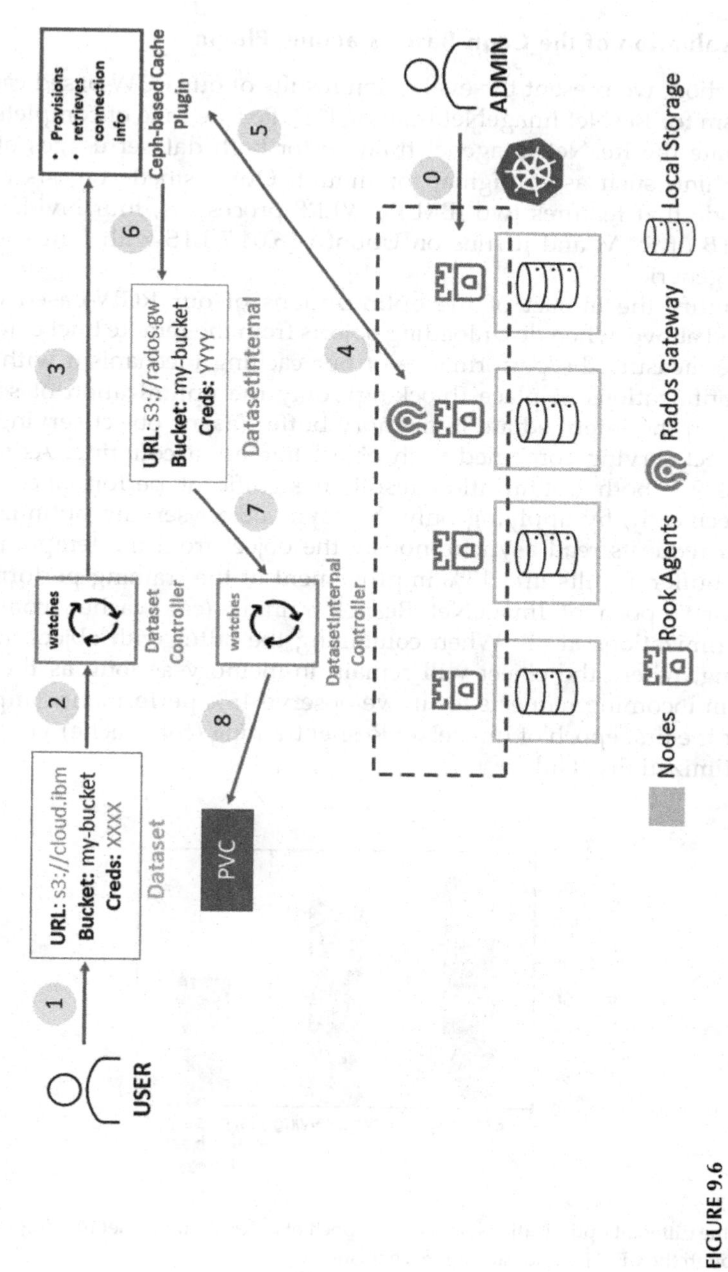

FIGURE 9.6
End-to-end flow of the caching framework.

mount point of the PVC would reflect the cached version provided by the RGW instance we provisioned, so the user's pods will benefit in terms of data access.

9.5.8 Evaluation of the Ceph-Based Caching Plugin

In this section, we present the evaluation results of our RGW-based caching mechanism for ResNet ImageNet training [11]. For the sake of completeness, we evaluate the ResNet Imagenet training for both dataset usages offered by DataShim, such as configmap or mount. Our testbed consists of one server node that features two IBM POWER8 processors, four Nvidia P100 GPUs, 1TB of RAM and it runs on Ubuntu 16.04.7 LTS with Linux Kernel 4.4.0-116-generic.

To quantify the impact of the optimizations of our RGW-based cache, mentioned above, when downloading objects from the remote bucket (cache-miss), we measure the performance of our caching mechanism with none of the optimizations in place (blocking), only the optimization of serving objects from the intermediate in-memory buffer (async object serving) and async object serving combined with object hotness accounting. As shown in Figure 9.7, both optimization result in significant performance gains. More specifically, by applying only the async object serving optimization, where all requests read asynchronously the object from the temporary in-memory buffer, results in a 13% improvement of the training performance for the first epoch of ImageNet Resnet training (cold-cache), compared to no optimizations at all. When combining the latter with object hotness accounting, where the object will remain in-memory as long as there are short-term incoming requests for it, we observe 45% performance improvement, for the first epoch of ImageNet Resnet training (cold-cache), compared to no optimizations at all.

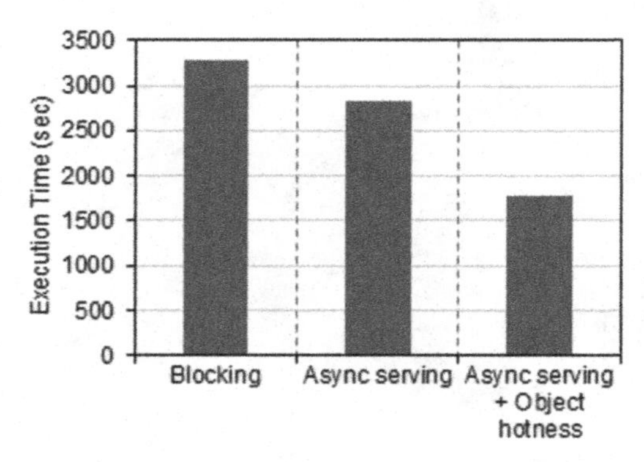

FIGURE 9.7
Impact of the different optimizations on the first epoch of a ResNet Imagenet training (cold cache) through the CEPH-based caching mechanism.

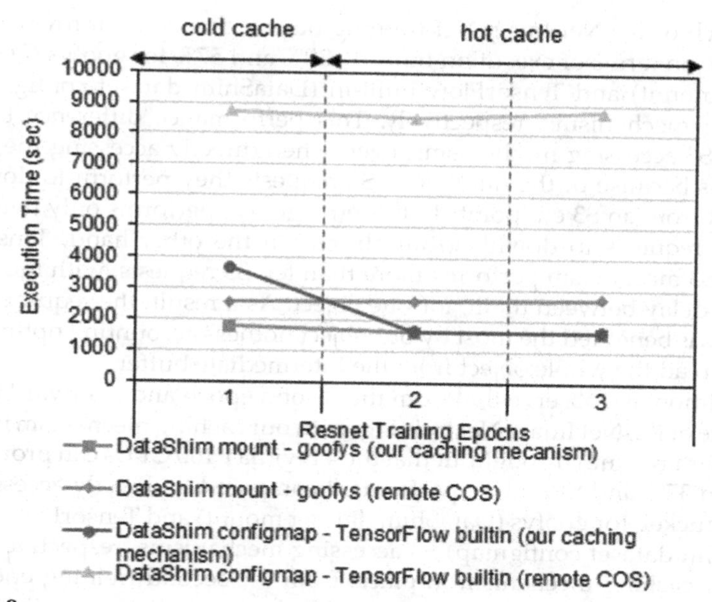

FIGURE 9.8
Performance of training three epochs of ResNet Imagenet on 4x Nvidia P100 GPUs. Data accessed via the CEPH-based cache and by directly accessing the remote buckets.

As mentioned, depending on how the workload at hand uses DataShim's dataset (configmap or mount), under the hood, this translates to different S3 accessing mechanisms, such as TensorFlow built-in or goofys [12] mechanism. Every one of these mechanisms, depending on the size of the requested object, splits up the corresponding downloading process to multiple individual requests that will read only a chunk of the object. Note that, every S3 request is coupled with an extra overhead that includes connection initialization, HTTP headers parsing and dumping, object data downloading, and merging the different data chunks of the object. However, in the case of our mechanism, as soon as a cache-miss occurs, it will start downloading the whole object in an intermediate in-memory buffer, by doing a single request to the remote bucket. In this way, any upcoming short-term request of the same object will collect the corresponding data from the intermediate buffer and it will be served locally to the application client. As a result, any overhead coupled to the S3 requests, accessing directly the remote bucket, is minimized.

Figure 9.8 depicts the performance, in terms of execution time, for the first three epochs of ResNet ImageNet training. Note that during the first epoch our caching mechanism is cold; every ImageNet object request from the client results in a cache-miss. For the reasons presented above, we observe performance gains even for the first training epoch. More specifically, for the first epoch of ResNet ImageNet training our mechanism improves the performance, in terms of execution time, by 30% and 57% for goofys (DataShim dataset mount) and TensorFlow built-in (DataShim dataset configmap) S3 accessing mechanisms, respectively. This performance difference between the two S3 accessing mechanisms, even when directly accessing the remote bucket, is because of the number of S3 requests they perform to download an object from an S3 endpoint. To this end, goofys performs only two timely close S3 requests to download an object. On the other hand, TensorFlow built-in S3 mechanism performs more than ten S3 requests, with multiple of seconds delay between them, for one object. As a result, the request pattern of goofys is benefited the most by our object hotness accounting optimization and can read the whole object from the intermediate buffer.

In addition, we observe that from the second epoch and afterward the performance of ResNet ImageNet training, with our caching mechanism in-place, reach the maximum throughput that the 4 Nvidia P100 GPUs can provide and results in 37% and 81% more performant compared to directly accessing the remote bucket, for goofys (DataShim dataset mount) and TensorFlow built-in (DataShim dataset configmap) S3 accessing mechanisms, respectively. Also, with our caching mechanism in-place, from the second training epoch and afterward using the DataShim dataset either as mount or as configmap, we observe the same performance because the whole ImageNet is downloaded and, thus, any object is rapidly served to the application client.

9.6 H3

H3 is an embedded object store, backed by a high-performance key–value store.

H3 provides a user-friendly, S3-like object API but without the REST layer commonly used in object stores – a design choice that boosts the performance of data operations significantly, especially for smaller-sized objects. Additionally, STORE's pluggable back-end architecture allows adapting the object store's scale and performance to a variety of deployment requirements. H3 supports several key–value storage implementations, ranging from in-memory, to distributed, RDMA-based services. In the USL, H3 is integrated with DataShim as a "local" dataset storage facility, providing the means to transparently utilize local devices and RAM in nodes, for high-speed access to intermediate workflow "artifacts" (data exchanged between workflow stages).

9.6.1 Overview

Cloud-based object stores currently support the majority of Internet-based services and applications, with Amazon's S3 (Simple Storage Service) [13] being by far the most popular. The term "S3" is often used both in referring to the Amazon service, as well as the underlying communications protocol. Many other service providers strive to provide S3 compatibility (like DigitalOcean Spaces [14] or Wasabi [15]), as the amount of applications already using the protocol through respective libraries is vast. MinIO [16] is an open-source EVOLVE that provides an S3-compatible server for use on local premises, and S3proxy [17] implements an S3 API over a selection of other storage services or a local filesystem. However, in local setups, the HTTP-based RESTful web service API, used by S3 and most object store services, greatly inhibits performance and especially latency of storage operations.

H3 is a thin object-to-key–value translation layer (Figure 9.9), forwarding storage operations to a key–value storage engine. A key–value store is a type of non-relational database (NoSQL) that stores data as a collection of key–value pairs, similar to a common dictionary data structure. Keys and values can range from simple strings to complex compound binary large objects (blobs), requiring no upfront data modeling or schema definition. Similar to objects stores, key–value stores lack an overall structure and offer a simple interface of *get*, *put*, and *delete* operations. As keys are independent of each other, key–value stores are highly partitionable and can be easily scaled on multiple underlying storage segments or even across a distributed setup of nodes. In H3 we use a plugin architecture in the back end to support different key–value implementations, like Redis [18], RocksDB [19], and Kreon [20], each with different characteristics: Redis stores data in memory, RocksDB in disk, while Kreon is optimized for flash-based storage, where CPU overhead and I/O amplification are more significant bottlenecks compared to I/O randomness. As H3 is stateless, it requires no synchronization among peer

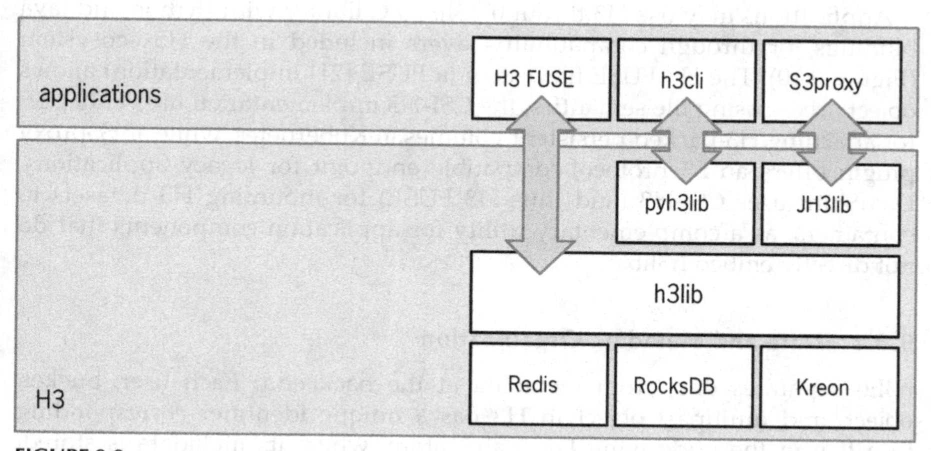

FIGURE 9.9
Example of H3 object to key mapping.

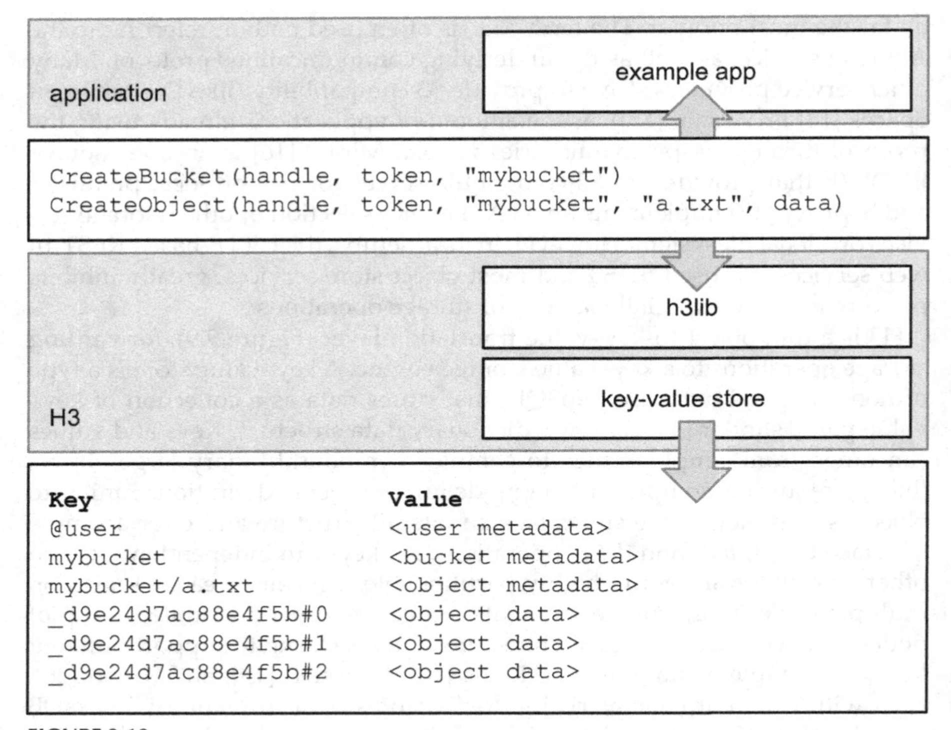

FIGURE 9.10
Compatibility layers of the H3 stack.

applications that operate on the same back end. Thus, if the key–value store is distributed, H3 can easily scale out across nodes.

Applications may use H3 through h3lib, a C library with Python and Java bindings, or through compatibility layers included in the H3 ecosystem (Figure 9.10). The H3 FUSE filesystem (a FUSE [21] implementation) allows object access using file semantics, the CSI-H3 implementation uses H3 FUSE for attaching H3-backed persistent volumes in Kubernetes, while an S3proxy plugin offers an S3 protocol-compatible endpoint for legacy applications. DataShim uses CSI-H3 (and thus H3 FUSE) for mounting H3 datasets to containers, as a complementary utility for application components that do not directly embed h3lib.

9.6.2 Data and Metadata Organization

h3lib separates data from metadata at the back end. Each user, bucket, object, and multipart object in H3 has a unique identifier corresponding to a key in the underlying key–value store, where its metadata is stored. "Multipart"' objects are objects that are created by uploading different data chunks in separate operations, while specifying their ordering in the object. The multipart object can then be finalized using a *complete* operation or cancelled with an *abort*. This allows parallel uploading of large files. Object and multipart object data may consist of a single or multiple parts of variable size, which are stored under different keys. As key–value back ends usually have a size limit on values, we select a part size in H3 and automatically break oversized values into an appropriate number of smaller parts.

h3lib provides a simple back-end interface that needs to be implemented by plugins, which includes basic *read*, *write*, *delete*, and *list* operations. The interface separates data and metadata operations, which allows a plugin to serve data and metadata from different stores. Our currently implemented back ends for Redis, RocksDB, and Kreon, store data and metadata in the same key space and serve them both from the same key–value store.

The complete h3lib API currently includes 32 calls, which translate to corresponding key–value read. write, and scan (return all keys starting with a prefix) operations. h3lib takes advantage of the rich set of data query and manipulation primitive provided by the back end, in order to minimize code complexity and exploit any optimizations done in the key–value layer. In addition to common *create*, *read*, *write*, *list*, and *delete* methods for buckets, objects, and multipart objects, we provide functions to perform data operations directly in the back end: create an object from part of another object, write in an object offset with partial data from another, copy and move objects, and truncate an object to a specific size. h3lib also supports sparse objects and provides functions to read objects directly to local files and vice versa.

9.6.3 The H3 Ecosystem

h3lib is written in C, and we implement wrapper functions in Java and Python, for embedding h3lib to a larger set of applications. Java wrapper functions access the store natively, with the use of Java Native Access (JNA) [22], while the Python wrapper functions are implemented as a Python extension module. Furthermore, we provide client libraries on top of the wrapper functions for each programming language, as the wrappers expose internal details that are not required to the end user. In addition to the language bindings, we implement H3 FUSE – a filesystem layer on top of h3lib with the use of Filesystem in Userspace (FUSE), an interface for user-space programs to export a filesystem to the Linux kernel. To support H3 FUSE, we include in h3lib necessary filesystem attributes as part of the object's metadata, such as access permissions and ownership. Directories are implemented by H3 FUSE as empty objects ending with a forward slash character.

Moreover, through a jclouds plugin [23] (the library used by S3proxy to perform data operations), we provide a custom version of S3proxy that acts as a bridge between the S3 protocol and a key–value store through H3. Both the filesystem and S3-compatibility layers aim to provide applications with the ability to utilize h3lib, without the need to alter their application code.

Considering that the FUSE framework imposes overheads that greatly affect performance [24] and S3 adds back protocol and serialization delays, the user is required to take into consideration the trade-off between compatibility and performance. The primary purpose of these layers is to support legacy applications that are no longer maintained and, consequently, cannot be altered, as well as to enable integrating H3 to existing systems.

We use H3 FUSE in a CSI plugin to integrate h3lib in the USL, for transparent data exchange between H3-compatible applications and applications supporting only filesystem semantics. This is of particular use in "hybrid" workflows, where one stage fetches data directly to a mounted filesystem path, another uses h3lib to access and transform the data, while a third is implemented with some ready-made utility that expects an S3 endpoint to interact with. The diverse set of H3 interfaces allows such heterogeneity in data access protocols.

9.7 Integration

Figure 9.11 shows how datasets are managed in the combined deployment of all USL components. The life cycle of a dataset starts when a new dataset CRD is created through Karvdash. Then, the Datashim controller converts the dataset definition to a Persistent Volume and Persistent Volume Claim in

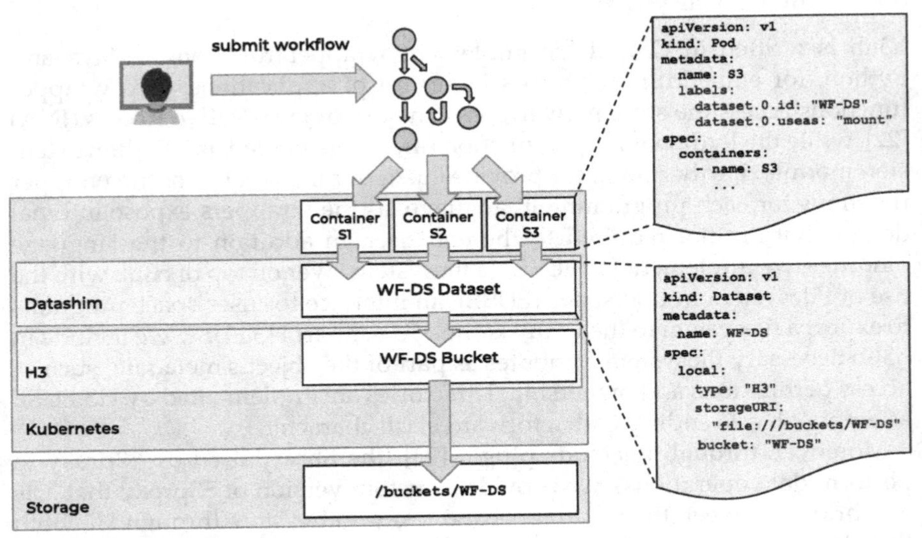

FIGURE 9.11
All components of the USL in action.

Kubernetes, which are in turn handled by the appropriate CSI plugin. Here, the user creates an H3 dataset, which results in an H3 volume, supported by a bucket in actual storage. Users can directly use volumes created by Datashim in their pods, however, to make this even easier, Datashim monitors specific labels in pod definitions and if they are found, automatically adds the necessary configuration to mount the volumes. In the example shown in the figure, the user submits a workflow, consisting of several containers annotated with the appropriate labels (i.e. "dataset.0.id") for Datashim to automatically mount the corresponding H3 volume.

The internal mechanics of the integration of all USL components is shown in Figure 9.12. In this example, all services are running in containers: Karvdash, Datashim, H3 (embedded in the CSI-H3 deployment), as well as Argo (the workflow controller) and some key–value store, like Redis. First, the user logs into the dashboard and creates an H3 dataset using the Redis service, which is stored in Kubernetes as a dataset CRD instance (1). Then, a workflow is submitted through Argo (2), which triggers the deployment of a pod as part of some step (3). This request is intercepted by Karvdash as it refers to a container running in the user's namespace (4). Karvdash adds the user's datasets as labels in the pod definition. Datashim intervenes, as there is a now a dataset label (5). Datashim has already prepared the volume in Kubernetes and attaches it to the pod. Assuming no other controllers intercept the call to create the pod, it's now up to Kubernetes to decide where to run it. Kubernetes selects the second node and contacts the CSI-H3 controller to ask for a volume mountpoint, as the pod requires an H3 volume. The controller, communicates with the nodeplugin that runs at the same node as the pod (7), which in turn creates an H3 bucket at the endpoint provided (the Redis server) and uses H3 FUSE to mount it (8). A new volume is created (9). Actually, FUSE runs in the CSI-H3 nodeplugin container, but the mountpoint is propagated to the host, so it can be shared among containers running in the same machine. Finally, the workflow step can start, with the H3 FUSE volume attached at a specific folder (10). Note, that the user only creates a dataset and launches a workflow – all the rest happens automatically (and almost instantaneously) by the USL.

This way, USL completely abstracts all storage complications from users. Workflow developers first utilize Karvdash to specify their data assets through the web interface and then submit a set of stages to the Argo workflow engine. The USL components intercept the calls, convert datasets to volume attachments, and mount the volumes to pods. Consecutive stages may communicate with each other by accessing data as local files. Thus, the workflow relies on the existence of some mountpoint, but its type is defined by the dataset used. The same workflow can run with a different dataset, without internal changes. In EVOLVE, one dataset definition may instantiate H3 with a key–value store using memory-only scratch space or persistent storage in the underlying Lustre file system. Another dataset may result in a workflow operating on objects in a remote S3 bucket residing in the cloud.

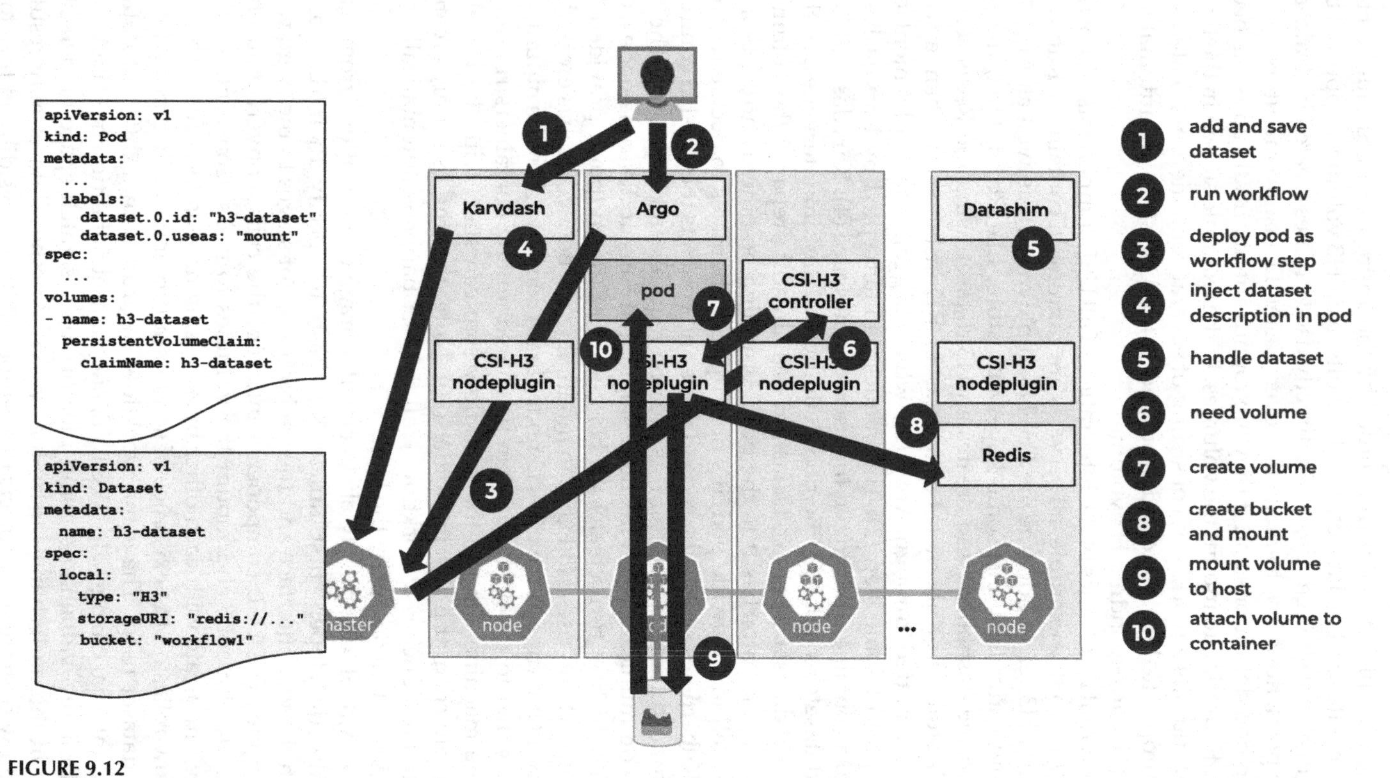

FIGURE 9.12
USL integration internals.

9.8 Related Work

Management of storage and data is not a new topic in the cloud world, with many solutions already existing targeting Kubernetes.

Rook [10] is a cloud-native framework for orchestration, provisioning, and monitoring of distributed storage on Kubernetes. Through Rook, cluster administrators can automatically deploy and scale the actual storage back end on a Kubernetes cluster. As an example, a fully functional Ceph [8] cluster can be deployed on a Kubernetes cluster, including the resources for provisioning volumes to pods. In contrast, USL takes care of abstracting the notion of volumes, and via dataset CRDs users can connect any storage source to their pods. USL, however, does not take care of the management of the storage back end and could be complemented by frameworks such as Rook for delivering a full experience from the user interface to storage back-end management.

OpenEBS [25] is a volume management framework for Kubernetes that implements the Container Attached Storage (CAS) paradigm. Each volume has a dedicated containerized storage controller and storage replicas for high availability of data. Having a per-volume manager enables setting per-volume storage-related policies and increase resilience to failures. OpenEBS also offers disaster recovery and monitoring via Prometheus. In contrast with the abstraction proposed in this chapter, OpenEBS focuses on management of volumes and does not take care of interfacing with users and their applications in a transparent way. In addition, in OpenEBS there is no support for object storage-based volumes.

Pachyderm [26] tries to increase the abstraction level between data and applications focusing on data science, especially in relation to generation of datasets and tracking of their life cycle. It offers tools for versioning of data, allowing access to a previous versions of data while it is being generated and transformed. In addition, Pachyderm also offers provenance monitoring to help users track where their data is coming from, as well as supporting the deployment of massively parallel data transformation pipelines. Pachyderm has a different target than that of the USL, but it presents another attempt at simplifying data access in Kubernetes.

NooBaa [27] targets users relying on multiple object storage solutions, enabling encryption, de-duplication of data and compression. Users do not need to know credentials and details for all the object storage solutions they rely on, because the NooBaa framework exposes a common endpoint through which data can be accessed. Compared to USL, NooBaa simplifies dealing with multiple cloud storage solutions, but does not remove the necessity for the user to deal with the details of the object storage endpoint.

All of the above technologies attempt simplifying the way cloud applications access storage, however many of them are targeting only one storage solution, like NooBaa, or focus more on the storage back end and volumes, as in the case of Rook and OpenEBS. Instead, USL focuses on handling the heterogeneity of many possible storage solutions, by providing a single interface and abstraction for applications. In addition, the aforementioned EVOLVEs are sometimes orthogonal to USL and can be integrated for delivering an even better experience. As an example, USL could be used for providing access to data in Pachyderm, allowing Pachyderm data pipelines seamless access to data from multiple storage back ends.

9.9 Conclusions

In this chapter we have presented the USL, a new approach that aims to bridge the gap between HPC and cloud-based storage offerings, by focusing on the increasing demand for performance and portability in HPDA workloads that use a diverse set of storage services. USL provides a turnkey solution, which includes a user-facing dashboard (Karvdash), an integrated data abstraction called *Dataset* (DataShim), as well as a novel, embedded object store, leveraging key–value storage technologies for providing high-performance access to data (H3). These components are stand-alone open-source software frameworks, whose integration creates a unique software stack for data access in Kubernetes environments.

USL has been developed in the context of EVOLVE, and is crucial to support the overall effort of creating a converged platform that minimizes the gap between big data, cloud, and HPC technologies. The USL provides an abstract interface for deployment of large data analytics workflows that benefit at the same time from big data software frameworks, cloud services, as well as "classic" HPC libraries and storage facilities, all running on a highly heterogeneous HPC cluster. Our experience with EVOLVE's industrial partners indicates that the USL has dramatically improved the way applications access multiple data sources, without sacrificing performance because of the added abstraction layers.

Note

1 A *pod* is a set of one or more container instances running as a single entity in Kubernetes.

References

[1] Antony Chazapis, Christian Pinto, Yiannis Gkoufas, Christos Kozanitis, and Angelos Bilas. 2021. A Unified Storage Layer for Supporting Distributed Workflows in Kubernetes. In Workshop on Challenges and Opportunities of Efficient and Performant Storage Systems (CHEOPS '21), April 26, 2021, Online, United Kingdom. https://doi.org/10.1145/3439839.3458735
[2] Kubernetes: Production-Grade Container Orchestration, https://kubernetes.io
[3] Argo Workflows, https://argoproj.github.io/projects/argo
[4] Karvdash: A dashboard service for facilitating data science on Kubernetes, https://github.com/CARV-ICS-FORTH/karvdash
[5] DataShim, https://datashim.io/
[6] H3: An embedded object store, https://github.com/CARV-ICS-FORTH/H3
[7] Kubernetes Container Storage Interface (CSI) Documentation, https://kubernetes-csi.github.io/docs/
[8] Sage A. Weil, Scott A. Brandt, Ethan L. Miller, Darrell D. E. Long, and Carlos Maltzahn. 2006. Ceph: a scalable, high-performance distributed file system. In Proceedings of the 7th symposium on Operating systems design and implementation (OSDI '06). USENIX Association, USA, 307–320.
[9] "Ceph Object Gateway," [Online]. Available: https://docs.ceph.com/en/latest/radosgw/
[10] "Rook," [Online]. Available: https://rook.io/
[11] ResNet50. [Online]. Available: https://github.com/tensorflow/models/tree/master/official/vision/image_classification#resnet50
[12] "Goofys," [Online]. Available: https://github.com/kahing/goofys.
[13] Amazon S3, https://aws.amazon.com/s3/
[14] DigitalOcean Spaces, www.digitalocean.com/products/spaces/
[15] Wasabi, https://wasabi.com
[16] MinIO, https://min.io/
[17] S3Proxy, https://github.com/gaul/s3proxy
[18] Redis, https://redis.io/
[19] RocksDB, https://rocksdb.org/
[20] Anastasios Papagiannis, Giorgos Saloustros, Pilar González-Férez, and Angelos Bilas. 2018. An Efficient Memory-Mapped Key–Value Store for Flash Storage. In Proceedings of the ACM Symposium on Cloud ComputingSoCC '18. Association for Computing Machinery, New York, 490–502. https://doi.org/10.1145/3267809.3267824
[21] FUSE (Filesystem in USErspace), www.kernel.org/doc/html/latest/filesystems/fuse.html
[22] Java Native Access, https://github.com/java-native-access/jna
[23] Apache jclouds, http://jclouds.apache.org
[24] Bharath Kumar Reddy Vangoor, Vasily Tarasov, and Erez Zadok (2017). To FUSE or Not to FUSE: Performance of User-Space File Systems. In 15th

USENIX Conference on File and Storage Technologies (FAST 17) (pp. 59–72). USENIX Association.
[25] OpenEBS: Kubernetes storage simplified, https://openebs.io/
[26] Pachyderm, www.pachyderm.com/
[27] NooBaa, www.noobaa.io/

10

The DeepHealth HPC Infrastructure: Leveraging Heterogenous HPC and Cloud-Computing Infrastructures for IA-Based Medical Solutions

Eduardo Quiñones, Jesus Perales, Jorge Ejarque, Asaf Badouh, Santiago Marco, Fabrice Auzanneau, François Galea, David González, José Ramón Hervás, Tatiana Silva, Iacopo Colonnelli, Barbara Cantalupo, Marco Aldinucci, Enzo Tartaglione, Rafael Tornero, José Flich, Jose Maria Martínez, David Rodriguez, Izan Catalán, Jorge García, and Carles Hernández

CONTENTS

DOI: 10.1201/9781003176664-10

10.1 Introduction

Deep learning (DL) is increasingly being considered to effectively support medical diagnosis of complex diseases. DL includes two main operations: (1) *training*, which refers to the process of generating a predictive model, in the form of a deep neural network (DNN), based on large datasets composed of biomedical information (e.g. medical images); and (2) *inference*, which refers to the process of predicting a diagnosis based on a reduced dataset.

DL training is the most computationally intensive operation, requiring very large memory and computing power. The training operation is an iterative process, which means it is required to iterate many times over all the dataset samples to properly adjust the model, that is, the weights of the DNN. In general, larger datasets allows users to obtain predictive models with higher accuracy, which, in turn, increases the running time needed for the training procedure.

This makes it unfeasible to run training operations on general-purpose computing systems, such as those available in hospitals, for large medical datasets, even when such computers are equipped with powerful processors featuring many-core or GPU acceleration devices.

The exploitation of the *parallel capabilities of HPC infrastructures*, composed of dozens or hundreds of computing nodes, allows the training operation to be split into smaller datasets upon which parallel training operations can be applied. These methods are mandatory when the samples do not fit into a single computing node. As an example, the dataset provided by the FISABIO Foundation (a partner of the DeepHealth project) composed of Covid-19 images is 141 GB which does not fit into the memory of a single computing node (e.g. the Marenostrum 4 Supercomputer, features 96 GB of memory per node).

Moreover, training operations include data augmentation processes to transform original images to mitigate the problem of overfitting, that are applied on-the-fly, because each unique image from the dataset may require to be transformed in a different way at every iteration, that is, epoch. Applying a sequence of image transformations on-the-fly is also computationally expensive. It is important to note that medical imaging datasets are typically composed of extremely large images, further increasing the computing requirements.

DeepHealth has developed an HPC toolkit capable of efficiently exploiting the computing capabilities of HPC infrastructures to execute DL training operations, in a *fully transparent way*. To do so, the data/computer scientists only need to describe in a file (CSV or JSON or XML or similar) the set of computing nodes to be used during the training process and launch it.

HPC facilities are not well suited for every kind of application. Queue-based workload managers that commonly orchestrate HPC centers cannot satisfy the strict time-to-solution requirements of the inference phase, and air-gapped worker nodes are not able to expose user-friendly web interfaces for data visualization. To address these issues, the DeepHealth HPC toolkit also supports cloud environments, with innovative hybrid solutions to support private and public clouds and a new workflow management system (WMS) capable of scheduling and coordinating different steps of a DL pipeline on hybrid HPC/cloud architectures.

Overall, the DeepHealth HPC toolkit offers vision and DL functionalities included in the European Computer Vision Library (ECVL) and the European Distributed Deep Learning Library (EDDL),[1] the HPC capabilities needed to efficiently exploit the computing capabilities of HPC and cloud infrastructures. The two libraries are also developed within the context of the DeepHealth project.

10.2 The Parallel Execution of EDDL Operations

EDDL is a general-purpose DL library initially developed to cover DL needs in healthcare use cases within the DeepHealth project. EDDL provides hardware-agnostic tensor operations to facilitate the development of hardware-accelerated DL functionalities and the implementation of the necessary tensor operators, activation functions, regularization functions, optimization methods, as well as all layer types (dense, convolutional, and recurrent) to implement state-of-the-art neural network topologies.

In order to be compatible with existing developments and other DL toolkits, the EDDL uses ONNX,[2] the standard format for neural network interchange, to import and export neural networks including both weights and topology. As part of its design to run on distributed environments, the EDDL includes specific functions to simplify the distribution of batches when training and inference processes are run on distributed computing infrastructures. The EDDL serializes networks using ONNX to transfer weights and gradients between the master node and worker nodes. The serialization includes the network topology, the weights, and the bias. To facilitate distributed learning, the serialization functions implemented in the EDDL allow users to select whether to include weights or gradients.

The next sections present the parallel strategies implemented to execute the EDDL operations and so leverage HPC and cloud architectures.

10.2.1 COMPSs

COMPSs is a portable programming environment based on a *task model*, whose main objective is to facilitate the parallelization of sequential source code written in Java or Python programming languages, in a distributed and heterogeneous computing environment. In COMPSs, the programmer is responsible for identifying the units of parallelism (named *COMPSs tasks*) and the synchronization data dependencies existing among them by annotating the sequential source code (using annotations in case of Java or standard decorators in case of Python).

Figure 10.1 shows a snipped (simplified for readability purposes) of the parallelization of the EDDLL training operation with COMPSs. COMPSs tasks are identified with a standard Python decorator @task (lines 1 and 5). The IN, OUT and INOUT arguments define the data directionality of function parameters. By default, parameters are IN, and so there is no need to explicitly specify IN parameters. Moreover, when a task is marked with is_replicated=True, the COMPSs task is executed in all the available computing

```
@task (is_replicated = True)
def build (model):
  # The model is created at each worker
  [...]
@task(INOUT = weights)
def train_batch(model, dataset):
  # Executed at each worker
  [...]
def main():
  # A new model is created
  net = eddl.model([...])
  build(net)
  for i in range(num_epochs):
    for j in range(num_batches):
      weight[j] =
train_batch(net,dataset)
    # Synchronize all weights from
workers
    compss_wait_on(weight)
    # Update weights on the model
    update_gradients(net,weight)
```

FIGURE 10.1
A (simplified) snippet of EDDLL training operation parallelized with COMPSs.

nodes for initialization purposes; otherwise, it executes on the available computing resources.

The train iterates over num_epochs epochs (line 13). At every epoch, num_batches batches are executed (line 14), each instantiating a new COMPSs task (line 15) with an EDDLL train batch operation. All COMPSs tasks are synchronized at line 17 with compss_wait_on, and the partial weights are collected. The gradients of the model are then updated with the partial weights at line 19.

The task-based programming model of COMPSs is then supported by its runtime system, which manages several aspects of the application execution and keeps the underlying infrastructure transparent to the programmer. The COMPSs runtime is organized as a master–worker structure:

- The *master*, executed in the computing resource where the application is launched, is responsible for steering the distribution of the application and data management.
- The *worker(s)*, co-located with the master or in remote computing resources, are in charge of responding to task execution requests coming from the master.

One key aspect is that the master maintains the internal representation of a COMPSs application as a Direct Acyclic Graph (DAG) to express the parallelism. Each node corresponds to a COMPSs task and edges represent data dependencies (and so potential data transfers). As an example, Figure 10.2

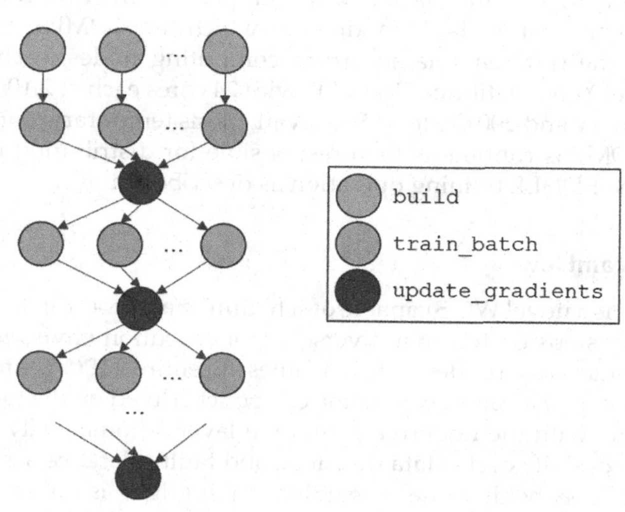

FIGURE 10.2
DAG representation of the EDDLL training operation presented in Figure 10.1.

presents the DAG representation of the EDDLL training operation presented in Figure 10.1.

Based on this DAG, the runtime can automatically detect data dependencies between COMPSs tasks: as soon as a task becomes ready (i.e. when all its data dependencies are honored), the master is in charge of distributing it among the available workers, transferring the input parameters before starting the execution. When the COMPSs task is completed, the result is either transferred to the worker in which the destination COMPSs tasks execute (as indicated in the DAG), or transferred to the master if a compss_wait_on call is invoked.

One of the main features of the COMPSs framework is that it abstracts the parallel execution model from the underlying distributed infrastructure. Hence, COMPSs programs do not include any detail that would tie them to a particular platform, boosting portability among diverse infrastructures and so enabling its execution in both a classical HPC environment and a cloud-based environment. To do so, COMPSs abstracts the underlying infrastructure by creating a set of execution environments, named COMPSs workers, in which COMPSs tasks execute. Internally, the COMPSs runtime implements different adapters to support the execution of COMPSs tasks in a given resource. Through a set of configuration files, the user specifies the available computing resources, which may reside in a computing cluster or in the cloud.

Figure 10.3 shows an example of this deployment. The execution starts in the *Computing Resource 1*, where the COMPSs master executes. Then four workers are deployed in four different resources to distribute the workload, where the EDDLL training operations can be distributed.

The COMPSs runtime is already supported in the Marenostrum Supercomputer[3] as a loadable module, in which the COMPSs workers are executed in the different Marenostrum computing nodes, each equipped with two Intel Xeon Platinum 8160 CPU with 24 cores each at 2.10GHz, 96 GB of main memory and 200 GB local SSD available as temporary storage during jobs. The COMPSs runtime is then responsible for distributing the parallel version of the EDDLL training operation as described above.

10.2.2 StreamFlow

StreamFlow[4] is a novel WMS capable of scheduling and coordinating different DL workflow steps on top of a diverse set of execution environments, ranging from practitioners' desktop machines to entire HPC centers. In particular, each step of a complex pipeline can be scheduled on the most efficient infrastructure, with the underlying runtime layer automatically taking care of worker nodes' life cycle, data transfers, and fault-tolerance aspects.

The basic idea behind the StreamFlow paradigm is to easily express correspondences between the description of a coarse-grain application

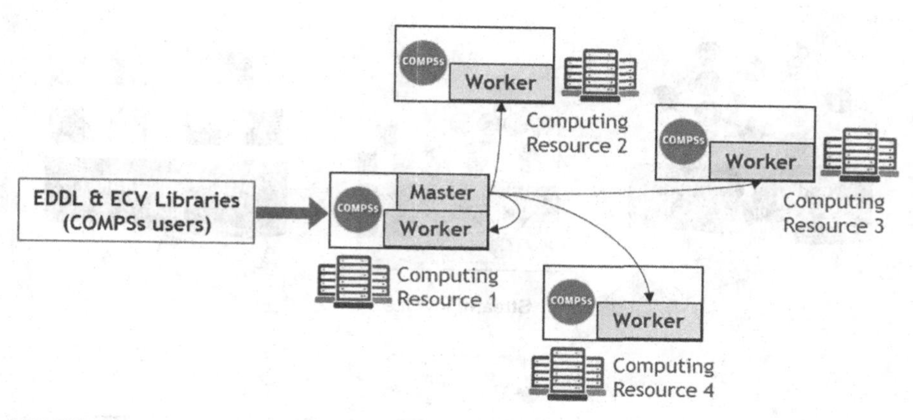

FIGURE 10.3
Example of a COMPSs deployment in a HPC infrastructure.

workflow, that is, a graph containing the application steps with the related data dependencies, and the description of an execution environment, that is, a manifest defining the capabilities of a target infrastructure. Starting from such description, the StreamFlow runtime layer is then able to orchestrate both the worker nodes' life cycle and the execution of the application on top of them.

With respect to the majority of WMSs on the market, StreamFlow gets rid of two common design constraints:

- There is *no need for a single shared data space* accessible from all the workers involved in a workflow execution. This allows supporting complex and hybrid execution infrastructures, including hybrid HPC/cloud architectures.
- Steps can be offloaded to *multi-agent execution environments*, ensuring the co-allocation of multiple and potentially heterogeneous worker nodes. This allows steps involving distributed architectures, for example, a multi-node DNN training driven by a COMPSs' master–worker infrastructure to be offloaded.

To provide enough flexibility, StreamFlow adopts a three-layered hierarchical representation of execution environments. A complex, multi-agent environment is called *model* and constitutes the unit of deployment, that is, all its components are always co-allocated when executing a step. Each agent in a model, called *service*, constitutes the unit of binding, that is, each step of a workflow can be bound to a single service for execution. Finally, a *resource* is a single instance of a potentially replicated service and constitutes the unit of scheduling, that is, each workflow step is offloaded to a configurable number of resources to be processed. As an example, a Helm chart describing

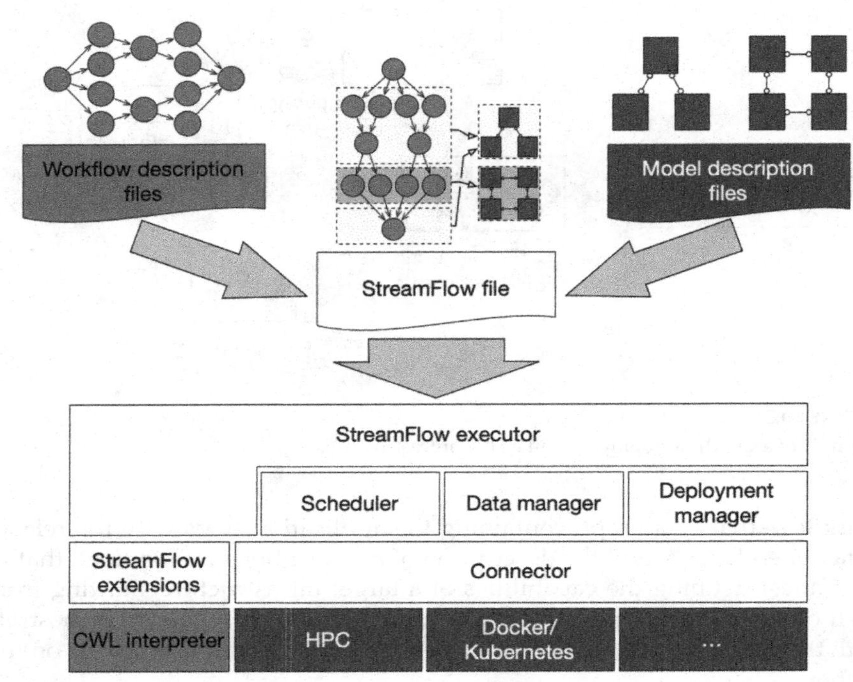

FIGURE 10.4
StreamFlow logical stack.

a COMPSs master pod and four COMPSs worker pods constitutes a model with two services, the former with one resource and the latter with four resources.

Figure 10.4 shows StreamFlow's logical stack. The deployment manager is the component in charge of creating and destroying models when needed. To do that, it mainly relies on external orchestration technologies (e.g. Slurm, Kubernetes, or Docker Compose) by means of pluggable implementations of the connector interface. After a model is deployed, the scheduler component is in charge of selecting the best resource on which each workflow step should be executed while guaranteeing that all requirements are satisfied. Finally, the data manager, which knows where each step's input and output data reside, must ensure that each service has access to all the data dependencies required to complete the assigned workflow step, performing data transfers only when necessary.

Rather than defining a new format to describe workflow models, StreamFlow relies on an existing coordination format, called Common Workflow Language (CWL). CWL is an open standard for describing analysis workflows, following a declarative JSON or YAML syntax. As CWL is a fully declarative language, it is far simpler to understand for domain experts than its Make-like or dataflow-oriented alternatives. Moreover, the fact that many products offer support for CWL, either alongside a proprietary coordination language or at higher-level semantics on top of low-level APIs, fosters portability and interoperability.

It is also worth noting that StreamFlow does not need any specific package or library to be installed on the target execution architecture other than the software dependencies required by the host application (i.e. the involved workflow step). This agentless nature allows virtually any target architecture reachable by a practitioner a potential target model for StreamFlow executions, as long as a compatible connector implementation is available.

10.3 Cloud Infrastructures

10.3.1 Hybrid Cloud

The hybrid cloud is the combination of private cloud and public cloud, allowing the exploitation of the best of both types. There are several reasons why it may be interesting to have a hybrid cloud. The three most typical cases are:

- Security: Sensitive information that cannot be at risk, – banking or health information, is kept in the private part of the cloud to minimize potential risks.
- Cost: The presence of sporadic peaks on the computing workload can be served with public cloud resources, without requiring the private cloud infrastructure to dimension to compensate for the peaks. By doing so, the hybrid cloud part is used for the usual load, and the public part is used to compensate for the peaks while keeping the service alive.
- Availability: In the same way as with a content delivery network (CDN), a hybrid private–multi-public cloud can be created. Multi-public clouds can be chosen to be close to customers to reduce latency and maximize available bandwidth, to reduce cost, or to improve availability.

The *DeepHealth HPC toolkit* includes an on-premises private cloud based on a Kubernetes cluster (and hosted by the DeepHealth partner TREE technology), and a public cloud in AWS computing resources, as illustrated in Figure 10.5.

By doing so, the DeepHealth hybrid cloud allows computing resources to be provisioned in a flexible way to accommodate project requests. GPU computing capabilities can be added – both through provisioning on the on-premises Kubernetes cluster and via nodes provisioned using Amazon Elastic

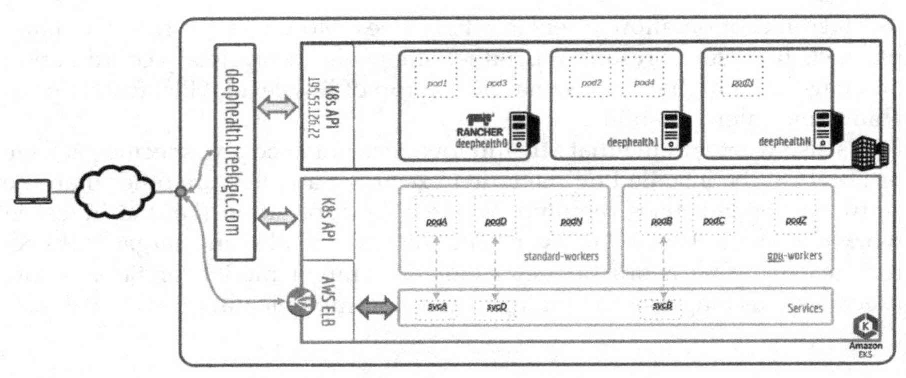

FIGURE 10.5
Hybrid cloud environment included in the DeepHealth HPC toolkit.

Kubernetes Service (EKS) technology on Amazon Web Services (AWS). This hybrid infrastructure has two main parts: Rancher and API services.

1. The open-source Rancher multi-cluster orchestration platform is used[5] to support the management of the clusters. Rancher allows the operational and security challenges to be managed on multiple Kubernetes clusters across any infrastructure. Moreover, its web interface provides control over deployments, jobs, pipelines, including an app catalog for fast deployment using the Helm software package manager.[6]
2. An API has been developed to facilitate the use and/or integration of the libraries with the Kubernetes platform. Moreover, a high-level REST API has been developed on top of the Kubernetes API to help abstract the user from the infrastructure itself, simplifying the deployment and management of the workflows. It provides functions of varying complexity, from simple ones, like list of pods, to more complex ones such as expose pods – which implements functionality abstracting the user from the potentially complex configuration of the clusters (e.g. multi-cloud, hybrid cloud). The API itself can support the addition of new Kubernetes clusters both on-premises and in the cloud from any provider. To guarantee properly authenticated and authorized access to the DeepHealth cloud, a connection through a VPN is required to access the API.

10.3.2 Parallel Execution on Cloud Environments

DeepHealth exploits the parallelization of DL operations on cloud environments by the usage of the COMPSs and StreamFlow parallel frameworks presented in Section 10.2. The following sections go on to describe these.

10.3.2.1 Parallel Cloud Execution Based on COMPSs

Unlike the Linux-based infrastructure, there is no need for setting up the execution environment in all the computing resources, but a docker image must be available, for example, Docker Hub.[7] Figure 10.6 shows an example of this deployment. The execution starts in the *computing resource 1*, where the COMPSs master executes. Then three workers are deployed in three different containers in the cloud infrastructure (in our case the cloud is provided by TREE), where the COMPSs application is distributed (in our case, EDDLL training operations).

Moreover, the COMPSs runtime is being adapted to support the cloud infrastructure provided by TREE. The cloud is based on Kubernetes (K8S),[8] and allows applications to be managed in a container technology environment and the manual processes to deploy and scale containerized applications to be automated. Moreover, an API is being developed by TREE to help abstract the user from the infrastructure itself, speeding up the processes of deployment and management of the workflows. COMPSs runtime interacts with this API to deploy workers and distribute the workload.

The DeepHealth hybrid cloud has been integrated with the COMPSs framework (see Section 10.2.1), which allows the DL training operations to be accelerated by dividing the training datasets across a large set of computing nodes available on cloud infrastructures, and upon which partial training operations are then performed. This combination is done through the REST API, allowing COMPSs to abstract from the infrastructure and perform their workflow regardless of where they are deployed.

Figure 10.6 shows a possible distribution of the execution of the parallel version of the EDDLL training operation presented in Section 10.2.1, in the

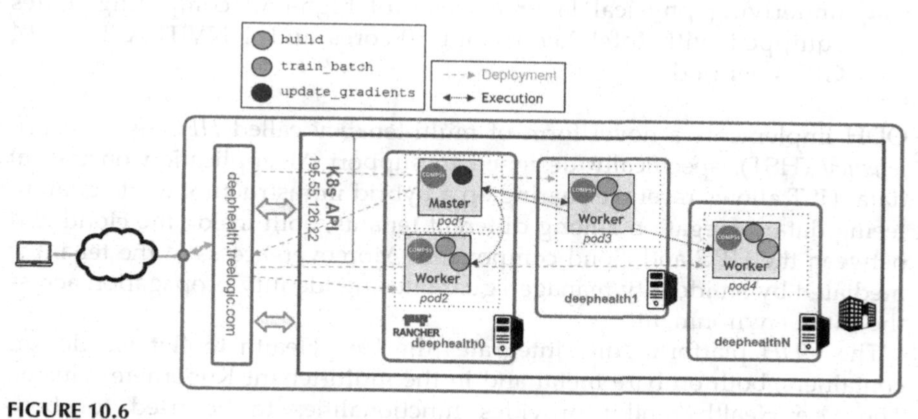

FIGURE 10.6
Distributed EDDLL training operation presented in Figure 10.2 on the DeepHealth hybrid cloud.

DeepHealth hybrid considering three COMPSs workers, setting the number of replicas to 3. The COMPSs runtime use the DeepHealth cloud API to automatically deploy the master and the three replicas in which the COMPSs workers will be executed. Once the deployment is completed, the parallel execution of the training operation is initiated, and so the COMPSs runtime starts the distribution of the different COMPSs tasks (i.e. the build and train_batch tasks shown in Figure 10.2), guaranteeing the data dependencies among tasks. In this case, the update_gradients function is executed in the COMPSs master to aggregate the partial computed weights at the end of each epoch.

10.3.2.2 Parallel Cloud Execution Based on StreamFlow

StreamFlow relies on the University of Torino's OpenDeepHealth (ODH), which implements a hybrid HPC/cloud infrastructure to effectively support the training and inference of AI models. The WMS introduced in Section 10.2.2 is the key technology enabling a transparent offloading of AI tasks to heterogeneous sets of worker nodes, hiding all deployment and communication complexities to the expert user.

As shown in Figure 10.7, ODH includes two different infrastructures:

- The HPC component is the C3S OCCAM (Open Computing Cluster for Advanced data Manipulation)[9] cluster, with 46 heterogeneous nodes also including NVIDIA Tesla K40 and V100 GPUs. Workloads are orchestrated through an elastic virtual farm of hardened docker containers, running directly on top of the bare-metal layer.
- The cloud component is the HPC4AI[10] private cloud, a multi-tenant Kubernetes cluster running on top of an OpenStack cloud. The underlying physical layer consists of high-end computing nodes equipped with Intel Xeon Gold 80-cores and 4 NVIDIA Tesla T4 GPUs per node.

ODH implements a novel form of multi-tenancy called *HPC secure multi-tenancy* (HST), specifically designed to support AI application on critical data. HST allows resource sharing on a hybrid infrastructure while guaranteeing data segregation among different tenants, both inside the cloud and between the HPC and cloud components. Moreover, access to the tenant is mediated by an identity manager, guaranteeing identity propagation across the entire environment.

The ODH platform fully integrates the DeepHealth toolkit via docker containers, both on bare metal and in the multi-tenant Kubernetes cluster. The DeepHealth toolkit provides functionalities to be used for both training and inference, addressing the complexity of the different available

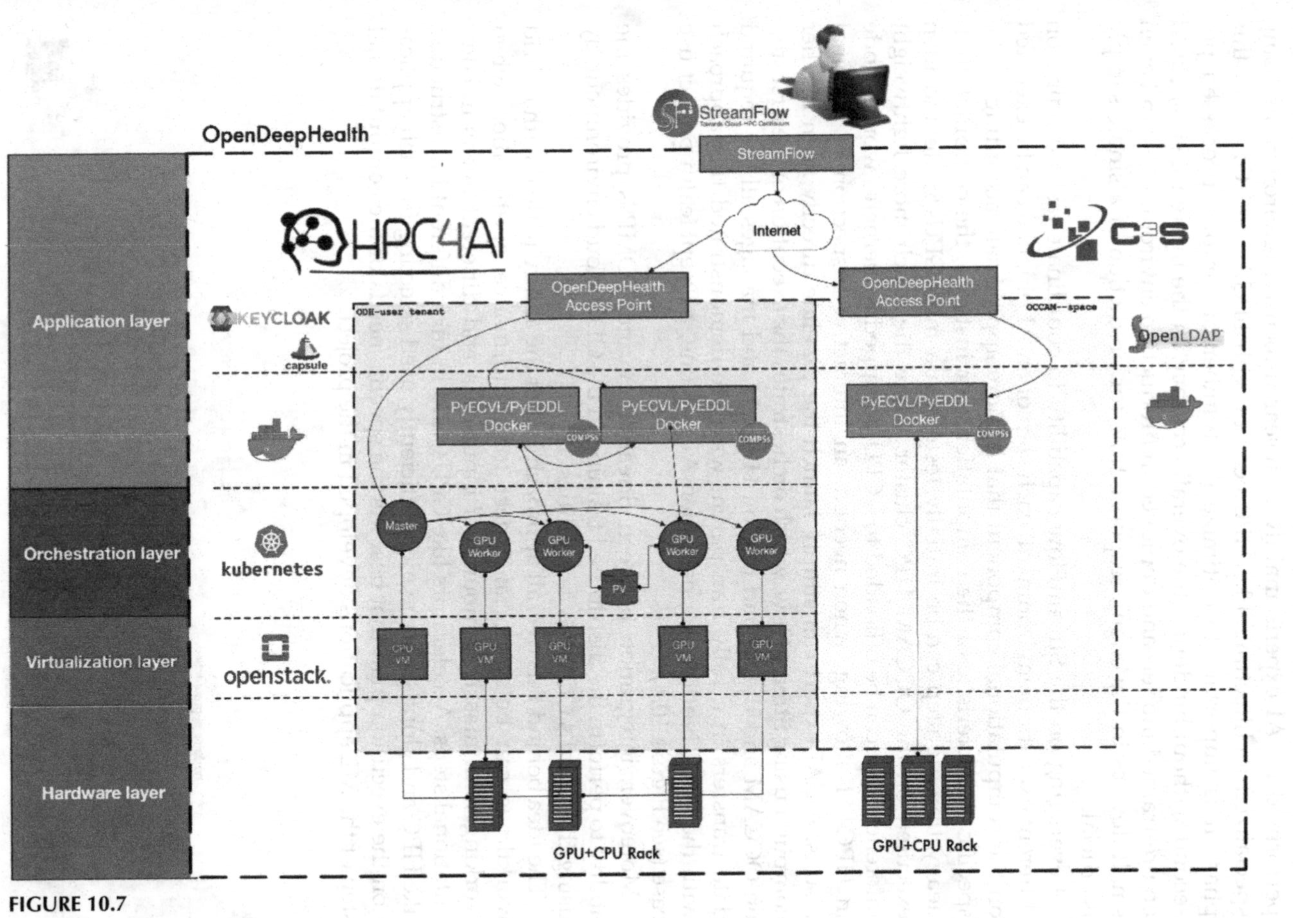

FIGURE 10.7
The OpenDeepHealth Architecture.

computational resources and target architectures. The training phase is performed by AI experts mainly in research-focused environments using specialized architectures like HPC centers with FPGAs and GPUs. In this phase it is important to optimize the number of samples processed per second without hindering the overall accuracy. In the inference step, based on pre-trained models and deployed in production environments (and even small devices in the edge), response time for prediction of a single sample is crucial.

Leveraging on the StreamFlow capability to coordinate tasks running on different execution environment, each step of an AI pipeline can be executed on the computational component that is more appropriate according to the specific characteristic of the computation. For instance, the computational-heavy training step can be initially tested on a multiGPUs node and then executed on the OCCAM HPC cluster, while the much more lightweight inference step, can be offloaded to a CPU-equipped Kubernetes worker node in HPC4AI allowing, when needed, an interactive inspection of the final results. An AI expert can simply launch the pipeline directly from his/her computer using StreamFlow, which orchestrates the execution of the first step on OCCAM and the second one on HPC4AI and manages all the required data transfers in a fully transparent way. We demonstrated this approach with the Lung Nodule Segmentation AI pipeline, a DeepHealth project use case (see Figure 10.8).

Moreover, integration between StreamFlow and COMPSs provides the ability to perform the distributed training in the ODH cloud environment, by using the cloud's COMPSs capabilities.

The idea behind this overall approach is that the ability to deal with hybrid workflows can be a crucial aspect for performance optimization when working with massive amounts of input data and different needs in computational steps. Accelerators like GPUs, and in turn different infrastructure like HPC and clouds, can be more efficiently used selecting for each application the execution plan that best fits the specific needs of the computational step of the ML applications developed in the project.

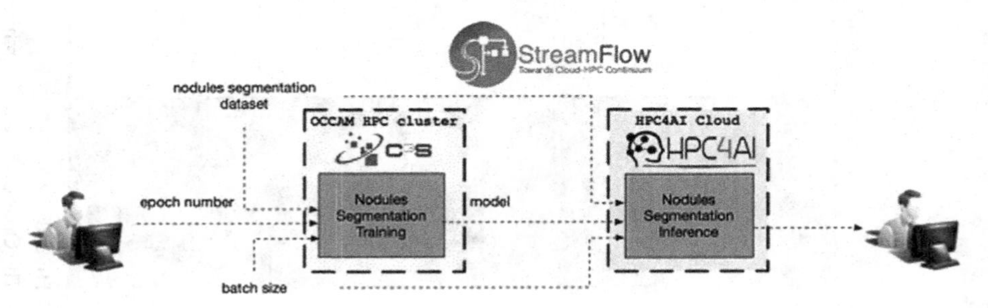

FIGURE 10.8
A DeepHealth use case AI pipeline executed using StreamFlow.

10.4 Acceleration Devices: GPU and FPGAs

10.4.1 FPGA Acceleration

FPGAs are devices with reconfigurable logic that can be combined and interconnected in order to process a specific algorithm. Contrary to CPUs and GPUs, in an FPGA the architecture (design) is adapted to the algorithm, thus offering an opportunity to optimize them. On the other hand, the resources of an FPGA device are limited and usually the device works at a lower frequency. Therefore, a careful and custom design is needed in order to benefit from it. Although FPGAs are well suited for inference in DL, they can also be used in specific situations for training processes.

FPGAs are supported in the DeepHealth project in two orthogonal although complementary and needed directions. First, large FPGA infrastructures are being adapted to the project by suitable interfaces and protocols specific to FPGAs. Second, specific and optimized FPGA algorithms are developed for specific use cases within the project. In the next sections we describe the adaptations and developments performed for the infrastructure and then the optimizations at algorithmic level.

10.4.1.1 The DeepHealth FPGA Infrastructure

10.4.1.1.1 The MANGO FPGA Platform

The DeepHealth toolkit includes the MANGO FPGA platform, developed within the European MANGO project. As part of DeepHealth activities we are evolving this cluster of FPGAs from a hardware prototyping platform to a high-performance and low-energy compute platform. The platform consists of two clearly differentiated subsystems: the general-purpose nodes (GNs) and the heterogeneous nodes (HNs). The former executes the host applications, as well as the low-level communication libraries; the latter represent the computational part of the system built upon the FGPA modules:

- A GN (see Figure 10.9) consists of a Supermicro SuperBlade module SBI-7128RG-F equipped with an Intel Xeon processor E5-2600 v3, 64 GB of RAM memory and 1 TB of SSD storage. Each GN is connected via PCIe to two HNs, so it can use both HN subsystems. GNs are also connected to the HNs via Ethernet and USB for cluster programming and management purposes.
- An HN (see Figure 10.9 and Figure 10.10) consists of 12 FPGA modules mounted on top of four proFPGA motherboards and placed in an FPGA cluster. This setup is extended with a total amount of 22 GB of RAM memory split in several DDR3 and DDR4 modules among different FPGAs. The HN is a heterogeneous subsystem since it contains different

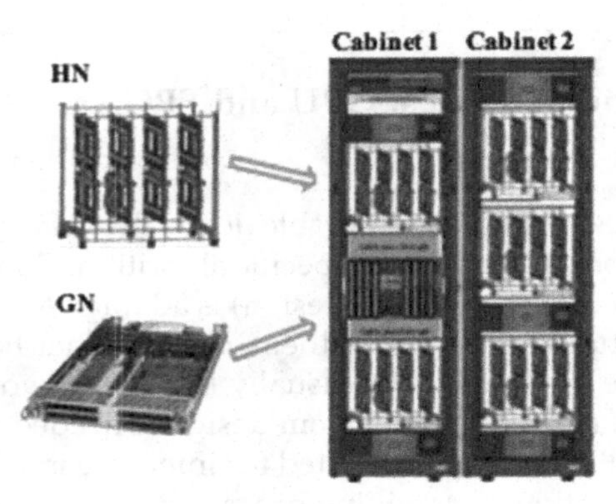

FIGURE 10.9
MANGO hardware: GNs and HNs.

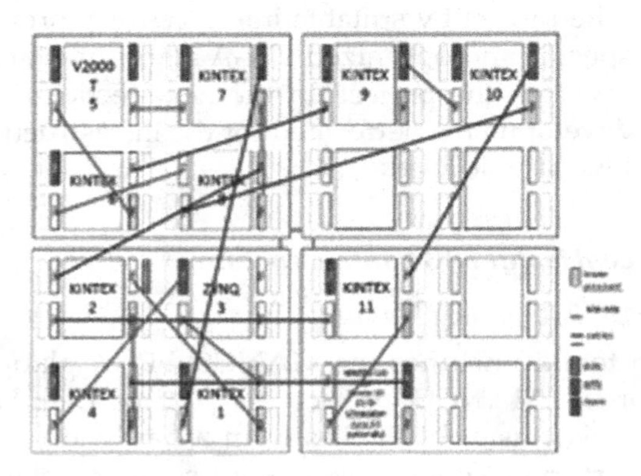

FIGURE 10.10
HN modules and interconnection.

types of FPGAs, it is composed of Xilinx Kintex Ultrascacle XCKU-115, Xilinx Virtex 7 V2000T, Xilinx Zynq 7000 SoC Z100, and Intel Stratix 10 SG280H FPGA modules. The HN cluster is also equipped with FPGA interconnection cables to enable direct communications between its FPGAs. The cables are arranged in such a way that maximize communication bandwidth among the overall system but keeping the throughput balance between the different FPGA modules. The HN also includes one PCIe extension board that enables PCIe communications with the GN. Figure 10.10 shows the positioning and interconnections of the different FPGA/memories and cables between the different modules.

Overall, the complete DeepHealth FPGA-based infrastructure consists of a total of four GNs and eight HNs arranged in two cabinets as shown in Figure 10.9.

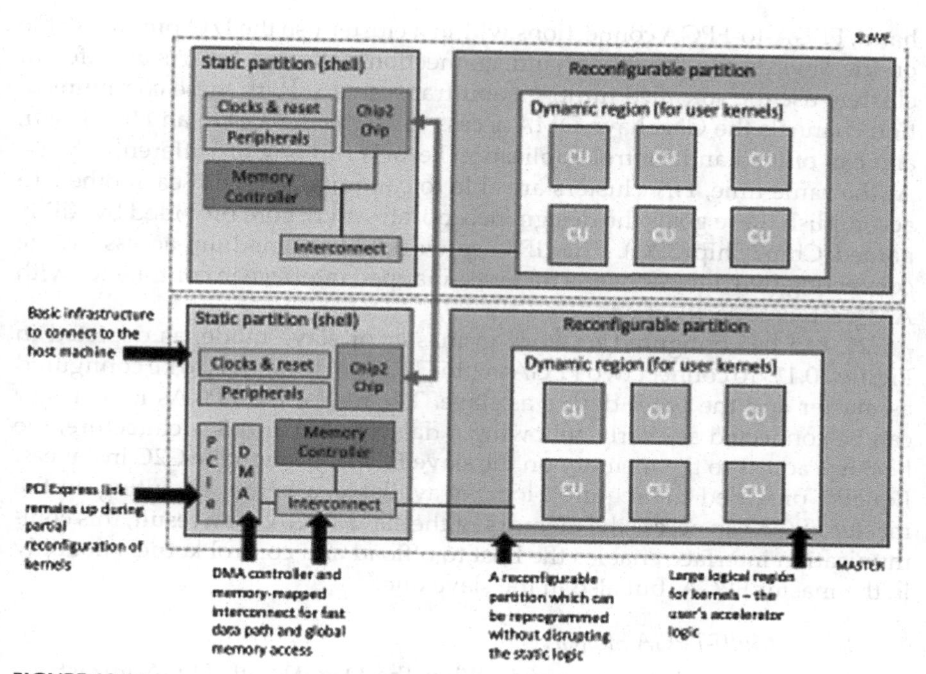

FIGURE 10.11
FPGA logical design of two FPGAs in the same cluster connected with Chip2Chip.

With the objective of facilitating programmability, the DeepHealth FPGA implements two partitions: a static partition, known as *shell*, and a reconfigurable partition (see Figure 10.11). The former remains unchanged while the host is up and running. When the host is booting this partition is loaded into the FPGA from an external memory drive. Once uploaded, it provides the required interfaces to communicate with attached peripherals and the host where the FPGA is connected to in an efficient way. In the MANGO cluster, the connection to the host is accomplished through a PCIe Gen3 x8 bus, offering a bidirectional raw bandwidth of 16GB/s, approximately. In addition, the static partition provides the clocks and reset networks to the rest of the elements in the FPGA, such as kernels. The reconfigurable partition consists of a placeholder, inside the FPGA, which can be changed at runtime using the features that the *shell* provides. This partition contains the resources in which application kernels can be deployed.

The MANGO cluster is compatible with the OpenCL application programming interface (API)[11] for FPGA initialization, data transfer, and kernel offloading, supporting both Xilinx and Intel FPGAs. Kernels are the part of the application that will run on the FPGA, so as to provide acceleration capabilities to the applications.

Furthermore, the FPGA design has been instrumented to support communication between the different FPGAs an HN is composed of. On the one hand, FPGA-to-FPGA connections within a cluster use the I/O pins available on the devices. On the other hand, connections between FPGAs at different clusters use the so-called multi-gigabit transceivers. With these communication channels the GNs have DMA access to all the devices in an HN cluster, and can offload and control application kernels running on different FPGAs. At the same time, HN clusters are able to communicate with each other. To accomplish these goals the design incorporates an IP core provided by Xilinx named Chip2Chip(C2C). This IP core works like a medium-access bridge connecting two devices over a memory-mapped interface in compliance with AXI protocol specifications.

C2C can be configured to work in master or slave mode, as depicted in Figure 10.11. To connect two FPGAs with C2C one device has to be configured as master and the other device as slave. The rest of the FPGAs in a cluster can be connected similarly, following a daisy chain. In this architecture, the host has access to the memory on the slave FPGA through the C2C instances. Equally, provided the required logic is available, a kernel executing on the master FPGA can access the memory of the slave FPGA. As a result, this communication interface enables the host to offload and control kernels not only in the master FPGA, but also in the slave one.

10.4.1.1.2 Multi-FPGA Support

Another feature being researched within the DeepHealth FPGA infrastructure is the analysis on the use of multiple FPGAs for the implementation for a single inference network. Uniting the compute and memory resources of all used FPGAs makes it possible to implement neural networks with higher requirements of memory (weights) and/or with a higher level of parallelism, reducing the latency.

The N2D2 (Neural Network Design & Deployment) DL framework made by DeepHealth partner CEA[12] (see Section 10.4.1.3 for further details), features a technology called dNeuro,[13] which is able to export an inference network as a hardware description in RTL language, suitable for synthesis and implementation on an FPGA. dNeuro is optimized to use the available compute resources (mainly, DSPs) as efficiently as possible, and to use only the embedded memory resources (block RAMs), also in the scope of efficiency. dNeuro generates the network according to specified constraints, allowing more or less parallelism depending on the number of available DSP units. Specifying a number of DSPs higher than in a single FPGA results in a network suitable for multi-FPGA execution.

CEA is developing technology to map netlists automatically onto a multi-FPGA platform, as if this platform were a single, larger FPGA. For this, the netlist has to be split into multiple netlists, one for each FPGA. The partitioning must be as efficient as possible, in order to maintain efficient communication between the FPGAs, both in terms of critical path preservation and control of the number of inter-FPGA signals. To achieve this goal, specifically adapted state-of-the-art hypergraph partitioning techniques are being used to ensure the overall performance improvement of the application.

10.4.1.2 An Optimized FPGA Board Design for DL

DeepHealth is also investigating the development of an FPGA board (see Figure 10.12) optimized for inference. The key component of the board is an INTEL Stratix-10 MX1650 or MX2100 FPGA. Both types are package and pin compatible and have FPGA internal high bandwidth memory (HBM) embedded. The MX2100 provides 16GByte of HBM memory and the MX1650 provides 8GByte of HMB memory. The MX2100 is the preferred choice and will be used for the first board designs.

Furthermore, SODIMM connectors are used for memories and peripherals which are connected to regular FPGA I/Os. In DeepHealth these SODIMM extension board sites can be used to attach additional memories to the FPGA depending on the application's requirements. Examples of such memories are DDR4 memory or high-speed SRAM memories to support random memory access with a minimal latency.

For direct host communication a PCIe interface (Gen3 x 16) is implemented. To communicate with external devices via high-speed interfaces four QSFP28 interfaces are available at the bracket. Each of these interfaces provides a full-duplex bandwidth of 100GBit/s. To interconnect multiple of these FPGA boards or to further extend the capabilities for external communication general-purpose connectors are available at the backside of the board.

Finally, a board support package is provided for using OpenCL/HLS tools for programming and integration of the hardware into the DeepHealth infrastructure.

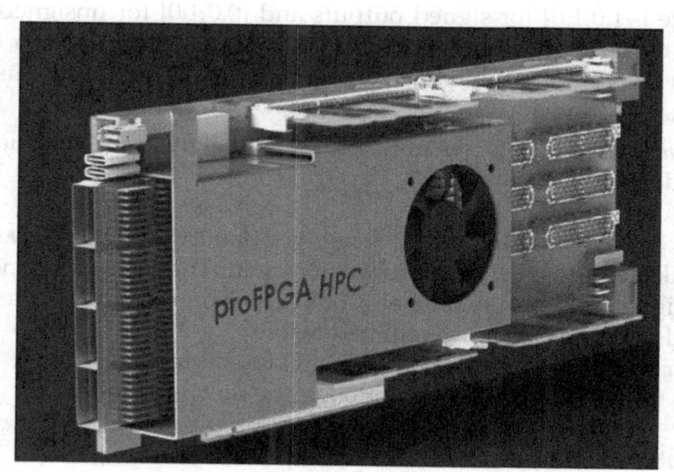

FIGURE 10.12
The new DeepHealth FPGA board optimized for inference operations.

10.4.1.3 FPGA-Based Algorithms

Pruned and quantized models enable the use of FPGA devices for energy-efficient inference processes when compared to GPUs or CPUs. This section presents the use of these two techniques in the DeepHealth FPGA infrastructure.

10.4.1.3.1 Quantization and N2D2 and Interface with DeepHealth via ONNX

N2D2 is a comprehensive solution for fast and accurate DNN simulation and full and automated DNN-based applications building. It integrates database construction, data preprocessing, network building, benchmarking, and hardware export to various targets. It is particularly useful for DNN design and exploration, allowing simple and fast prototyping of DNN with different topologies.

Once the training DNN performances are satisfactory, an optimized version of the network can be automatically exported for various embedded targets. An automated network computation performances benchmarking can also be performed among different hardware targets. Various targets are currently supported by the tool-flow: from plain C code to C code tailored for high-level synthesis (HLS) with Xilinx Vivado HLS and code optimized for GPU. Various optimizations are possible in the exports: (1) DNN weights and signal data precision reduction (down to 8-bit integers); (2) nonlinear network activation functions approximations; and (3) different weights discretization methods.

The post-training quantization algorithm is done in three steps:

1. *Weights normalization:* all weights are rescaled in the range [−1.0,1.0].
2. *Activations normalization:* activations at each layer are rescaled in the range [−1.0,1.0] for signed outputs and [0.0,1.0] for unsigned outputs. The optimal quantization threshold value of the activation output of each layer is determined using the dataset, and implies the use of additional shifting and clipping layers.
3. *Quantization:* inputs, weights, biases, and activations are quantized to the desired precision.

Interworking between the N2D2 framework and the EDDL library is possible thanks to the ONNX file exchange format. This allows any neural network designer to take advantage of both environments. This integration is illustrated with the two flows outlined below.

In order to obtain an inference code making an intensive usage of integer operators rather than floating-point ones, the flow is as follows:

1. regular training performed in EDDL;
2. EDDL exports networks to ONNX format;
3. N2D2 imports networks from ONNX format;

4. 8-bit quantization of the network using N2D2;
5. exporting the quantized network to ONNX;
6. importing the ONNX into EDDL; and
7. generating the inference code using EDDL facilities.

This flow is schematized in Figure 10.13. It can be adjusted to take advantage of the various automated inference code generators proposed by N2D2.

FPGA-based HPC oriented flows are as follows:

1. N2D2 users access the FPGA-based HPC inference code generation provided by the EDDLL library;
2. EDDL users benefit from N2D2 training capabilities not yet available in EDDL;
3. regular training performed in N2D2;
4. 8-bit quantization of the network within N2D2;
5. export the quantized network to ONNX;
6. import the ONNX neural network into EDLL; and
7. generation of the inference code using EDDLL facilities.

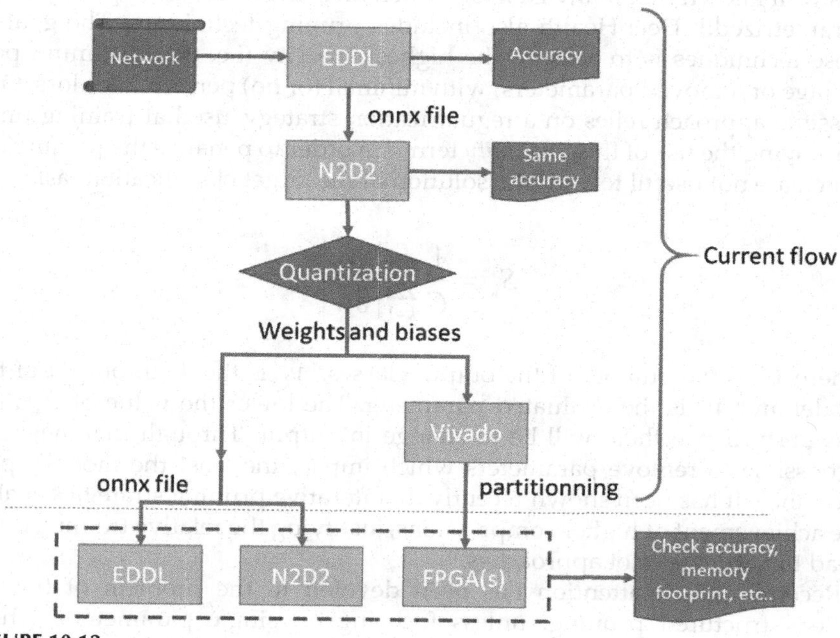

FIGURE 10.13
Quantization flow with EDDL.

The performance of the ISIC classification use case generated through this flow are detailed below.

With 63% classification accuracy, it shows up the same accuracy as the floating-point version with a memory footprint divided by four. The storage memory size for the weights and biases of the standard VGG16 network is estimated to 512.28 MB. When performing quantization, the weights are stored in signed 8-bit integers, and biases in signed 16-bit integers. Each layer requires an additional scaling factor, stored in an unsigned 8-bit integer. Coming from 32-bit values, the required storage size will be divided by almost four. The storage memory size for the weights and biases of the quantized VGG16 network is estimated to 1,074,445,952 bits, which is 128.08 MB.

Similar results were obtained with different versions of the MobileNet network.[14] The original version of MobileNet was trained on the ISIC dataset: recognition rates ranged from 73% to 79% depending on the alpha factor (from 0.25 to 1) which drastically impacts the required memory size. Eight-bit quantization and export times were in the range of 150 to 500 seconds, and the recognition rates of the quantized networks remained in the same range as before. The estimated memory footprint was 440 kB (alpha = 0.25), 1250 kB (0.5), and 4050 kB (1) with estimated frame rate on FPGA ranging from 250 to 1780 frames per second.

10.4.1.3.2 DNN Compression Methods

It is well known that many DNNs, trained on some tasks, are typically overparametrized.[15] DeepHealth also includes pruning techniques. The goal of these techniques is to achieve the highest sparsity (i.e. the maximum percentage of removed parameters) with minimal (or no) performance loss. One possible approach relies on a regularization strategy, used at training time, employing the use of the *sensitivity* term[16] in order to penalize the parameters which are not useful toward the solution of the target classification task:

$$S_w = \frac{1}{C}\sum_{k=1}^{C}\left|\frac{\partial y_k}{\partial w}\right|$$

where C is the number of the output classes, y_k is the k-th output of the model and w is the evaluated parameter. The lower the value of S_w, the less perturbance there will be to change in output. Through this metric, it is possible to remove parameters which impact the least the model's performance. It has been shown recently that iterative pruning strategies enable the achievement of higher compressions, justifying the relative training overhead toward one-shot approaches.[17]

Recently, much attention has been devoted to the problem of the so-called structured pruning: unless focusing on single parameters, which enable a reduced gain in terms of FLOPs and memory footprint at training time, pruning algorithms should be focused on removing entire neurons. Comparing LOBSTER,[18] which is a state-of-the-art unstructured pruning algorithm which removes a very large quantity of parameters, to SeReNe,[19] which is a sensitivity-based structured pruning algorithm, results that structured pruning strategies, despite removing fewer parameters than the unstructured ones, bring significant advantages in terms of memory footprint and FLOPs reduction, achieving up to $2\times$ footprint saving and $2\times$ FLOPs reduction, evaluated at inference time on resource-constrained devices, for complex architectures like ResNet and VGG-16, trained on state-of-the-art tasks, with no performance loss.

10.4.1.3.3 Development of DL Kernel on the DeepHealth FPGA Infrastructure

DL kernels are specialized on the most frequent neural network layers used in the specific domain of the project (image-related classification and segmentation processes for health), therefore convolutions, resizing, and pooling operations are the focus of the kernel. However, the design of the kernel is performed using HLS. With HLS an algorithm described in a high-level language such as C++ is transformed into an implementation on a reconfigurable device or even into an ASIC.

The kernel design follows the dataflow model by Xilinx combined with the use of streams. With this model a pipelined design between modules can be created and data can be processed concurrently on all the connected modules. The use of streams enables concurrency.

Figure 10.14 shows the baseline design for the kernel targeting convolutional operations. Each box represents a different module and arrows represent streams. The kernel reads data (activations, bias, and weights) from a DDR memory attached to the FPGA and produces features being written back to the DDR memory. The images (activations) are pipelined through all the modules. The padding module provides padding support to the input images read in a streamflow fashion and then forwards the padded image to the next module. The cvt module converts the input stream into frames

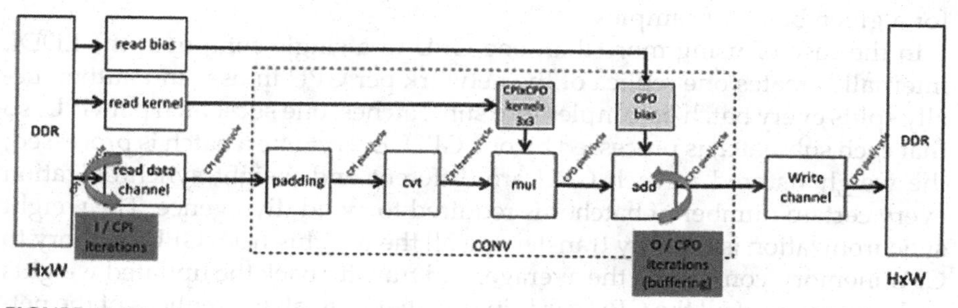

FIGURE 10.14
Baseline kernel design for convolutional operations on the EDDL.

of pixels that will be convolved in the mul module and reduced in the add module. Finally, the produced features are written back to memory.

The design supports parallel access to memory in order to processes a defined number of input (CPI) and output (CPO) channels. Therefore, the design can be customized to different granularities of the convolution operation. Indeed, the number of convolutions operations performed in each cycle is CPI x CPO as each CPI channel is used for each CPO channel (direct convolutions supported).

10.4.2 Many Core and GPU Acceleration

The EDDL implementation is optimized to efficiently execute on a single computing node featuring many-core fabrics or a set of GPU cards accelerators. In that regard, the EDDL includes an API to build the neural network and the associated data structures (according to the network topology) on the following acceleration technologies (named *computing service* in EDDL nomenclature): CPU, GPU, and FPGA. Tensor operations are then performed using the hardware devices specified by means of the computing service provided as a parameter to the build function. Moreover, the number of CPU cores, GPU cards, or FPGA cards to be used are indicated by the computing service. This section presents the many-core and GPU acceleration included in the EDDL. See Section 10.4.1 for FPGA acceleration.

On many-core CPUs, tensor operations are performed by using the Eigen[20] library which relies on parallelization on OpenMP.[21] When using GPU cards, the forward and backward algorithms are designed to minimize the number of memory transfers between the CPU and the GPU cards in use, according to the configuration of the computing service. EDDL incorporates three modes of memory management to address the lack of memory when a given batch size does not fit the memory of a GPU. The most efficient one tries to allocate the whole batch in the GPU memory to keep memory transfers to a minimum, the intermediate and least efficient modes allow working with larger batch sizes at the cost of increasing the number of memory transfers to perform the forward and backward steps for a given batch of samples.

In the case of using more than one GPU in a single computer, the EDDL internally creates one replica of the network per GPU in use, and automatically splits every batch of samples into sub-batches, one sub-batch per GPU, so that each sub-batch is processed by one GPU. Every time a batch is processed, the weights stored in each GPU are different and weight synchronization every certain number of batches is required to avoid divergence. The weight synchronization is done by transferring all the weights from GPU memory to CPU memory, computing the average, and transfer back the updated weights to the memory of all the GPU cards in use, that is, to all the replicas of the network. The computing service used for defining the use of GPU cards has an attribute to indicate the number of batches between weight synchronizations. As memory transfers between CPU and GPU must be reduced as much as possible, a trade-off between performance and divergence must be reached by means of this attribute, whose optimal value will vary depending on the dataset used for training and the topology of the neural network.

EDDL support for GPUs has been implemented twice, by means of CUDA kernels developed as part of the EDDL code, and by integrating the NVIDIA cuDNN library. The use of different hardware accelerators is completely transparent to developers and programmers who use the EDDL; they only need to create the corresponding computing service to use all or a subset of the computational resources.

10.5 Conclusions

DeepHealth HPC toolkit allows the EDDL to be efficiently executed on HPC and cloud infrastructures. On one side, it includes HPC and cloud workflow managers to parallelize the execution of EDDL operations, including the parallelization of the costly training operations; on the other side, it supports the most common hardware acceleration technologies, that is, many cores, GPUs, and FPGAs, to further accelerate the training and inference operations in single computing nodes. Moreover, the DeepHealth HPC toolkit provides the data/computer scientists with the level of abstraction needed to describe the underlying computing infrastructure in a fully transparent way.

Notes

1 https://github.com/deephealthproject
2 "Open Neural Network Exchange. The open standard for machine learning interoperability," [online]. Available: https://onnx.ai/
3 www.bsc.es/marenostrum/marenostrum
4 I. Colonnelli, B. Cantalupo, I. Merelli, and M. Aldinucci, "Streamflow: cross-breeding cloud with HPC," *IEEE Transactions on Emerging Topics in Computing*, 2020. doi:10.1109/TETC.2020.3019202
5 https://rancher.com/docs/rancher/v2.x/en/
6 https://helm.sh/
7 www.docker.com/products/docker-hub
8 https://kubernetes.io/
9 M. Aldinucci et al. "OCCAM: a flexible, multi-purpose and extendable HPC cluster," *Journal of Physics: Conference Series*, vol. 898, 2017

10 M. Aldinucci, S. Rabellino, et al. "HPC4AI, an AI-on-demand federated platform endeavour," in *ACM Computing Frontiers*, Ischia, Italy, 2018. doi:10.1145/3203217.3205340
11 Khronos.org/opencl
12 https://github.com/CEA-LIST/N2D2
13 www.cea.fr/cea-tech/leti/Documents/démonstrateurs/Flyer_DNEURO.pdf
14 https://arxiv.org/abs/1704.04861
15 H. N. Mhaskar, and T. Poggio, "Deep vs. shallow networks: An approximation theory perspective," *Analysis and Applications* 14 (06) (2016) 829–848.
16 E. Tartaglione, S. Lepsøy, A. Fiandrotti, and G. Francini, "Learning sparse neural networks via sensitivity-driven regularization," in *Advances in Neural Information Processing Systems*, 2018, pp. 3878–3888.
17 Tartaglione, Enzo, Andrea Bragagnolo, and Marco Grangetto. "Pruning artificial neural networks: a way to find well-generalizing, high-entropy sharp minima." *International Conference on Artificial Neural Networks*. Springer, Cham, 2020.
18 Tartaglione, Enzo, et al. "LOss-Based SensiTivity rEgulaRization: towards deep sparse neural networks." arXiv preprint arXiv:2011.09905 (2020).
19 Tartaglione, Enzo, et al. "SeReNe: Sensitivity based Regularization of Neurons for Structured Sparsity in Neural Networks." arXiv preprint arXiv:2102.03773 (2021).
20 G. Guennebaud, B. Jacob, et al. "Eigen v3.," 2010. [online]. Available: http://eigen.tuxfamily.org
21 https://openmp.org

11

Applications of AI and HPC in the Health Domain

Dana Oniga, Barbara Cantalupo, Enzo Tartaglione, Daniele Perlo, Marco Grangetto, Marco Aldinucci, Federico Bolelli, Federico Pollastri, Michele Cancilla, Laura Canalini, Costantino Grana, Cristina Muñoz Alcalde, Franco Alberto Cardillo, and Monica Florea

CONTENTS

11.1 Introduction

This chapter aims to show how the convergence of high-performance computing (HPC), big data, and artificial intelligence (AI) (and deep learning (DL) in particular) can support improvements in the health domain, by providing an overview of a few use cases from the DeepHealth project. The next sections are organized as follows. Section 11.2 presents the latest technical progresses in the health field. Section 11.3 briefly describes the DeepHealth concept and introduces the DeepHealth toolkit. Section 11.4 presents general info (typical workflow, key performances indicators (KPIs)) about the

DOI: 10.1201/9781003176664-11

DeepHealth use cases. Section 11.5 provides more details of five DeepHealth use cases. Section 11.6 highlights the value offered by the DeepHealth toolkit to healthcare providers. Section 11.7 concludes emphasizing the impact of the DeepHealth project.

11.2 AI and HPC in the Health Domain in 2020

The European Commission (EC) focuses on innovation in health and on technical areas addressing current health issues in relation to people's well-being, as well as on increasing the sustainability of health systems, as presented in the EU eHealth Action Plan (2014–2020).

In April 2018, the European Union (EU) signed a Declaration on Cooperation on Artificial Intelligence (AI) that emphasized the commitment of EU states toward boosting Europe's technology and industrial capacity in AI. Linking AI technology developments with health was an inherent step to take, as shown by the innovation calls launched by H2020 -> https://cordis.eur opa.eu/programme/id/H2020_ICT-11-2018-2019: HPC and Big Data enabled Large-scale Test-beds and Applications.

In 2020, the Next Generation EU recovery instrument and the Annual Sustainable Growth Strategy 2021 were launched by the EC to fight the effects of the COVID-19 pandemic, providing €750 billion funding. The EU4Health 2021–2027 programme, EU's response to COVID-19, is one of the main pillars of this strategy and has a budget of €9.4 billion, which among other objectives, will strengthen health systems so that they can face epidemics as well as long-term challenges by stimulating their digital transformation. In this context, big data, AI, and supercomputers, with their analytical power, are major assets in detecting patterns in the spread of the virus or potential treatments. The EC will continue investing in the use of AI to speed up the diagnosis of COVID-19 and improve future treatment of patients.

Not only public health authorities or institutions, but also private healthcare providers are incorporating AI technologies in their activities, as a way to gain competitiveness, optimize diagnostic processes, and improve treatment monitoring; and, lately, to join efforts to fight against COVID-19 pandemics.

11.3 DeepHealth Concept

Two of the most known areas of healthcare that can benefit from the advances in AI are medical imaging and electronic health records (EHR), as highlighted in "AI in Healthcare Whitepaper" (BDVA Task Force7 – Subgroup Healthcare – November 2020). The main goal of the DeepHealth project is to put HPC computing power at the service of biomedical applications, to apply DL and computer vision techniques on large and complex biomedical datasets to support a new and more efficient way of diagnosis and to generate insights into complex diseases in a scalable and efficient way.

The DeepHealth concept is based on the scenario, depicted in Figure 11.1, where processing large quantities of images is needed for diagnosis.

Health professionals (doctors and medical staff) are end users – they are experts in medical areas and diseases.

ICT experts involved in the project will provide medical users with the software applications necessary for obtaining results according to their needs and requirements.

The training environment represents the context where ICT experts work with datasets of images for training predictive models.

In the production environment the medical personnel insert images coming from scanning sessions into a software platform or biomedical application that uses predictive models to get clues that can help them to make decisions during diagnosis. Doctors have the knowledge to label and annotate images, define objectives, and provide related metadata. ICT staff are in charge of processing the labeled and annotated images, organizing the datasets, performing image transformations when required, training the predictive models, and loading such models into software platforms once tested and validated.

To perform all these operations, ICT staff will use the DeepHealth toolkit and the DeepHealth HPC and cloud infrastructure, as it can be seen in Figure 11.1.

As described also in Chapter 10, the DeepHealth toolkit is a general-purpose DL framework, including image processing and computer vision functionalities, enabled to exploit HPC and cloud infrastructures for running parallel/distributed training and inference processes.

The core of the toolkit consists of two libraries, namely the European Distributed Deep Learning Library (EDDL) and the European Computer Vision Library (ECVL) that are ready to be integrated in any software application. Additionally, the DeepHealth toolkit also has two complementary software components, the back end, and the front end. The back end exposes a RESTful API to allow the use of all the functionalities provided by the two libraries. The front end makes accessible the functionalities of the two libraries via a web-based graphical user interface.

The DeepHealth toolkit incorporates the most advanced parallel programming models to exploit the parallel performance capabilities of HPC and cloud infrastructures, featuring different acceleration technologies such as symmetric multiprocessors (SMPs), graphics processing units (GPUs), and field-programmable gate arrays (FPGAs), as shown in Chapter 10.

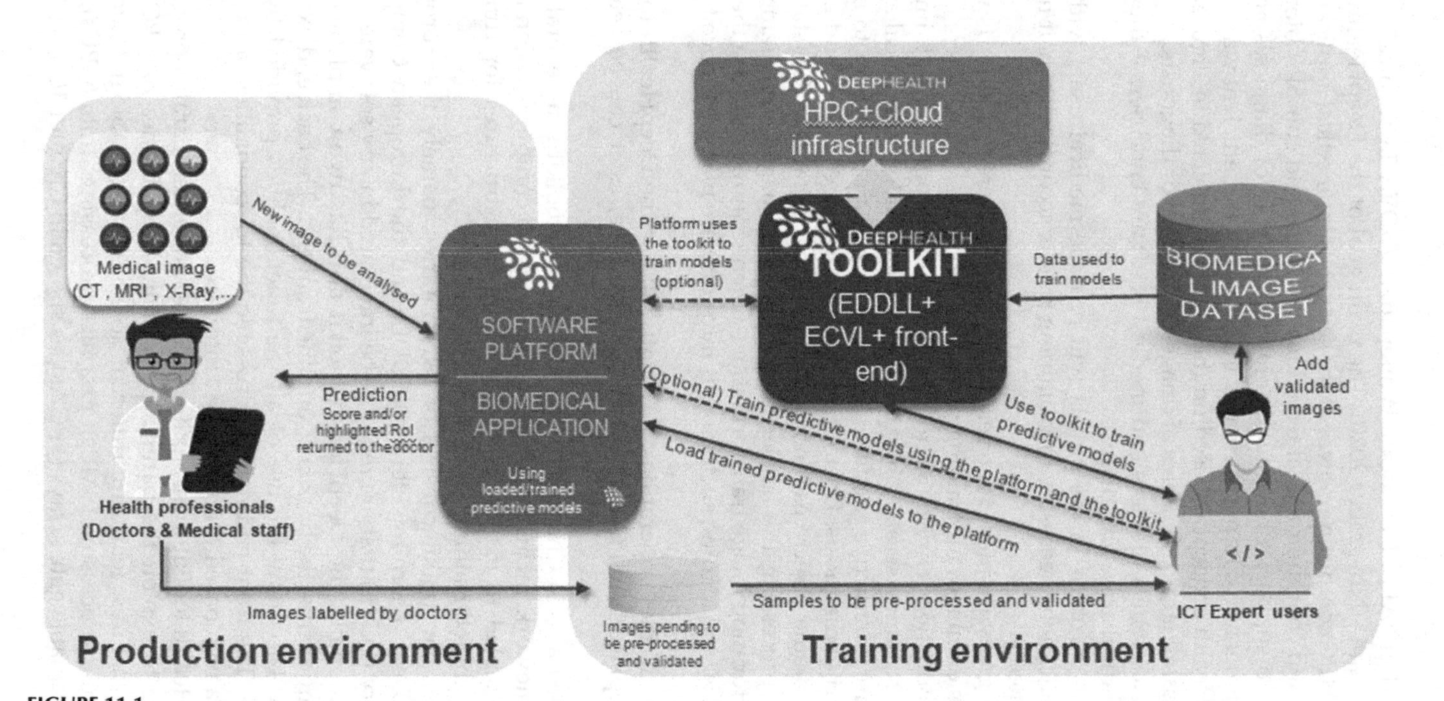

FIGURE 11.1
DeepHealth concept

Moreover, the toolkit provides functionalities to be used for both training and inference, addressing the complexity of the different available computational resources and target architectures at both the training and inference stages.

This process is transparent to doctors; they just provide images to the system and get predictions such as indicators, biomarkers, or the semantic segmentation of an image for identifying tissues, bones, nerves, blood vessels, etc.

To demonstrate and validate the concept of the project, the DeepHealth open-source toolkit and the HPC and cloud infrastructure are tested and validated in real environments thanks to the 15 pilot use cases (test beds) and on seven biomedical commercial or research platforms, raising the innovation potential of European companies.

11.4 DeepHealth Use Cases

This section will summarize the main aspects that are common to all DeepHealth uses cases, that is, the testing and validation workflow and KPIs.

A typical use case in the DeepHealth project is based on a medical imaging dataset, which is trained and tested on the DeepHealth toolkit and one or more commercial health platforms, taking advantage of hybrid and heterogeneous HPC and big data clusters. First, data scientists and members of the team preprocess (labeling, annotation, anonymization) and prepare the dataset by splitting it into three subsets, namely, training, validation, and testing subsets. Next the development team designs several artificial neural networks (ANNs) and launches the training processes on HPC and cloud architectures by means of the runtime of the toolkit adapted to HPC frameworks like the ones described in Chapter 10. The team evaluates the models using the validation subset, and redesigns some models if necessary. Sometimes, the team would reconsider the dataset itself in view of the knowledge learned in previous iterations. The model that gets the best accuracy using the testing subset is selected; then computer scientists, members of the same team, configure an instance of the application with the best model and deploy the solution in a production environment.

The most important KPIs to be validated and tested in DeepHealth pilots are the time to model in production (ttmip), time of preprocessing images (toppi), and time-of-training-models (totm).

- Time of preprocessing images (toppi). This KPI is the sum of T1 and T2, where T1 is the time a data scientist (or ICT expert) needs to design the pipeline of transformations to be applied to the images and the time he/she needs to implement that pipeline of transformations using the ECVL. T2 is the time to run the pipeline over all the images in the dataset.

- Time of training models (totm). This KPI is calculated by measuring the execution time of the training procedure.
- Time to model in production (ttmip). This KPI is the sum of various variables, the definitive model used in production will commonly be the result of choosing one among several tested models, so the time of designing and testing each model must be taken into account (totm is part of the total time necessary for testing each single model). Additionally, if each model needs a particular pipeline of image operations, the value of toppi for each defined image-preprocessing procedure must be included as part of the ttmip.

Other KPIs that will measure the performance of the algorithms for training DNNs on distributed architectures are speedup and efficiency of parallelism.

Each use case also has specific KPIs to assess that the obtained predictive models provide accuracies equivalent to those obtained with the same DNN topologies when the models used are trained using other toolkits. These KPIs are accuracy, precision, recall, and F1-score for classification tasks and intersection over union (IoU) in semantic segmentation tasks. IoU can be computed in average for all the classes to be detected, and individually for each one in order to know in more detail which tissue types or skin spots or tumor types are more difficult to detect by the DNN-based models.

The 15 use cases validate the DeepHealth toolkit performances in the following medical fields (1) neurological diseases, (2) tumor detection and early cancer prediction, and (3) digital analysis of brain pathologies and automated image annotation, exemplifying how the toolkit can create specific biomedical applications. At the end of 2020 a new use case was added with a COVID-19+[1] related dataset released by the partner FISABIO.

11.5 Use of HPC and Cloud in Medical Pilots

As discussed in previous chapters, medical datasets are in a state of constant growth and are required to be as large as possible to train robust and accurate enough models based on CNNs. However, due to the fact that the process of training neural networks requires to iterate over all the samples of the training subset, and the number of iterations/epochs ranges from 100 to 1000 or more, depending on the use case and the architecture, the use of HPC and cloud-computing infrastructures are becoming indispensable for distributing the workload by splitting the training subset. In this way each worker node in the computing infrastructure is in charge of performing the back propagation algorithm on a data partition. Chapter 10 provides details on how to distribute the workload on different computing architectures supported by the DeepHealth toolkit. As Figure 11.1 illustrates, leveraging all the computer power offered by hybrid HPC and cloud infrastructures equipped with hardware accelerators and many-core CPUs is completely transparent to medical personnel, and ICT experts do not need to have deep knowledge of programming on distributed environments. Thanks to the DeepHealth toolkit, ICT experts can run the training processes on hybrid HPC and cloud infrastructures, test and validate the trained models, and, finally, update the models in production with the new ones trained with updated datasets after including recently acquired medical images.

Five out of the 15 use cases of the DeepHealth project are described in this chapter to show the joint use of ECVL and EDDL. These five uses cases are:

- UC2 UNITOPatho, based on whole-slide colorectal images obtained from colonoscopies;
- UC3 UNITOBrain, based on CT scans of the brain;
- UC4 Chest, based on CT scans of lungs;
- UC5 UNITO Deep Image Annotation, based on X-ray chest images; and
- UC12 Skin Cancer Melanoma Detection, based on dermoscopic images.

11.5.1 UC2 – UNITOPatho

The expansion of cancer screening programs and the demand of colonoscopy surveillance routines are leading the gastrointestinal histopathology to grow.

Predictive signals of possible gastrointestinal cancer development are colorectal polyps, which are pre-cancerous lesions located in the lining of the colon.[2]

Colorectal tissue samples are collected by biopsies and colonoscopies. Gastrointestinal pathologists examine them to find signs that predict tissue neoplastic process and invasive carcinomas.

In this use case, we focus our effort to develop a neural network-based pipeline to automatically diagnose colorectal cancers from whole-slide images (WSIs).

WSIs are super high-resolution images retrieved by scanning biopsy material, more specifically hematoxylin and eosin-stained slices of tissue, from patients undergoing cancer screening. These images can reach more than 100.000×100.000 pixels. The WSI corresponds to the whole specimen collected from the patient: it is provided with a label (the diagnosis) and the annotation on the subset of tissue taken into consideration by the pathologist to elaborate the diagnosis (the so-called diagnostic tissue).

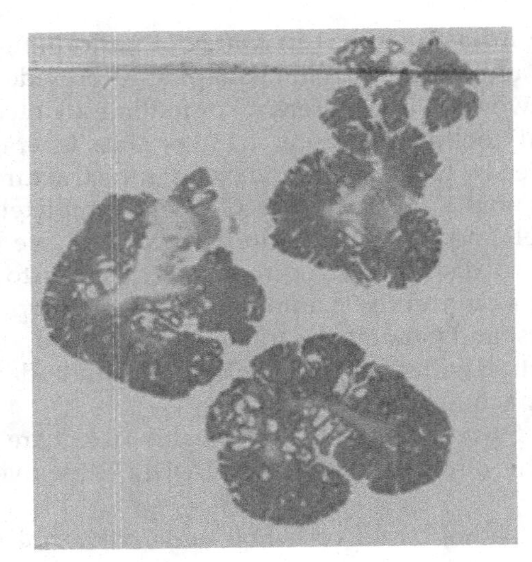

FIGURE 11.2
An example of WSIs containing tubulovillous adenoma tissue.

Given the nature of the problem, learning from huge images is a hard task which requires HPC infrastructure in order to successfully perform training and, most importantly, inference at diagnosis time. The DeepHealth toolkit is also designed to efficiently handle large-scale images and to train DL models efficiently.

The experts collected six different types (or classes) of tissue: normal tissue, hyperplastic polyp, tubular adenoma with high- and low-grade dysplasia, tubulovillous adenoma with high- and low-grade dysplasia.

A first version of this collection is the open-access dataset UNITOPATHO.[3] It consists of 9536 hematoxylin and eosin-stained patches extracted from 292 WSIs.

In order to solve the classification task, we propose the use of a residual neural network model (ResNet). The neural network model predicts the class for each tissue patch from the WSI and, finally, a label for the WSI itself.[4,5] This UC relies on the University of Torino's OpenDeepHealth (ODH) platform, which implements a hybrid HPC/cloud infrastructure to effectively support the training and inference of AI models. As detailed in Chapter 10, ODH integrates the DeepHealth toolkit via docker containers, both on bare metal and in the multitenant Kubernetes cluster and it was specifically designed to support AI application on critical data, such as biomedical images.

11.5.2 UC3 – UNITOBrain

The occlusion of a cerebral vessel causes a sudden decrease of the blood perfusion of the corresponding vascular territory. Identifying such an occlusion in a fast and reliable way is critical under emergency scenarios.[6] The so-called

CT perfusion, with a sample time of roughly 1 Hz, measures the passage of a contrast media bolus into the brain, on a pixel-by-pixel basis. Serial low-dose scans are acquired; time–density curves, corresponding to the contrast media passage in brain tissue, are calculated; parametric maps are calculated. The most relevant parameters used in clinical practice are cerebral blood volume and cerebral blood flow (CBF and CBV).[7]

Given the nature of the task, generating the aforementioned parametric masks in the least time possible is crucial. To this end, HPC infrastructure offers parallel computation capabilities which can be successfully exploited through the use of the DeepHealth toolkit in order to train and to infer from an AI-based model. Data from 115 patients were collected: a subset of 100 patients was used to train the deep model, while the remaining 15 were held to validate the results.

In order to generate ground-truth (GT) maps, we relied on a state-of-the-art deconvolution-based algorithm.[8] The validation step was carried out evaluating concordance between more expert medical evaluators among the segmented lesions using the GT maps and those produced by the ANN model.

As an ANN, we have taken inspiration from the state-of-the-art U-Net architecture.[9] Since the model was originally thought to segment medical images, and in our case we aim at generating parametric images, we introduced some changes to the standard model, like reducing the granularity of the convolutional operations and using average pool layers instead of maxpool.

Contrary to the other state-of-the-art approaches, no extra information (like the arterial input function) was provided to the U-Net model, favoring its use in an emergency scenario where the time to obtain these maps is critical.

Overall, on the generated maps, we achieve an average Dice score on the lesions above 0.70, resulting in a generation of good quality perfusion maps from the U-Net model. Inter-rater concordance was measured, finding a very strong correlation between lesion volumes of CNN maps and GT maps, achieving above 0.98 Pearson correlation.[10] This UC leverage on the HPC features of ODH platform, previously described.

11.5.3 UC4 – Chest

Lung nodules are small focal lesions in the lung parenchyma that can be solitary or multiple and in many cases are accidentally found in CT scans. Their identification is time-consuming in the current clinical activity for the radiologist and, since these small lesions are difficult to spot, patients often need to perform follow-up CT scans in order to assess their benignity/malignancy, resulting in increased radiation exposure and anxiety for the patient and increased work amount for doctors. Lung nodules are quite common incidental findings in CT scans and can be defined as small focal lesions (ranging from 5 to 30 mm) that can be solitary or multiple. DL models outperform traditional computer vision techniques in various tasks. Typically, features of

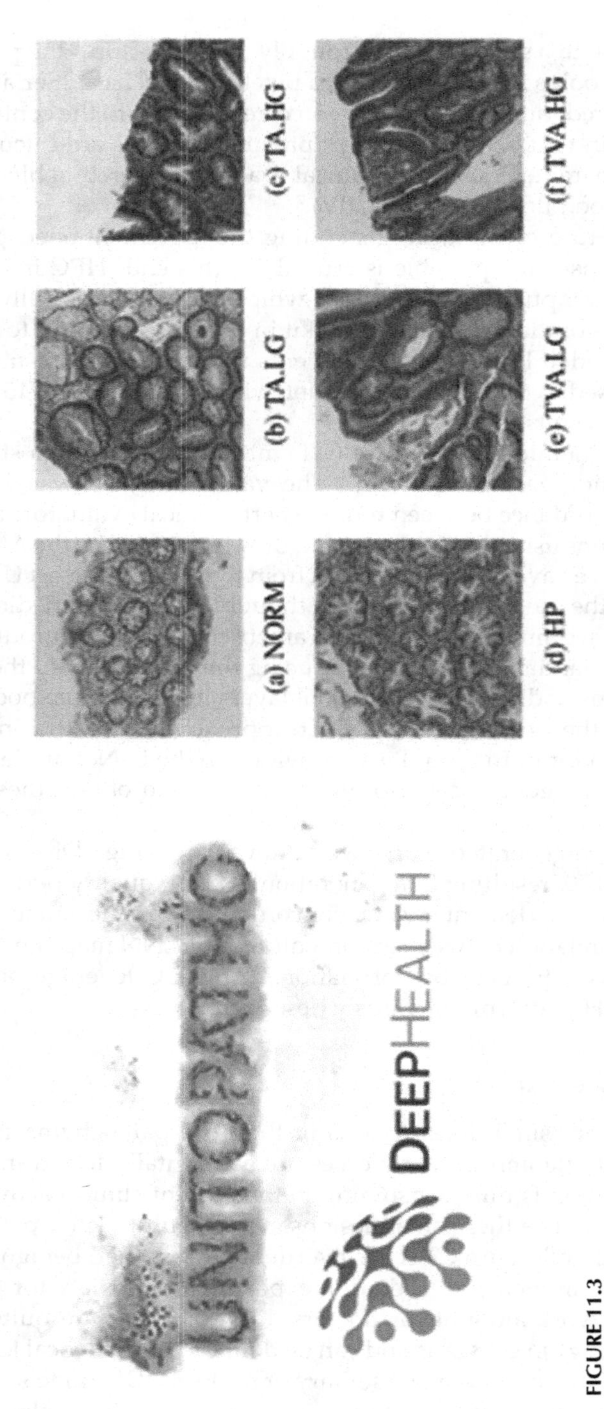

FIGURE 11.3
UniToPatho logo, followed by a collection of type-representative tissue samples.

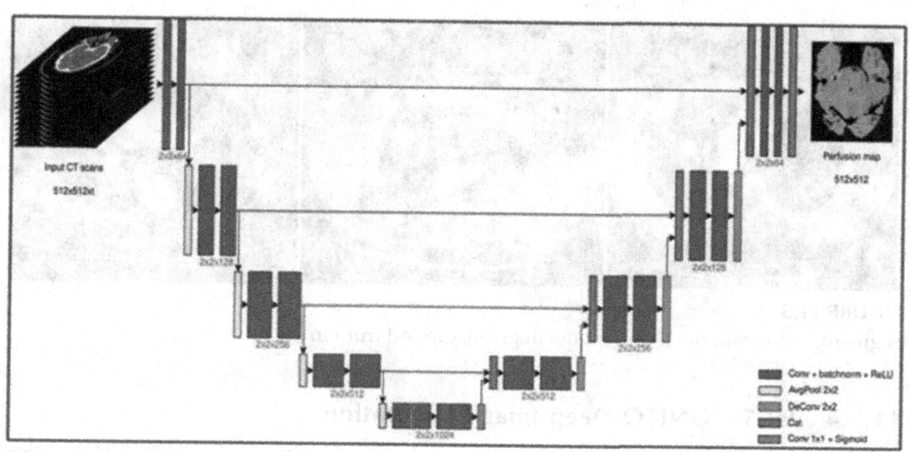

FIGURE 11.4
Artificial neural network model deployed for UC3.

TABLE 11.1

	Patients	Images
Train	247	13589
Validation	61	1699
Test	109	1708

the input data are hand-crafted, but DL features are learned in an end-to-end fashion. Convolutional neural networks (CNNs) are one of the most popular models that are also employed in medical imaging.[11] More specifically, the U-Net model is used to identify lung nodules in CT scans. The dataset used to train and evaluate the network is provided by Citta' della Salute e della Scienza di Torino. See Table 11.1 given below.

In order to get preliminary results, splits were created with only images with a ground truth mask, such that the training set contains 80% of images, validation 10% and test 10% CT scans are in DICOM format.

CT scans are very large images: every exam may contain more than 200 acquisitions which need to be properly processed. This use case exploits HPC by means of the ODH platform to reduce the computation time. Handling these image format and training the DL model to perform the segmentation task efficiently is crucial for the success of this task: the DeepHealth toolkit offers efficient capability of training segmentation models on medical images in an efficient way and is therefore crucial toward the success of this use case.

The model currently reaches an intersection over union of 0.62 with only rotations as data augmentation.

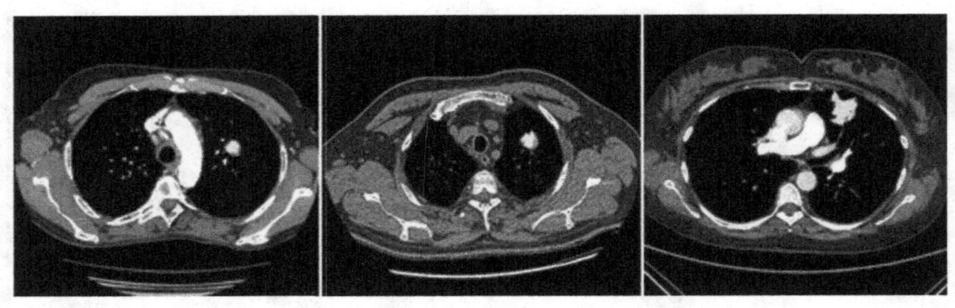

FIGURE 11.5
Segmentations from the U-Net model in green, ground truth in yellow.

11.5.4 UC5 – UNITO Deep Image Annotation

Statistics prepared by the EU clearly show that the number of imaging devices and, consequently, the number of daily scans is steeply increasing all over Europe.[12] The large amount of digital images collected daily by the national health systems or even by single hospitals poses new challenges and a huge burden on the clinical community. Indeed, radiologists need to inspect every single image and write detailed reports; often taking into account also information stored in EHR. Such a heavy workload may easily lead the operators to make mistakes not so much in providing the correct diagnosis but rather in the repetitive and tedious data entry and reporting procedures. Information technology (IT) and, specifically, the innovative solutions based on AI implemented in several use cases of the DeepHealth project might help also in reducing the mistakes caused by habituation and fatigue.

UC5 "Deep Image Annotation" sets its main goal to the automatic reporting of medical images, that is, given one or more input images, UC5 aims at automatically generating a description of the image content using natural language.

The multimodal processing in UC5 is based on a combination of two different ANNs set in a processing pipeline.[13] In the first stage, a CNN classifies and encodes the input images; in the second stage, a recurrent neural network (RNN) generates sentences starting from the encoding coming from the CNN. UC5 uses the public and anonymized dataset named "Indiana University chest X-ray Collection"[14] containing 7470 X-ray images and 3955 semi-structured reports (see Figure 11.6). Each report, in XML format, is paired with two views of the chest and contains multiple textual annotations, including sections with the "indication," the "findings," and the "impression," that correspond to the actual reports written by the radiologists. The reports contain also MeSH and MTI (respectively, medical subject headings and medical text indexer, both by the National Library of Medicine, USA) terms (or tags).

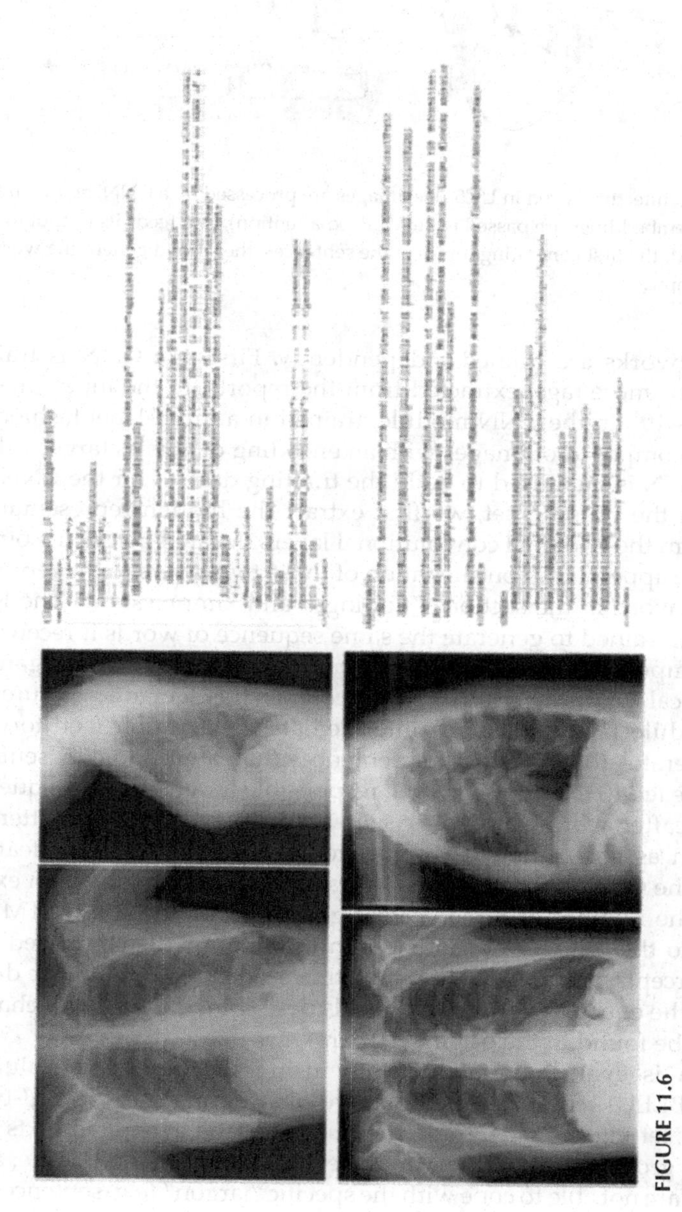

FIGURE 11.6
Two examinations composed of two views of the chest. Upper row: normal examination and related report with the sections used in UC5. Lower row: anomalous examination and fragment of the related report with the diagnosis and the MeSH/MTI terms (labeled as "automatic").

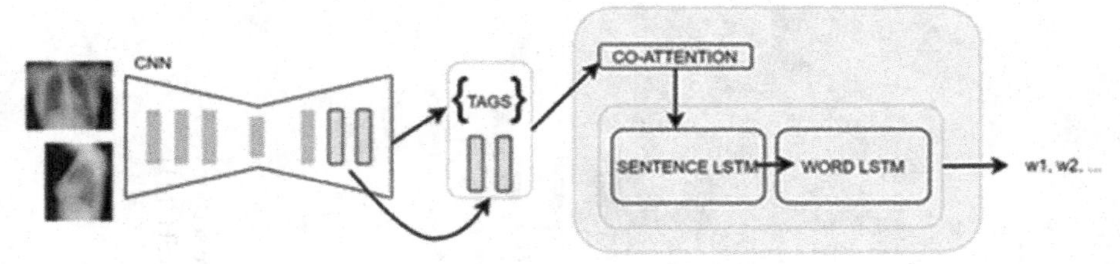

FIGURE 11.7
Neural network architectures used in UC5. The images are processed by a CNN, and then their classification and embeddings are passed to an MLP (co-attention), that feeds its output to the cascaded LSTM, the first generating topics of the sentences, the second generating words according to the topic.

The two networks are trained independently. First, the CNN is trained to assign one or more tags, extracted from the reports, to the input images. We use a VGG-19[15] in the CNN module, trained in a multi-label fashion: its training set is composed of images with an encoding of tags as target values. The trained CNN is then used to build the training dataset for the RNN: for each image in the training set, we first extract the hidden representations (encoding) from the two final convolutional layers (located before the output layer), then we append the concatenation of those two layers to the representation of each word in the sections "findings" and "impression." The RNN module is then trained to generate the same sequence of words it receives as input. It is composed of two long-short term memory networks[16] organized in a hierarchical way: the first LSTM, receiving the semantic features by the CNN module (actually, by an intermediate co-attention feed-forward module), generates topic vectors for sentences (hereafter, topic or sentence LSTM) that are fed to the second LSTM, responsible for generating sequences of words (hereafter, word LSTM). The sentence LSTM receives a co-attention vector[17] which assigns different weights to parts of the semantic features extracted by the CNN, mimicking the visual exploration made by an expert radiologist. The number of topic vectors generated by the first LSTM that corresponds to the number of sentences in the final text is controlled by a multilayer perceptron, trained at the same time as the two LSTMs. We do not provide here the equations or the full details describing the system behavior, but these can be found in the original paper.[18]

The system is evaluated automatically using the bilingual evaluation understudy (BLEU) score. Furthermore, we plan to make a "crowd"-based evaluation by letting humans grade the quality of the generated texts with respect to the ground truth. However, the evaluation is an open issue: automatic metrics are not able to cope with the specific "jargon" (e.g. sentences like "no anomalies detected" and "anomalies detected" have opposite meanings but receive a high similarity score) and it is not easy to enroll many domain experts in a crowd-based evaluation.

Based on the preliminary results, UC5 will likely not be able to produce reports as reliable as human radiologists do (as expected). Nevertheless, UC5 can still have a positive and measurable impact in the daily clinical routine, even if limited, for example, in speeding up the reporting activity while potentially reducing the number of errors as well (e.g. mistakes in copying and pasting from previously written reports).[19] UC5 "Deep Image Annotation" offers several interesting opportunities. It is computationally challenging, thus providing a good test bed for the two ECVL and EDDLL libraries under development in DeepHealth. If well engineered, it can be a valuable support tool in current commercial systems. And, with more data becoming available, the reports that it will be able to generate will represent a valuable step forward for both experienced and junior radiologists.

11.5.5 UC12 – Skin Cancer Melanoma Detection

Skin cancer represents a major public health issue, being the most common form of human cancer worldwide, with an increasing trend.[20] In the last decades, many efforts were made to improve skin cancer treatments; however, the early detection remains a key factor in preventing cancer progression to advanced stages and ensuring a lower mortality rate.[21]

For epithelial skin cancer, such as basal cell carcinoma and squamous cell carcinoma, delayed diagnosis is mainly responsible for a larger and more disfiguring surgery, which could also affect relevant functional structures (e.g. mouth, ears, eyelids, nose). For melanoma, a delayed diagnosis may mean death due to potential tumor aggressiveness. Early diagnosis represents the ideal and cheap solution to fight the skin tumor consequences.

Dermoscopy is a form of in vivo skin surface microscopy performed using high-quality magnifying lenses and a powerful light source to mitigate the surface reflection of the skin, to enhance the visibility of the pigmentation of the lesion. This technique is broadly employed by dermatologists since it allows for a fast diagnosis, and significantly increased the diagnosis accuracy, sensitivity, and specificity with respect to the naked eye examination.

However, this kind of noninvasive imaging approaches requires the eye of expert clinicians to successfully diagnose skin cancer. Therefore, many efforts were given toward the development of automatic tools for supporting and training physicians in the analysis of dermoscopic images.[22] Due to its outstanding results in many research areas including image understanding and image classification,[23,24] DL has become the main option for analyzing medical images in several fields, including skin lesion classification.[25,26,27]

Unfortunately, these algorithms require a huge amount of annotated data to ensure the correct learning process. When dealing with medical imaging, collecting and annotating data can be cumbersome and expensive. This mainly relates to the nature of the data and to the need for well-trained expert technicians. Moreover, such algorithms require a significant amount of (expensive) computational resources, which are often inaccessible to medical personnel.

As introduced in the previous sections, the main goal of the DeepHealth project is compensating the second issue, boosting the productivity of data scientists operating in the medical field by providing a unified framework for the distributed training of neural networks, which is able to leverage hybrid HPC and cloud environments in a transparent way for the user, without requiring a deep understanding of DNNs and distributed HPC.

On the other hand, UC12 has the main goal of collecting dermoscopic images and design advanced DL algorithms for the segmentation and classification of skin lesion, providing clinicians with Computer-Aided Diagnosis (CAD) systems for the automated melanoma recognition. Indeed, segmentation of images and extraction of features can lead to a rapid and automatic identification of diagnostic clues which can facilitate image interpretation and diffusion of technologies among other doctors.

In this sense, the UC12 combines and exploits existing publicly available datasets with a huge internal one, completed of clinical, dermoscopy and confocal microscopy images, annotated with conclusive diagnosis (histologic or clinically confirmed), and relevant patient's data. Samples of such a kind of images are provided in Fig. 8.

This dataset consists of 25,849 dermoscopy images, collected between 2003 and 2019 using several distinct acquisition tools. The dataset presents a different category distribution compared to the public datasets, with a higher percentage of melanoma cases. Visual artefacts which could be considered source of biases, such as rulers, ink markings/staining, and colored patches are almost completely absent.

For what concerns public datasets, the International Skin Imaging Collaboration (ISIC) began to aggregate a large-scale collection of dermoscopic skin lesion images in 2016.[28] The 2019 version of the ISIC archive contains a total amount of 25,331 labeled dermoscopic images belonging to nine different classes, which represent eight types of skin lesion plus an additional category named *none of the others*, which contains samples of different natures that do not belong to any of the other eight classes.

Differently from ISIC 2019, the 2020 dataset is focused on a binary classification problem. In this case images are divided in only two classes: melanoma or non-melanoma. Moreover, this set of dermoscopic images contains patient-level contextual information, providing for each image an identifier which allows lesions from the same patient to be mapped to one another.

This additional knowledge is frequently used by clinicians to diagnose melanoma and is especially useful in ruling out false positives in patients with many atypical nevi.

This new dataset is composed of 33,126 images and collected from 2,056 patients (21% of them with at least one melanoma, 79% with zero melanomas) gathered from 1998 to 2020, with an average of 16 lesions per patient.

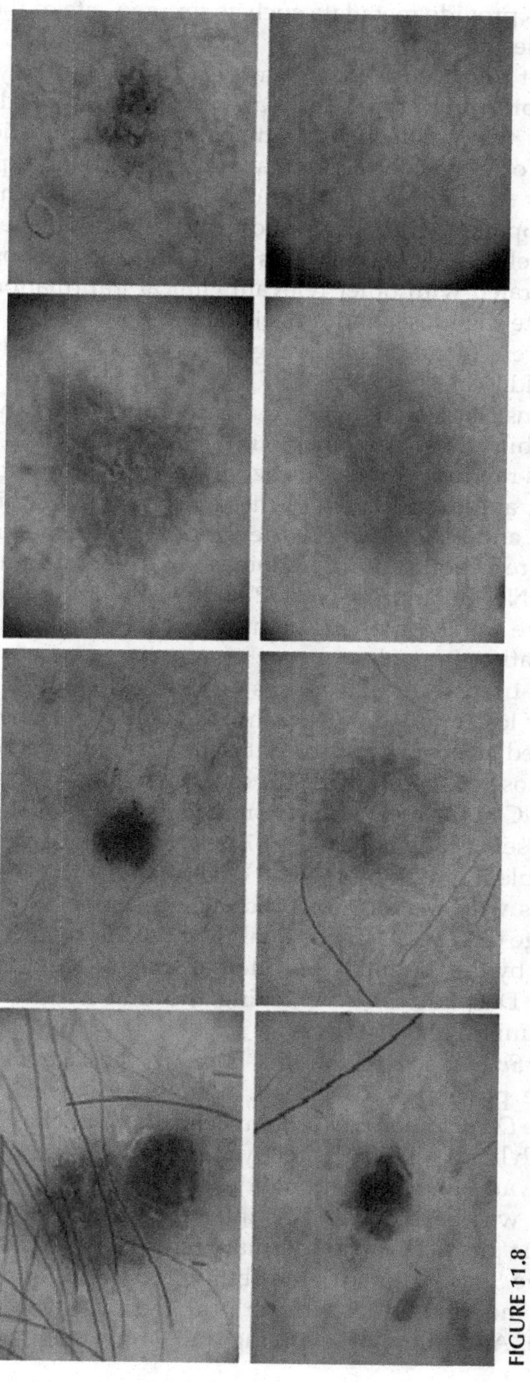

FIGURE 11.8
Samples of internal dataset. From upper left to right: melanoma (MEL); melanocytic nevus (NV); basal cell carcinoma (BCC); actinic keratosis (AK); benign keratosis (BKL); dermatofibroma (DF); vascular lesion (VASC); squamous cell carcinoma (SCC).

All three datasets provide metadata such as sex, age, site of the lesion, and number of patient lesions.

Image analysis of this use case focus on the extraction of key parameters (such as pagetoid spreading, atypical cells at the junction, atypical melanocytic nests, alteration of the architecture for melanoma, identification of tumor islands and cords for basal cell carcinoma, identification, and quantification of dyskeratosis for squamous cell carcinoma) useful to obtain an accurate description of histopathologic background from confocal images. This information will be correlated with parameters obtained from dermoscopic image analysis and integrated with other relevant clinical information in order to provide an accurate diagnosis and a decisional protocol, able to reproduce the analytical process of differential diagnosis of doctors with the highest expertise in the field.

For what concerns classification, this use case implements an ensemble of state-of-the-art architectures, combined with several techniques for data augmentation, such as random vertical/horizontal flips, random rotation, additive Poisson noise, and dropout. All of these techniques were implemented by means of ECVL and EDDL libraries, developed within the project.

The ensemble prediction output combines six different neural networks based on EfficientNet-Bx and ResNet152 state-of-the-art architectures. The single models were trained with different input and batch sizes and using different augmentation strategies.

Each network is trained for ~20 epochs on ISIC 2019 and ISIC 2020 using the Cross-Entropy loss and Adam[29] optimizer, with a learning rate of 3e-5. Two fully connected layers process the dataset metadata and then the output is concatenated those of the networks to be assessed as well. The ensemble achieves a 0.94 AUC in the melanoma/non melanoma classification problem on the official test set of ISIC 2020 and 0.87 on the internal dataset.

The second problem issued by the use case concerns the segmentation of skin lesion images with the aims of producing a segmentation mask for a certain input image (Figure 11.9). The quality of output segmentation results is, again, ensured by the ensemble of different state-of-the-art architectures, mostly based on DeepLabv3+,[30] achieving the outstanding mIoU (mean intersection over union) of 0.850.

The *DeepHealth Service* acts as the link between the DeepHealth libraries implementing the previously described models and the HPC and cloud systems. Using the *DeepHealth Service*, data scientists do not have to write code for the DeepHealth library APIs or directly manage computing resources, but directly use ECVL and EDDL functionalities through a RESTful web service and, optionally, a web-based GUI. This interface provides clinicians useful tools that can assist physicians with dermatologist-grade decision support.

Additionally, the *Service* provides the ability to design, train and test additional predictive models and to perform pre- and post-processing without writing any code. And allowing evolving and improving the existing use

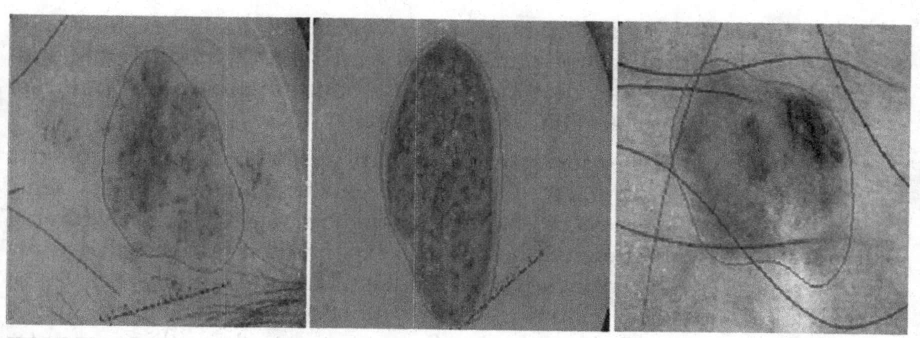

FIGURE 11.9
Example of segmented (red contour) skin lesion images from the ISIC dataset.

case whenever new state-of-the-art architectures will be released published in literature.

Instead, the REST interface enables managed service usage scenarios, where a potentially complex and powerful computing infrastructure (e.g. HPC, cloud computing, or even heterogeneous hardware) could be transparently used to run DL jobs without the users needing to directly interface with it.

A specific instance of the *service* was deployed for the UC12 tasks and it was configured for the asynchronous and distributed training (and test) of aforementioned classification and segmentation neural networks. With this goal, the *service* spreads the jobs among different cloud nodes which have several GPUs each.

11.6 DeepHealth Value Proposition

As described in Section 11.3, DeepHealth addresses two kinds of users: ICT expert users who train AI models and healthcare professionals who use the trained models to predict on images.

DeepHealth provides a multi-step value proposition addressing all the potential users of these two categories. Each value proposition is adapted to each of the target users and presented in the format: For – who needs – That – Unlike and, finally, the positioning statement:

- For **research institutions** who need state-of-the-art, open-source DL&CV technologies to advance in their research concerning AI for imaging with a reduced time-to-model that allow for an optimized use of the latest hardware technologies in a flexible and easy way unlike non-European tools, not adapted to European HPC and hardware accelerators, nor integrated in European platforms, which are less innovative or not open source. DeepHealth Toolkit is positioned as "The European, open-source DL&CV framework, modular and scalable, that streamlines development of models, thanks to an optimised use of state-of-the-art HPC technologies."

- For **independent expert users and start-ups** who need technologies to offer highly accurate AI models for medical images in less time, with higher accuracy and better diagnostic functionalities, by having a technological differential in order to expand their business that allow them to generate innovative AI models thanks to transparent use of HPC technologies; including specific functions for biomedical images and being a unified easy-to-configure framework, with an easy-to-install back-end and user-friendly front-end unlike non-open source tools which imply an additional cost; other OSS tools non-optimized for HPC architectures & hardware accelerators; other OSS tools difficult to configure, deploy and orchestrate among them and which do not include an integrated back end and front end; or other OSS tools not specifically designed for biomedical imaging, which may lack specific functionalities. DeepHealth Toolkit is positioned as an "An all-in-one, open-source DL&CV framework specialized for developing innovative medical imaging AI solutions, with higher accuracy and for new use cases thanks to the optimised use of HPC technologies."
- For **healthcare technology vendors** who need new, easy to integrate open-source tools to expand their offering in order to provide the latest technological innovations, with special focus on improving expert users' productivity and addressing new medical imaging use cases that allow them to stay competitive, by reducing the time-to-market of their models and increasing their accuracy thanks to multiplatform frameworks for transparent exploitation of different high performance hardware unlike non-multiplatform OSS tools, difficult to integrate and without associated professional services (consulting, support, etc.); OSS tools which do not accelerate time-to market; Non-innovative OSS tools, which do not include specific methods for distributed computing and adapted middleware or which are not transparent in the use of hardware. DeepHealth Toolkit is positioned as "An open-source DL&CV framework that makes transparent the optimised use of state-of-the-art HPC technologies with the flexibility of a multiplatform solution, increasing expert users' productivity, reducing time-to-market. Specialized for healthcare and validated by fifteen biomedical imaging Use Cases."
- For **healthcare professionals** who need support in diagnosis due to growing shortage of imaging specialists, increasing numbers of patients, and to decrease erroneous diagnosis and healthcare costs that allow them to leverage the vast amounts of datasets available and benefit from the high value provided by AI technologies, thanks to highly accurate algorithms validated by clinicians and developed in collaboration with

them and with an adapted interface for clinicians' needs unlike non-reliable algorithms, developed with few iterations to decrease development time; AI solutions which cannot address specific use cases due to the high computing capabilities necessary; long time-to-model AI solutions for urgent use cases, like COVID-19 lung predictive models; or complex software solutions which cannot be easily integrated in the clinical workflow. DeepHealth Solutions are positioned as "Reliable, highly accurate AI-for-imaging solutions, developed in collaboration with expert clinicians and validated by them, for high added-value use cases and integrated in user-friendly platforms adapted for healthcare professional needs."

11.7 Conclusions

In conclusion, the validation of the DeepHealth concept in large-scale pilots will support the impact and the benefits the project is expected to have.

Using the DeepHealth toolkit and taking advantage of HPC and cloud architectures will increase the productivity of ICT staff working in the health sector by allowing them to design, train, and test significantly more predictive models in the same time period.

This will facilitate the daily work of expert users that are managing large image or other types of datasets; AI systems used in radiology could outperform human experts or aid them by reducing their workload.

Knowledge about diseases and pathologies will be extended by applying the DeepHealth toolkit. Early diagnosis and improving treatments will be possible, finally impacting the welfare of people and saving direct and indirect healthcare costs. Outcomes of the project will be useful to other sectors: EDDLL is a general-purpose DL library and ECVL will be useful for image processing in general.

Other industries can easily adopt the DeepHealth toolkit, following the trend AI+HPC as a service for an increasing number of applications (other DL-based applications, graph-based applications: data discovery, digital twins, among others). The project thus contributes to increasing the impact of AI on European society.

Notes

1 https://bimcv.cipf.es/bimcv-projects/bimcv-covid19/

2 Bevan, R., and Rutter, M.D., "Colorectal cancer screening – who, how, and when?" *Clinical Endoscopy*, 51(1): 37 (2018).

3 Bertero, L., Barbano, C.A., Perlo, D., et al. UNITOPATHO. IEEE Dataport (2021). https://dx.doi.org/10.21227/9fsv-tm25
4 Perlo, Daniele, et al. "Dysplasia grading of colorectal polyps through CNN analysis of WSI." (2021).
5 Barbano, C.A. et al., "UniToPatho, a labeled histopathological dataset for colorectal polyps classification and adenoma dysplasia grading." (2021).
6 Donahue, J., and Wintermark, M., "Perfusion CT and acute stroke imaging: Foundations, applications, and literature review." *Journal of Neuroradiology* (2015). https://doi.org/10.1016/j.neurad.2014.11.003
7 Albers, G.W., Marks, M.P., Kemp, S., et al., "Thrombectomy for stroke at 6 to 16 hours with selection by perfusion imaging." *New England Journal of Medicine* (2018). https://doi.org/10.1056/NEJMoa1713973.
8 Bennink, E., Oosterbroek, J., Kudo K., et al., "Fast nonlinear regression method for CT brain perfusion analysis." *Journal of Medical Imaging* (2016). https://doi.org/10.1117/1.jmi.3.2.026003
9 Falk, T., Mai, D., Bensch, R., et al., "U-Net: deep learning for cell counting, detection, and morphometry." *Nature Methods* (2019). https://doi.org/10.1038/s41592-018-0261-2.
10 Gava, Umberto A. et al. "Neural network-derived perfusion maps: a model-free approach to computed tomography perfusion in patients with acute ischemic stroke." (2021).
11 G. Nam et al., "Development and validation of deep learning–based automatic detection algorithm for malignant pulmonary nodules on chest radiographs." *Radiology*, 290(1): 218–228 (2019).
12 EU healthcare resource statistics, http://ec.europa.eu/eurostat/statistics-explained/index.php?title=Healthcare_resource_statistics_-_technical_resources_and_medical_technology&oldid=280129
13 Jing, B., Xie, P., Xing, E. P., "On the Automatic Generation of Medical Imaging Reports." *Proceedings of the 56th Annual Meeting of the Association for Computational Linguistics*, pp. 2577–2586, Melbourne, Australia, July 15–20, 2018.
14 Demner-Fushman, D. et al. "Preparing a collection of radiology examinations for distribution and retrieval." *Journal of the American Medical Informatics Association*, 23(2): 304–310 (2016).
15 Simonyan, K., and Zisserman, A., "Very deep convolutional networks for large-scale image recognition." CoRR, abs/1409.1556, 2014.
16 Hochreiter, S., and Schmidhuber, J., "Long short-term memory." *Neural Computation* 9(8): 1735–1780 (1997).
17 Jing et al., "On the Automatic."
18 Ibid.
19 Tsou, A.Y. et al., "Safe practices for copy and paste in the EHR: systematic review, recommendations, and novel model for health IT collaboration." *Applied Clinical Informatics*, 8(1) (2017).
20 Bray, F., Ferlay, J., Soerjomataram, I., et al., "Global cancer statistics 2018: GLOBOCAN estimates of incidence and mortality worldwide for 36 cancers in 185 countries." *CA: A Cancer Journal for Clinicians*, 68(6): 394–424 (2018).
21 Rigel, D. S., Russak, J., and Friedman, R., "The evolution of melanoma diagnosis: 25 years beyond the ABCDs." *CA: A Cancer Journal for Clinicians*, 60(5): 301–316 (2010).

22 Allegretti, S., Bolelli, F., Pollastri, F., et al., "Supporting skin lesion diagnosis with content-based image retrieval." In *2020 25th International Conference on Pattern Recognition (ICPR)* (2020).
23 He, K., Zhang, X., Ren, S., and Sun, J., "Deep residual learning for image recognition." In *Proceedings of the IEEE Conference on Computer Vision and Pattern Recognition*, pp. 770–778 (2016).
24 Tan, M., and Le, Q., "Efficientnet: Rethinking model scaling for convolutional neural networks." In *International Conference on Machine Learning*, pp. 6105–6114, PMLR (May 2019).
25 Zhang, J., Xie, Y., Xia, Y., and Shen, C., "Attention residual learning for skin lesion classification." *IEEE Transactions on Medical Imaging*, 38(9): 2092–2103 (2019).
26 Pollastri, F., Parreño, M., Maroñas, J., et al., "A deep analysis on high resolution dermoscopic image classification." *IET Computer Vision* (2021).
27 Canalini, L., Pollastri, F., Bolelli, F., et al., "Skin lesion segmentation ensemble with diverse training strategies." In *International Conference on Computer Analysis of Images and Patterns*, pp. 89–101 (Cham: Springer, 2019).
28 Codella, N. C., Gutman, D., Celebi, et al., "Skin lesion analysis toward melanoma detection: A challenge." 2017 International Symposium on Biomedical Imaging (ISBI), hosted by the international skin imaging collaboration (ISIC). In *2018 IEEE 15th International Symposium on Biomedical Imaging*, pp. 168–172 (ISBI/ IEEE, 2018).
29 Kingma, D.P., and Ba, J., "Adam: A method for stochastic optimization." arXiv preprint arXiv:1412.6980 (2014).
30 Chen, L. C., Zhu, Y., Papandreou, et al., "Encoder–decoder with atrous separable convolution for semantic image segmentation." In *Proceedings of the European Conference on Computer Vision* (ECCV), pp. 801–818 (2018).

12

CYBELE: On the Convergence of HPC, Big Data Services, and AI Technologies

Sophia Karagiorgou, Aikaterini Papapostolou, Zeginis Dimitris, Yannis Georgiou, Eugene Frimpong, Ioannis Tsapelas, Spiros Mouzakitis, and Konstantinos Tarabanis

CONTENTS

12.1 Introduction: Background and Driving Forces

With the earth's population approaching 8 billion, the United Nations (UN) estimates that global food production will need to increase by at least 60% to feed the world by 2050; and therefore, it is currently considered a daunting target [1]. This demand will also rise because of increase in people's wealth resulting in higher meat consumption plus the increasing use of cropland for biofuels [2]. Precision agriculture (PA) and precision livestock farming (PLF) promise both high quantity and quality of the products with a minimum of resource usage, such as water, energy, fertilizers, and pesticides, promoting profitability, efficiency, and sustainability, while protecting the environment [3]. Despite the advancements in the field of applications, data

DOI: 10.1201/9781003176664-12

and technologies, the adoption of novel products, services, and tools by the farm operators has fallen short of expectations [4], while the agri- and aqua-food industry is turning to the ICT solution providers for the answer [5]. Toward this direction, big data and artificial intelligence (AI) have a pivotal role to play and, together with the high-performance computing (HPC) technology, they are already disrupting the agri- and aqua-food industry and pointing the way forward. In addition to this, as the data volumes and varieties increase with the expansion in sensor deployments, novel data engineering techniques are also instrumental in collecting, harmonizing, enriching, and processing distributed data from different sources [6]. These should be conducted in a way that the latency, performance, and precision requirements by the end users and applications are also satisfied.

The driving forces for empowering vegetable or livestock farmers, fish, and seafood producers with digital agricultural and aquacultural innovation tools are leveraging the need for low-cost solutions using easy-to-deploy sensors, drones, computer vision, and machine learning (ML) algorithms. Understanding how HPC, big data, and AI technologies could improve farm or seafood productivity, it could significantly increase the world's food production by 2050 in the face of constrained arable land and with the water levels receding.

While much has been written about digital agriculture's potential, little is known about the economic costs and benefits of these emergent systems. In particular, the on-site decision-making processes, both in terms of adoption and optimal implementation, have not been adequately addressed. Besides, there are important questions to be answered related to technical viability, preparedness, and training of end users in such tools, economic feasibility, and data protection.

The biggest promise of digital agri- and aquaculture technological advancements is the ability to evaluate the system on a holistic basis at multiple levels (individual, local, regional, global) and generate tools that allow for improved decision-making in every sub-process. Some of the applications of such tools target reduction of risks in agriculture and aquaculture production, such as predicting and detecting crop diseases early on in production [7]. For instance, the use of drones in arable frameworks or optimizations in fish feeding enables detailed maps to be created for damage control, prevent food wastage, and benefit the entire value chain [8]. Some other technologies target risks associated with extreme weather conditions and climate change, which impact the efficiency in yield production and society more broadly, including consumers and citizens [9].

In the CYBELE project (Fostering Precision Agriculture and Livestock Farming through Secure Access to Large-Scale HPC-Enabled Virtual Industrial Experimentation Environment Empowering Scalable Big Data Analytics)[1] [9], all these questions are addressed within a multimodal and combinatorial approach. CYBELE demonstrates how the convergence of HPC, big data, AI,

IoT, and cloud computing could revolutionize farming, reduce scarcity, and increase food supply, bringing social, economic, and environmental benefits. The CYBELE framework coordinates unmediated access to huge amounts of datasets and their metadata from diverse data types including sensor data, textual data, spatiotemporal data, satellite, and aerial image data, etc. from a multitude of distributed sources targeting the agriculture and aquaculture domain. The CYBELE approach is holistic in a sense that it guides end-to-end time-demanding and compute-hungry analytic pipelines by seamlessly converging HPC, big data, and AI technologies under the umbrella of high-performance e-infrastructures without requiring previous or extensive user experience and technical skills.

12.2 Identified Gaps: Motivating the CYBELE Vision

PA and PLF come to assist in optimizing agricultural and livestock production and minimizing the wastes and costs. PA is a technology-enabled, data-driven approach to farming management that observes, measures, and analyzes the needs of individual fields and crops. Typical factors affecting standard agriculture include among others soil, climate, seed, cultivation practices, irrigation facilities, fertilizers, pesticides, weeds, harvesting, and post-harvesting techniques. These key enablers hold the potential of providing the concepts stemming from real-world cases, the information, the mathematical models, and the computational power required in order to make well-informed, optimal choices in various real-world driven PA and PLF verticals, and to ensure that the gaps currently encountered in these verticals are either: (1) due to the lack of information sources; or (2) due to the lack of proper mathematical models capable of generating value and extracting insights out of these data sources; or (3) due to the lack of the infrastructure capacity capable of handling the execution of the computationally demanding mathematical models harnessing the power of massive amounts of diverse types of big data. In addition to these, one of the key bottlenecks in leveraging the promise of digital agri-/aquaculture is the lack of proven benefit from data shared by farmers or data owners in agri-/aqua- domains. There are also other factors that result in so-called identified gaps, namely: data ownership concerns, privacy issues, economics, financial incentives of data ownership and dissemination, anticipated and quantified return on investment, data aggregation and pipelining from the source(s) to the desired locations, to name a few. If even a subset of the gaps described can be circumvented using a mix of technologies, policies, and awareness, the possible outcomes of digital agri-/aquaculture can shine. Some of these include seed-variety mapping to performance characteristics resulting in better selection (e.g.

soya yield and protein content prediction) [10], [11]; better understanding of regional and temporal conditions leading to sustainable and localized modeling of nutrition and supplementation needs (e.g. climate services for organic food production); and capability for statistical modeling of the on-site conditions (e.g. sustainable pig production, open sea fishing), leading to increased efficient hyperparameter tuning and optimal machine selection (e.g. pig weighing optimization, aquaculture feeding optimization).

Therefore, in CYBELE we are interested in both gaps and challenges on the one hand and some possible technological and policy solutions on the other to transform some of the promises of digital agri-/aquaculture into a reality. This is the reason why we designed and implemented scalable testbeds for efficient scheduling of HPC, big data, and AI tasks with diverse capabilities including: (1) the flexible CYBELE Data Model; (2) data check-in, curation, alignment, and integration with real agri-/aquaculture multi datasets; (3) efficient information retrieval from multiple sources with advanced data querying and exploration functionalities; (4) data science and business intelligent algorithms; (5) HPC; (6) high scalable storage, consistency, availability, and partition tolerance; (7) security-by-design; (8) data governance and monitoring; (9) distributed and cloud deployments.

12.3 Materializing the Solution: Convergence of HPC, Big Data, and AI

On the convergence of HPC, big data, and AI technologies and in order to materialize the solution, the CYBELE framework follows a layered approach which aims at ensuring interoperability among all involved components, putting emphasis on the way that pipelining of information (from data queries to simulations formulation, to data analytics and to visualizations) is supported, safeguarding smooth interoperation among the different services. Figure 12.1 presents the high-level architecture of the CYBELE framework.

More analytical information per component of the framework, as well as the perceived information flows, the technical interfaces (APIs) and the interaction among them are presented in Figure 12.1.

12.3.1 Data and Infrastructure Access Security Layer

CYBELE implements an integrated *data and infrastructure access security layer* which is spread along the whole framework and e-infrastructure. Following this approach, for all the designed components and the described workflows, security mechanisms and protocols are used toward a framework of enhanced security capabilities. Taking this approach into consideration,

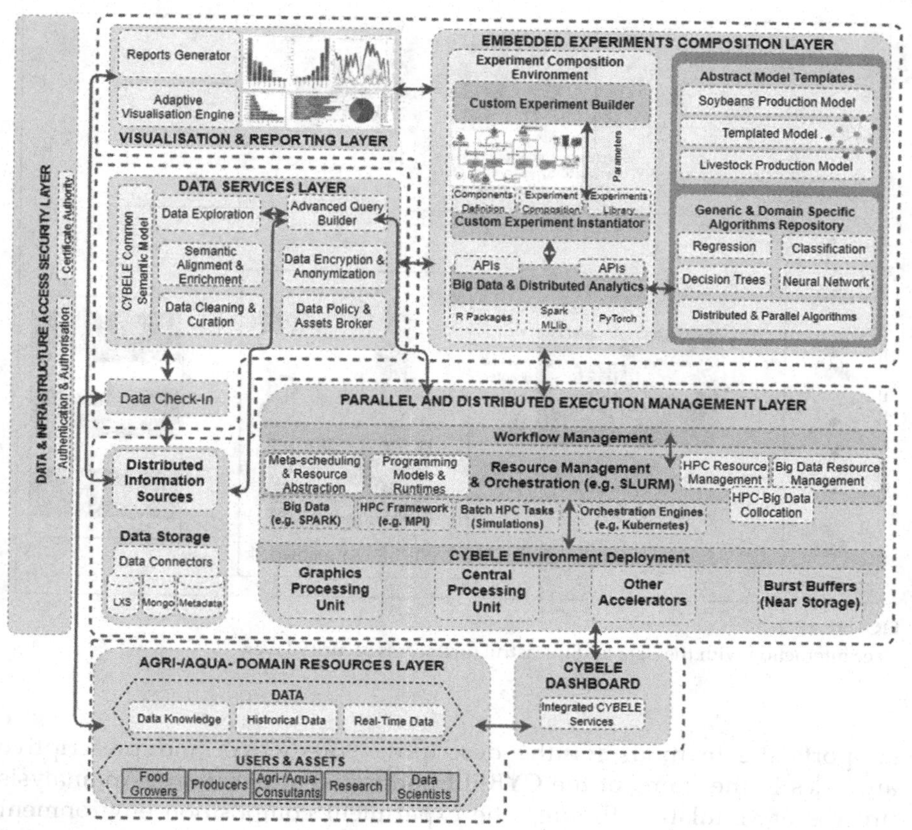

FIGURE 12.1
The CYBELE solution towards the convergence of HPC, big data, and AI technologies.

every component has been implemented with the appropriate security functionalities in mind. These include security of data at rest, data in motion, and data in use. The security layer consists of four (4) security modules: (i) certificate authority (CA); (ii) user authentication and authorization (UAA) server; (iii) vulnerability assessment (VA) toolkit; and (iv) anomaly detection service. Figure 12.2 presents user interactions with the data and infrastructure access security layer.

12.3.2 Embedded Experiments Composition Layer

The *embedded experiments composition layer* comprises two components: (a) the *experiment composition environment (ECE)* which automates the design, development, and execution of the big data analysis and simulation processes; and (b) the *generic and domain-specific analytic algorithms* which

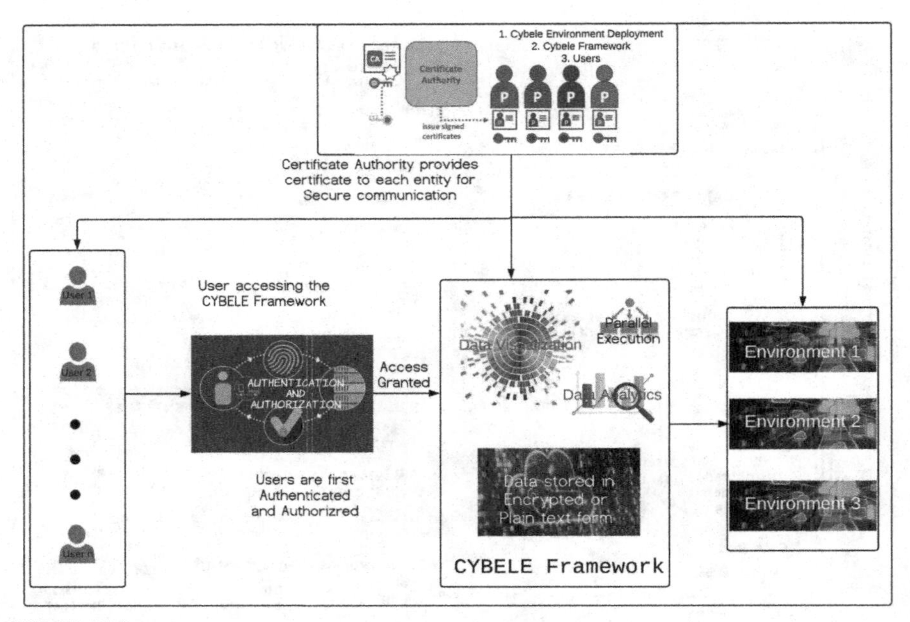

FIGURE 12.2
User interaction with the data and infrastructure access security layer.

supports the methods for the descriptive, predictive, and prescriptive analytics in the frame of the CYBELE industrial applications. An analysis process or simulation through the experiment composition environment includes retrieving as input the datasets from the *advanced query builder* and defining a new or selecting an existing analysis template from the *abstract model templates*. Then, the end user is able to (i) set end-to-end objectives for the experiment to be conducted (i.e. time performance, algorithm's accuracy, time constraints), (ii) select a specific algorithm with the associated software and execution endpoint, (iii) adjust the relevant configuration parameters, including input parameters for the algorithm along with their description and their default value, execution parameters that denote whether an analysis should be realized in a scheduled or automated way, as well as the periodicity factor for the latter case, parameters associated with networking, storage or computing resources, and (iv) adjust the output parameters along with their type (text, image, data, html). The implementation of the experiment composition environment is based on open source including custom UIs in React [12] and data pipelines engines including the Spring Cloud Data Flow [13] to facilitate workflows authoring as directed acyclic graphs (DAGs) of analytic processes. In this case, each process within a workflow is represented by a node which is fully parameterizable through the above-mentioned attributes (ii), (iii), and (iv) and the intercommunication of the processes is represented by an edge which is also parameterizable through the above-mentioned attribute (i). The user is able to save her end-to-end analyses by appending all the specifications of an experiment or simulation for future use and reuse in a YAML file [14]. The file can be stored to the *data storage* in a structure, namely the *experiments' library* for future reuse. Figure 12.3 presents user interaction with the embedded experiments composition layer.

Within CYBELE, the applications of ML for PA and crop management include yield prediction, disease or insect damage detection, weed detection crop quality, crop yield monitoring while the respective ones for PLF include animal welfare, livestock production, appetite detection, feed optimization, and biomass estimation. With regard to the aforementioned applications, the part of *domain-specific analytic algorithms* coupled with ECE are presented in Figure 12.4, where an analytic pipeline, based on user needs, also called *custom workflow*, is created consisting of both generic and custom algorithms. Most of the *domain-specific analytic algorithms* are built over *distributed processing and ML environments* (e.g. Apache Spark MLlib [15], Distributed TensorFlow [16]) exploiting their capabilities for scalable cluster computing on executing advanced analytics.

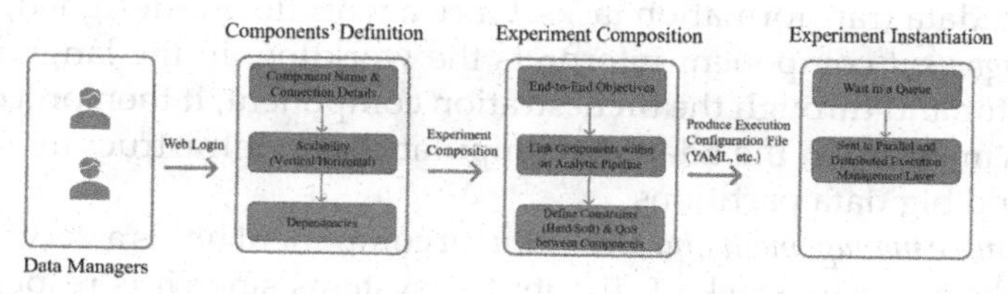

FIGURE 12.3
User interaction with the embedded experiments composition layer.

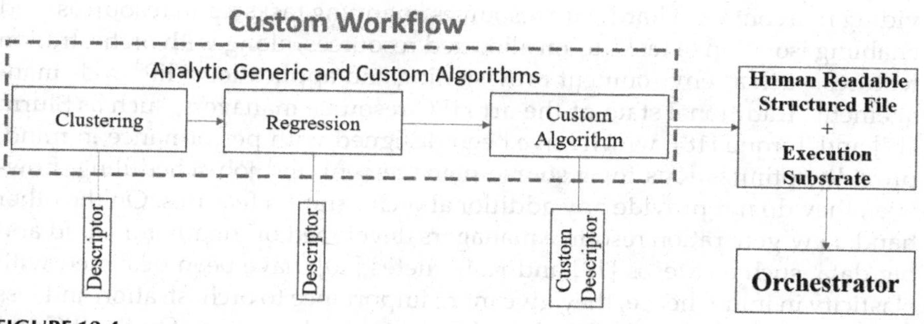

FIGURE 12.4
Algorithm implementation pipeline workflow diagram.

12.3.3 Parallel and Distributed Execution Management Layer

The data analysis workflows are deployed for execution to the *parallel and distributed execution management layer.* These workflows are instantiated on both HPC and big data resources abstracted to the end user, as depicted in the lower part of the CYBELE framework. The technological approach relies heavily on HPC e-infrastructure, to provide the compute power required to advance models and methods. The *workflow management* component is responsible for interpreting workflows designed with the experiment composition environment and forwarding them to the component responsible for orchestration, which performs the execution of the task collections upon the computational resources. A common pattern in scientific and cloud computing involves the execution of many computational and data manipulation tasks which are usually coupled, that is, the output of one task is used as an input for another task. Hence nontrivial coordination is required to satisfy data dependencies. The workload of task execution needs to be directed to the available distributed computational resources [13].

Therefore, a tight integration among ECE workflows and the orchestration component is required to guarantee execution and it is performed through the *workflow management* component. Individual tasks supported by CYBELE workflows are HPC simulations, big data analytics jobs, or even simple data transfer or data transformation tasks. Once a workflow is designed, the *workflow management* component interprets the workflow in the language of the orchestrator and through the orchestration component, it then proceeds with its deployment upon the adapted computational e-infrastructure consisting of HPC and big data partitions.

The *resource management and orchestration* component holds a very important place in the software stack of distributed systems since it is responsible for providing the necessary compute power to the executed tasks based on their requirements and the availability of resources. The component consists of five different modules. It combines resource management features, such as providing fine control of hardware resources, mapping tasks upon resources, and enabling isolation of tasks upon allocated resources, along with orchestration features, such as environment provisioning and applications' life-cycle management. Traditional state-of-the-art HPC resource managers, such as Slurm [17] and Torque [18], which have been designed with performance in mind, provide optimizations for resource management and job scheduling; however, they do not provide any additional orchestration features. On the other hand, new-generation resource managers developed originally for cloud and big data, such as Mesos [19] and Kubernetes [20], have been designed with elasticity in mind; hence, they give more importance to orchestration and less to performance. As the CYBELE workflows typically consist of both HPC and big data analytic tasks, as presented in Figure 12.5, the CYBELE framework

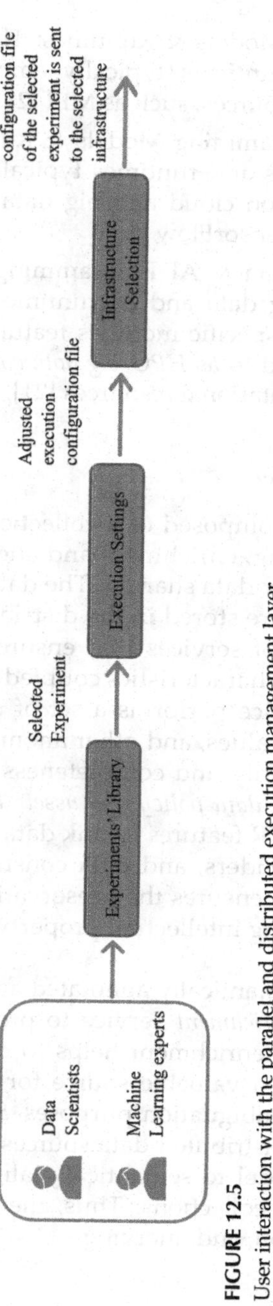

FIGURE 12.5
User interaction with the parallel and distributed execution management layer.

includes programming models and runtimes from both fields. Specifically, the *programming models and runtimes* component, consists of three modules:

- HPC Programming Models & Runtimes: These include the programming models and runtimes typically employed by scientific tasks executed on HPC resources, such as MPI [21].
- Big Data & AI Programming Models & Runtimes: These include the programming models and runtimes typically employed by data analytic tasks executed on cloud and big data e-infrastructures, such as Apache Spark [15], TensorFlow [16].
- HPC-enabled Big Data & AI Programming Models & Runtimes: To support multiple big data and AI runtimes to be deployed on HPC clusters, we deliver specific modules featuring optimized versions of the runtimes, referred to as *HPC–big data collocation*, tightly integrated with the HPC computational resources [21], [22].

12.3.4 Data Services Layer

The *data services layer* is composed of a collection of services which facilitate data check-in, cleaning, enrichment and alignment, storage, querying, and controlled proprietary data sharing. The data are ingested through the *data check-in* service and are stored in the distributed *data storage*. The data check-in is an umbrella of services that ensures the veracity, timeliness, transparency, and legacy characteristics coupled with the big data. The *data cleaning and curation* service performs a set of quality checks to discover inconsistencies, missing values, and other anomalies in the data and eventually ensures their integrity and completeness by following several data cleaning procedures. The *data policy and assets brokerage* service facilitates data sharing and offers IPR features to link data managers (i.e. agri-/aqua-tech providers, data providers, and data consumers). The *data encryption and anonymization* service ensures the preservation of the private information come with data having intellectual property rights and is integrated in the *data check-in*.

Checked-in data are semantically annotated and harmonized through the *semantic alignment and enrichment* service to promote data interoperability and reuse. Data-oriented enrichment helps to develop robust and flexible annotations and provide a valuable source for common representation of similar concepts for disambiguation purposes. Since the data come from a multitude of physically distributed data sources, the *common semantic model* serves as a reference model to semantically align, describe, annotate, and share these diverse data collections. Thus, the model enables on-demand data discovery, exploration, and querying.

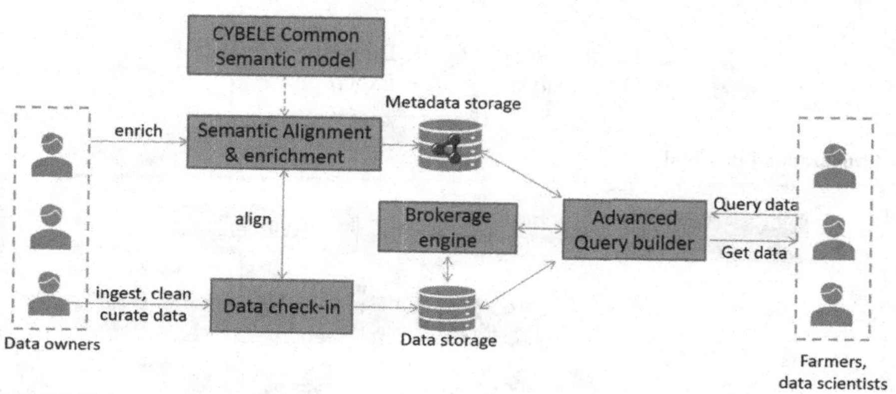

FIGURE 12.6
User interaction with the data services layer

The clean and semantically enriched data are stored in the *data storage* while the data annotations are stored at the *CYBELE metadata repository*. Both the data and annotations are made available to the *advanced query builder* for further exploration, analysis, and visualization. The *advanced query builder* provides an intuitive environment for end users to select preferred datasets, combine them, define and execute queries on the available/combined data in the distributed *data storage*. User interactions with the different data services are depicted in Figure 12.6.

12.3.5 Visualization and Reporting Layer

The *visualization and reporting layer* is responsible for the visual representation and reporting of the results produced from the other functional components of the CYBELE framework. This layer consists of an adaptive visualization tool which follows a user-centered design approach. End users have thus the facility to generate or use beautiful and appealing, as well as scientifically correct and relevant visualizations. Users are able to explore data, dynamics (i.e. evolving weather conditions, prices prediction), draw conclusions, and create reports. The *visualization and reporting layer* apart from applying the user interfaces, it is able to exploit large datasets resulting from computer simulations that use HPC resources provided that the data follow a machine-readable format. It is also capable of extracting insights from large and complex data coming from combined structured (e.g. simulations, sensors) or unstructured (free text, images) sources and presenting them in the most useful manner interacting with the *data storage, ECE,* and *advanced query builder*. User interaction with the *visualization and reporting layer* is presented in Figure 12.7.

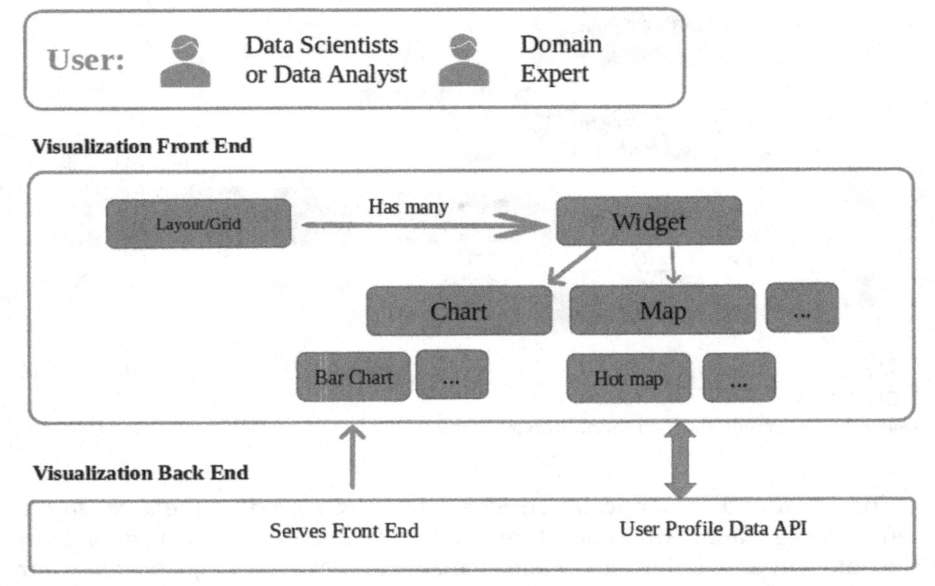

FIGURE 12.7
User interaction with the visualization and reporting layer.

12.4 Key Takeaways and Conclusions

The CYBELE framework facilitates the execution of different scenarios coming from the agri-/aquaculture domains by enabling the execution of batch, micro-batch, and streaming processes. Following a layered approach, each layer serves to abstract to the end user the technical details and eases the design, configuration, and enactment of complex big data, HPC, and AI applications. The *experiments composition environment* facilitates and automates the design, development, and execution of the big data, AI, and HPC tasks, supporting embedded scientific computing and reproducible frameworks. A set of *generic and domain-specific analytic algorithms* have been developed, stored, and fetched in the definition of data analysis workflows, consisted of a series of data analysis processes, interconnected among each other in terms of input/output data streams/objects. The *parallel and distributed execution management layer* focuses on tuning the HPC software stack to allow for efficient distributed execution of big data processing frameworks and AI algorithms on top of parallel HPC resources and enriches with programmable mechanisms the *resource management and orchestration* and its interface with big data processing frameworks and orchestration engines, thus bridging the gap between the HPC and big data worlds. The *data services layer* takes care of the entire data life cycle from ingestion and integration to semantic alignment and querying. The results of queries and analytics are exposed to the *visualization and reporting layer* with the ability to visually explore the different kinds of data, while discovering and addressing new patterns and insights. The analysis results use adaptive visualizations and user-friendly interfaces, improving the way in which information is presented.

Note

1 www.cybele-project.eu/

References

[1] Growing at a slower pace, world population is expected to reach 9.7 billion in 2050 and could peak at nearly 11 billion around 2100. Available Online: www.un.org/development/desa/en/news/population/world-population-prospects-2019.html

[2] Valin, Hugo, Daan Peters, Maarten Van den Berg, Stefan Frank, Petr Havlik, Nicklas Forsell, Carlo Hamelinck et al. "The land use change impact of biofuels consumed in the EU: Quantification of area and greenhouse gas impacts." (2015).

[3] Schrijver, R., et al. Precision agriculture and the future of farming in Europe. Report of STOA, Science Foresight Unit, European Union (2016).

[4] Schimmelpfennig, David. Farm profits and adoption of precision agriculture. No. 1477-2016-121190. 2016.

[5] Evans, Dean. Precision Farming with Big Data Analytics: The future of farming will rely on big data to give farmers new insights into how they can grow crops more efficiently and sustainably. Available Online: www.intel.co.uk/content/www/uk/en/it-management/cloud-analytic-hub/big-data-helps-farmers.html

[6] Eurobarometer team: Europeans, agriculture and the common agricultural policy. Special Eurobarometer 440, The European Commission (2016).

[7] Georgiou, Y., Zhou, N., Zhong, L., Hoppe, D., Pospieszny, M., Papadopoulou, N., Nikas, K., Nikolos, O.L., Kranas, P., Karagiorgou, S. and Pascolo, E., 2020, June. Converging HPC, Big Data and Cloud Technologies for Precision Agriculture Data Analytics on Supercomputers. In *International Conference on High Performance Computing* (pp. 368–379). Springer, Cham.

[8] Mouzakitis S., Tsapelas G., Pelekis S., Ntanopoulos S., Askounis D., Osinga S., Athanasiadis I.N. (2020). Investigation of common big data analytics and decision-making requirements across diverse precision agriculture and livestock farming use cases. Proceedings of 13th International Symposium

on Environmental Software Systems, February 5–7, 2020, Wageningen, The Netherlands. https://link.springer.com/book/10.1007/978-3-030-39815-6
[9] Perakis K., Lampathaki F., Nikas K., Georgiou Y., Marko O., Maselyne J. (2020). CYBELE – Fostering precision agriculture & livestock farming through secure access to large-scale HPC enabled virtual industrial experimentation environments fostering scalable big data analytics. Computer Networks, Vol. 168, 107035. https://doi.org/10.1016/j.comnet.2019.107035
[10] Fabiyi S. D., Vu H., Tachtatzis C., Murray P., Harle D., Dao T. K., Andonovic I., Ren J., Marshall S. (2020). Varietal Classification of Rice Seeds Using RGB and Hyperspectral Images. IEEE Access, Vol. 8, pp. 22493–22505. https://doi.org/10.1109/ACCESS.2020.2969847.
[11] Paudel D., Boogaard H., De Wit A., Janssen S., Osinga S., Pylianidis C., Athanasiadis I. N. (2020). Machine learning for large-scale crop yield forecasting, Agricultural Systems, 103016, ISSN 0308-521X, https://doi.org/10.1016/j.agsy.2020.103016.
[12] React. Available Online: https://reactjs.org/
[13] Spring Cloud Data Flow. Available Online: https://spring.io/projects/spring-cloud-dataflow
[14] YAML file. Available Online: https://en.wikipedia.org/wiki/YAML
[15] MLlib. Available Online: https://spark.apache.org/mllib/
[16] Distributed TensorFlow. Available Online: www.tensorflow.org/guide/distribute_strategy
[17] Slurm Workload Manager. Available Online: https://slurm.schedmd.com/documentation.html
[18] Torque Resource Manager. Available Online: www.adaptivecomputing.com/products/torque/
[19] Apache MESOS. Available Online: http://mesos.apache.org/
[20] Kubernetes (K8s). Available Online: https://kubernetes.io/
[21] Open MPI. Available Online: www.open-mpi.org/
[22] Horovod. Available Online: https://github.com/horovod/horovod

13

CYBELE: A Hybrid Architecture of HPC and Big Data for AI Applications in Agriculture

Naweiluo Zhou, Li Zhong, Dennis Hoppe, Branislav Pejak, Oskar Marko, Javier Cardona, Mikolaj Czerkawski, Ivan Andonovic, Craig Michie, Christos Tachtatzis, Emmanouil Alexakis, Philip Mavrepis, Dimosthenis Kyriazis, and Marcin Pospieszny

CONTENTS

DOI: 10.1201/9781003176664-13

13.1 Introduction: Vision and Challenges

Big-data analytics hosted by cloud clusters are becoming more data-intensive and computation-intensive lately, mainly due to development of artificial intelligence (AI) applications. High-performance computing (HPC) systems are often used to execute large-scale programs, such as the ones performing engineering, scientific or financial simulations that demand low latency and high throughput. By exploiting HPC resources, AI applications have the potential to achieve better performance compared to that on cloud. In general, an AI program incorporates a complex list of software and therefore its user needs flexibility to customize the working environment. However, HPC systems, supporting multi-tenant environments, typically provide complete stacks of software packages and often do not allow user customization in contrast to cloud systems. Containerization could offer a solution for provisioning AI applications with flexible working environments on HPC clusters.

Containerization is a virtualization technology. Rather than simulating a holistic operating system (OS) as in a virtual machine (VM), a container only shares its host OS and utilizes its dependencies. Typically, one container encapsulates one application with the corresponding libraries and configurations in an isolated environment thus enabling compatibility and portability. A container is booted by its host as a process, hence the start-up time of a container is similar to a native application [1–7]. Deployment of containers focus on environment compatibility on HPC clusters, where typically a heavy stack of software packages is encapsulated, making the size of containers relatively large. Per contra, containers on cloud, which are dedicated to run micro-services [8], are more lightweight.

Jobs on a cloud cluster are often managed by a container orchestrator, such as Kubernetes [9]. An HPC cluster is usually controlled by a workload manager, for example, TORQUE [10] or SLURM [11]. Typical HPC jobs are large workloads whose execution time can be ascertained. Whereas jobs on Kubernetes are often small programs that run continuously to provide services. An HPC workload manager lacks its efficiency in micro-service supports and deeply integrated container management in which a container orchestrator manifests its advantages. Nevertheless, a container orchestrator on its own cannot meet all the requirements of HPC systems, thereby cannot replace an existing workload manager.

The EU-funded research project CYBELE[1] enables convergence of HPC and cloud technologies being applied in the domain of agriculture that has become a trending field for AI applications. In this project, the cloud clusters host long-running service applications that supply straightforward graphic interfaces for designing application pipelines and workflows as presented in Chapter 12. The applications that require intensive computation or fast data access are scheduled for execution on HPC clusters where program performance can be significantly enhanced. In this chapter, a hybrid architecture that enables the synergy of TORQUE or SLURM managed HPC systems and Kubernetes managed cloud systems is given. Two use cases are detailed in order to illustrate how this architecture can be beneficial to AI applications and how AI applications should be adapted in order to scale on HPC systems.

13.2 Background

Agriculture is the basis of economy worldwide [12]. In recent years, AI technologies have been applied in various domains of agriculture, such as crop yield prediction, disease detection, soil content sensing and crop monitoring, to better understand growth of crops and improve production efficiency [13]. Monitoring data about crops automatically collected by sensors or satellites, such as the images of fields, can be continuously processed by AI applications, and when a problem pattern is recognized, the applications can recommend immediate actions to tackle the issue [14]. Pattern recognition [15] is one of the most important technologies in AI and it is the underlying technology of many decision-support systems that help farmers optimize their production.

13.2.1 AI in Big Data Analytics on Cloud

Advancement in machine learning (ML) and deep learning (DL) has brought more accurate solutions to applications and enhances application performance significantly. ML/DL models have been implemented by industry and academia in cloud clusters, where large datasets are crunched and algorithms are trained, thus allowing efficient scaling at a low cost.

Different parallelization algorithms and strategies have been developed in DL, where three predominant parallelization methods are commonly adopted, namely data parallelism, model parallelism and pipeline parallelism [16]. In *data parallelism*, each device in the cluster loads an identical copy of the DL model, and the whole training dataset is split into non-overlapping chunks which are fed into the model replicas of each device. Each model replica is trained on the data chunks, and the model parameters from different devices are synchronized after each step. However, when the model becomes too complicated and involves a large number of parameters, synchronization of the parameters on model replicas often becomes the bottleneck [17].

In contrast to data parallelism, *model parallelism* splits the model and each device loads a part of the model. The input is first fed to the device which holds the input layer, and subsequently its output is passed to the device that holds the next layer in the *forward pass* [18]. In the *backpropagation* [18] phase, the gradients are computed starting from the device which holds the output layer of the DL model, then this change is consequently passed to the previous layers. In theory, model parallelism could solve the problem raised in data parallelism. In practice, heavy communication among different devices is the bottleneck. *Pipeline parallelism* was thereby proposed. It combines both data parallelism and model parallelism, where not only the model is split into different devices, but also the dataset is fractured into chunks. This methodology is commonly adopted by most DL frameworks or libraries, for example, TensorFlow [19] and PyTorch [20]. In addition to the types of parallelism supported by the above DL parallel frameworks, Apache Spark [21], as unified analytic engine, is widely used for large-scale data processing and provides high-level APIs helping users create and tune practical ML pipelines.[2]

Computing resources and applications on cloud are managed by orchestrators. Kubernetes [9], as the trending container orchestrator, is based on a highly modular architecture that abstracts the underlying infrastructure and allows internal customization, such as deployment of different software defined network or storage solutions. It includes a powerful set of tools to control the life cycle of applications, for example, parameterized redeployment in case of failures, or state management. It also supports various bigdata frameworks (e.g. Apache Spark, Hadoop MapReduce [22] and Kafka [23]) and can be connected with Ansible [24] which is a popular software orchestration tool on cloud clusters.

13.2.2 AI on HPC Systems

Compared to cloud systems, HPC systems show the advantages in computational power, storage, and security [25]. Exploiting HPC infrastructures for ML/DL training is becoming a topic of increasing importance [26]. AI applications are usually developed with high-level scripting languages or frameworks, for example, TensorFlow [19] and PyTorch [20], which often require connections to external systems to download a list of opensource software packages during execution. For example, an AI application written in Python cannot be compiled into an executable that includes all the dependencies necessary for execution as in C/C++. Therefore, the developers need flexibility to customize their execution environments. HPC environments, especially HPC production systems, are often based on closed source applications and their users have restricted account privileges and security restrictions [27], for instance access to external systems is blocked. Containerization enables easy transition of AI workloads to HPC while exploiting HPC hardware and optimized libraries of AI applications, without compromising security of HPC systems.

HPC workload managers, such as TORQUE or SLURM, have been well designed to schedule conventional HPC parallel applications, for example, message passing interface (MPI) [28, 29] applications. Implementation and deployment of DL solutions on HPC systems are challenging in terms of parallelization, scheduling, elasticity, data management and portability. With a large number of DL models having complex structures and being trained with huge amounts of data, adapting DL/ML models in order to be efficiently scheduled by the workload managers to achieve high parallelization is the *sine qua non* of leveraging performance on HPC systems.

13.3 Hybrid Big Data and HPC Resource for AI Applications in CYBELE

The CYBELE project proposed and implemented a hybrid architecture that enables convergence of cloud and HPC clusters. This section only gives a brief introduction to this architecture and more details are described in [30–33].

The architecture illustrated in Figure 13.1 consists of an HPC cluster with either TORQUE or SLURM as its workload manager and a cloud cluster with Kubernetes serving as its container orchestrator. All the nodes on the left side are VM nodes. VM nodes rather than bare-metal nodes are more often utilized on cloud clusters. For simplicity, Figure 13.1 only shows a limited number of nodes and a single HPC partition. The important requirement of this architecture is the existence of one or more shared login node/nodes as highlighted in the dashed line in the middle of Figure 13.1. The login node is a Kubernetes worker node (or can be just a Kubernetes login node) and meanwhile functions as a login node for TORQUE or SLURM. Jobs are submitted via the login node in the form of *yaml* scripts, in which the TORQUE or SLURM job scripts are embedded. The embedded scripts are abstracted and scheduled to execute on the HPC cluster by a plugin named Torque-Operator [30, 31] or WLM-Operator [34]. The WLM-Operator implemented by Sylabs bridges Kubernetes with SLURM and the Torque-Operator developed within the CYBELE project connects Kubernetes with TORQUE (see [30] for more details). The normal Kubernetes micro-service jobs are still deployed via the login node. They are, however, scheduled to run on the cloud cluster. The advantages that this architecture bring are four-fold:

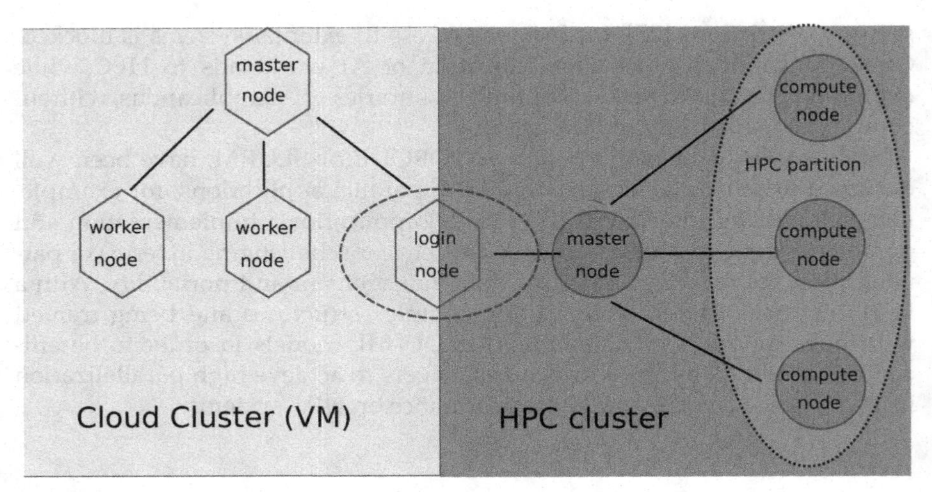

FIGURE 13.1
The hybrid architecture.

1. It provides a unified interface for users to access the cloud and HPC systems. All jobs are submitted in the form of *yaml* scripts.
2. It keeps the modification of HPC systems to the minimum. HPC systems only need to install Singularity on top of their current environments. Singularity [35] is the *de facto* standard container runtime for HPC systems.
3. Long-running micro-service applications stay on the cloud cluster and computation-intensive or data-demanding programs that have determined execution time are scheduled to the HPC cluster.
4. Users have the flexibility to customize their working environments on cloud clusters to develop their applications.

13.4 Parallelization and Deployment of AI Applications on HPC Systems

The key driving factor of success of AI, especially DL applications, is the optimization and scalability of training [26]. New breakthroughs in model accuracy were achieved by increasing the size and the complexity of models. Relatively simple model structures have been replaced with more

TABLE 13.1

Descriptions of Pilot Soybean Farming and Pilot Wheat Ear

Names	Descriptions	ML applications	Parallel models
Pilot Soybean Farming	Yield prediction for soybean farming	yes	Open MPI, Apache Spark
Pilot Wheat Ear	Provisioning of autonomous robotic systems within arable frameworks	yes	PyTorch

sophisticated architectures [36]. Moreover, the development of AI technologies has been pushed by an increasing amount of training data. DL models are data-intensive, and model accuracy can be improved significantly with larger-quantity of training data [37]. Therefore, massive compute resources available in HPC infrastructures enable training of complex models with vast amounts of data in a reasonable time. This section showcases two AI applications in the field of agriculture, which are developed within the CYBELE project, namely the Pilot Soybean Farming and Pilot Wheat Ear. Table 13.1 summarizes the descriptions of the two pilots and the details are presented in Section 13.4.1 and Section 13.4.2.

13.4.1 Pilot Soybean Farming

This section describes the Pilot Soybean Farming and its methods of parallelization and deployment on HPC systems.

13.4.1.1 Pilot Description

Pilot Soybean Farming focuses on the application of ML in soybean farming. Its goal is to develop a prediction algorithm that is able to infer hidden dependencies between the input parameters and the yield. The input dataset consists of three Sentinel-2 [38] images for each of the three years being analyzed. I.e. nine satellite images in total with each having 12 spectral bands with a resolution of 10m, 20m and 60m, depending on the channel. Additionally, yield monitor data from Austrian soybean farms is used as the ground truth for algorithm training (total size: 614 MB). The application is composed of two procedures, that is, the data preprocessing procedure followed by the training procedure. The input for the second procedure is a tabular file in the *xlsx* format (total size: 7.3 MB) yielded from the first procedure. Both procedures are written in Python scripts and containerized in Singularity.

13.4.1.2 Application Parallelization for HPC Systems

MPI is a conventional standard for parallel processing on HPC systems. The first procedure of Pilot Soybean Farming adopts MPI to speedup data processing. More specifically, mpi4py library, which is the Python support for MPI, is introduced to the program. In this case, the MPI launcher of the host system invokes the Singularity container and Singularity launches the MPI application within the container, which is the Python script with MPI support in this case.

The dataset produced by the first procedure is fed in as the input for the ML training model. The second procedure includes modules for image stacking, cropping to the field boundaries, yield map interpolation from point measurements and ML models for yield prediction. Its ML pipeline comprises the following steps:

1. The dataset is loaded into an Apache Spark data frame that can significantly decrement data access time. Since the ML model is coded in Python, Spark is implemented in PySpark library.
2. Filter out missing/erroneous records and split the training and testing datasets. In this step, the data is now ready to proceed further down the pipeline to feed the ML models, that is, Linear Regression [39], Decision Tree Regressor [40], Gradient Boosted Trees Regressor [41] and Random Forest Regressor [42]. The best results are selected from the four models.
3. The evaluation of the prediction models is performed over the test dataset providing a fair generalization approximation.

Description of the ML training procedure of Pilot Soybean herein is to demonstrate a method of scheduling a big-data framework to an HPC environment predominated by a workload manager. It is beyond the scope of this chapter to discuss prediction accuracy of various ML models. There are three ways of deploying Spark as standalone engine, though Hadoop YARN [43] and via MapReduce [44]. Standalone deployment is favored in the HPC environment. In this case, the HPC workload manager TORQUE/SLURM serves as the Spark scheduler in preference to Mesos [45] or YARN. The Spark standalone package is not required on an HPC system, and instead it is encapsulated inside Singularity. The master and workers of Spark are launched within the container processes rather than natively on the HPC cluster.

13.4.2 Pilot Wheat Ear

This section depicts the Pilot Wheat Ear and its methods of parallelization and deployment for HPC systems.

FIGURE 13.2
Input image of wheat ears and the output of ear counts generated by the DL model.

13.4.2.1 Pilot Description

Pilot Wheat Ear is targeted for provisioning of autonomous robotic systems within arable frameworks [46]. More specifically, the application provides a framework for automatic detection and count of wheat ear kernels in fields from the images collected by sensors on the ground so that it enables crop yield prediction and can suggest decisions for sales planning. A DL model, utilizing image segmentation based on U-Net architecture [47], is trained to automatically detect the wheat ears from a series of RGB images (138 images, 95 MB in total) captured in the crop fields in Serbia. Figure 13.2 illustrates an example of an input wheat-ear image and its output image. The output will serve as an input for yield prediction, which will relate the number of ears and their size to the actual yield.

13.4.2.2 Application Parallelization for HPC Systems

In recent years, a wide range of potential solutions has been brought for the challenge of image segmentation. The solution can encompass a specific neural network architecture and a set of techniques applied during training and inference. Typically, the decisions will be made based on a number of experiments. A high degree of parallelism can greatly speed up this process and allow more selection of the network parameters.

In Pilot Wheat Ear, its DL model is written in Python script based on Fastai/ PyTorch [20, 48] which are widely used DL libraries. PyTorch can get scale-up by using its method torch.set_num_threads() or the environment variable OMP_NUM_THREADS to adjust the number of threads. The library is containerized in Singularity, correspondingly each Singularity process boots one PyTorch thread. The workload manager schedules all the Singularity processes to the available HPC cores.

13.5 Performance Evaluation for Pilot Soybean Farming and Pilot Wheat Ear

The hybrid architecture elucidated in Section 13.3 has been implemented on the testbeds in different institutes.[3] This chapter showcases a testbed with a relatively small number of nodes. The testbed technical specifications are indicated in Table 13.2. The cluster comprises four VM nodes and the HPC cluster is composed of three bare-metal nodes. The HPC cluster has TORQUE installed as its workload manager while the cloud cluster is managed by Kubernetes. Docker [49], as one of the most widely used container runtime on cloud, is utilized to continuously run the CYBELE service programs as described in Chapter 12. Due to security concerns, Docker container runtime is restricted to the cloud cluster. AI applications are containerized in Singularity images. The core applications required for the cloud and HPC systems are listed in Table 13.3. Flannel[4] is deployed for the container network interface (CNI) of Kubernetes. Without this additional network model,

TABLE 13.2

Technique specifications of the testbed

Cluster name	HPC cluster (bare-metal)	Cloud cluster (VM)
Total number of nodes	3 (2 compute nodes)	4 (3 worker nodes)
Number of cores	20 (10 cores per CPU, 2 CPU per node)	2 cores per node
RAM per node (NUMA)	128 GB	8 GB
CPU frequency	Intel(R)Xeon(R) CPU E5-2630 v4, 2.20GHz	Intel i7 9xx (Nehalem Core i7, IBRS) 2.79GHz
Operating System	Ubuntu 18.04.3 LTS	Ubuntu 18.04.3 LTS
VM type	—	QEMU 2.11.1, KVM 2.11.1

TABLE 13.3

The list of core applications on the testbed

Cluster type	HPC cluster	Cloud cluster
Orchestrator	TORQUE	Kubernetes
Container runtime and interface	Singularity	Singularity, Docker, Singularity-CRI[1]
Plugin	—	Torque-Operator, Flannel
Compiler	Golang compiler	Golang compiler
Parallel model	Open MPI	—

[1] Singularity-CRI: A Singularity-specific implementation of Kubernetes CRI (CRI: container runtime interface).

containers located in separate nodes could not establish communications among each other.

The results are compared between the cloud cluster and the HPC cluster. In Pilot Soybean Farming, the size of the training dataset for the training procedure is relatively small, the performance evaluation of this pilot concentrates on the data preprocessing procedure which is computation-intensive and data-intensive. Figure 13.3 compares the execution time with different levels of parallelism for data preprocessing procedure of Pilot Soybean Farming and Pilot Wheat Ear. It is worth noting that the execution time of the two pilots can vary from minutes to days depending on the amount of input data sets that are used for processing. The details are given in Table 13.4 and Table 13.5.

The three-year data processed in Pilot Soybean Farming is organized in three independent groups. The number of processes is, therefore, set to be the multiple of 3. However, for the sake of completeness, the results of a single process are also presented for both. On the HPC cluster, the execution time is measured with the number of processes to be 3, 6, 9, and 12, respectively. For Pilot Wheat Ear, the performance ranges from the execution with two processes (the maximal number of cores on one cloud node) to 20 processes (the maximal number of the cores on one HPC node). The results of the two applications are calculated as the mean values of three executions, and the numbers are proved to be stable (as indicated by the variances in the table). The time spent in job scheduling from cloud to HPC is negligible compared to the total execution time. The performance improvement of both applications is significant when running on an HPC node. In Figure 13.3(a), the application exhibits optimal performance with nine processes, whereas deteriorating with 12 processes. Similarly, for Pilot Wheat Ear in Figure 13.3(b) 18 processes outperform 20 processes. The reasons could be due to the fact that the volume of the dataset does not require too many processes running simultaneously, and the process synchronization cost rises with the increment of process number. It is out of the scope of this chapter to discuss the algorithms of process/thread parallelism.

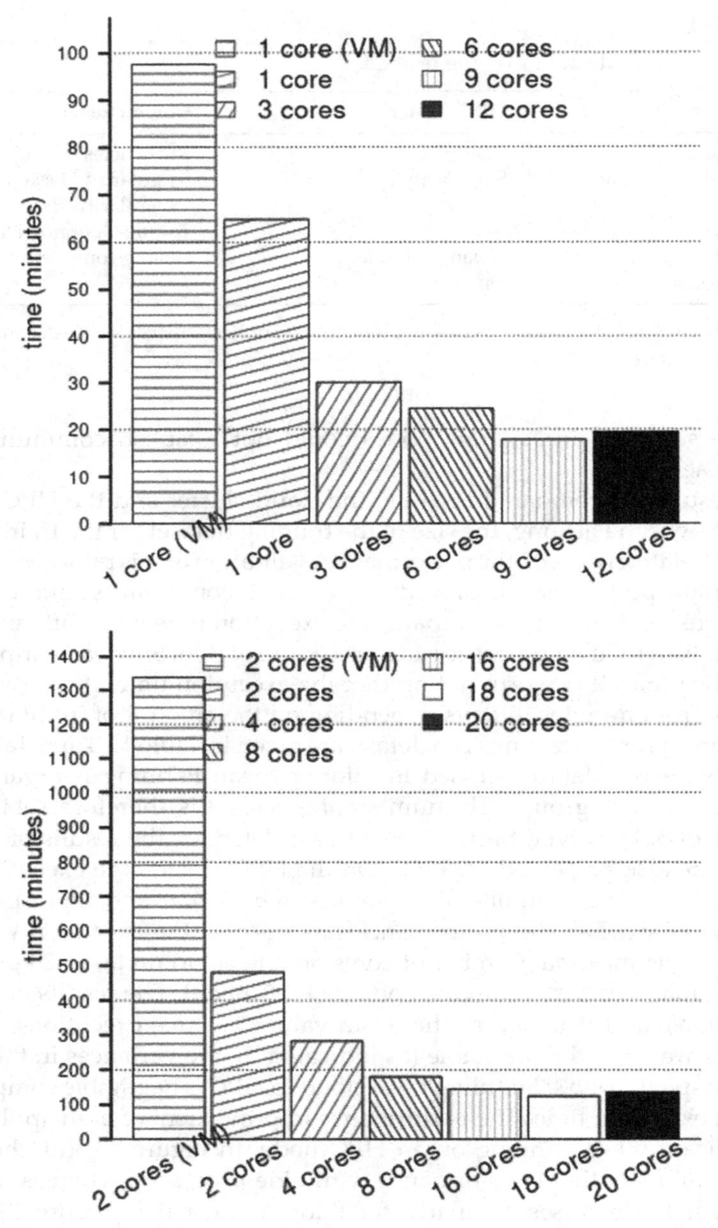

FIGURE 13.3
Time comparison for Pilot Soybean Farming (data preprocessing procedure) and Pilot Wheat Ear.

TABLE 13.4
Execution time on the cloud (VM) and HPC (bare-metal) cluster for data preprocessing procedure on a single node (unit: minutes). The numbers in brackets are the mathematical variances

Cores (on 1 node)	1 core	3 cores	6 cores	9 cores	12 cores
Cloud (VM)	97.60 (0.23)	—	—	—	—
HPC (bare-metal)	64.73 (0.16)	30.07 (0.01)	24.41 (0.01)	17.81 (0)	19.52 (0)

TABLE 13.5
Execution time on the cloud (VM) and HPC (bare-metal) cluster for Pilot Wheat Ear on a single node (unit: minutes). The numbers in brackets are the mathematical variances

Cores (on 1 node)	2 core	4 cores	8 cores	16 cores	18 cores	20 cores
Cloud (VM)	1337.13 (0.00)	—	—	—	—	—
HPC (bare-metal)	480.50 (4.97)	282.39 (0.28)	179.54 (1.02)	144.76 (0.12)	124.25 (0.06)	134.99 (0.55)

13.6 Discussion

The immense gain in performance, which comes out of the box, is one major reason for leveraging HPC for AI applications, while containerization provides compatibility and portability to ease deployment of AI applications onto HPC systems. The use cases presented in this chapter illustrates how AI applications can be adapted to utilize the conventional HPC parallelization models (e.g. Open MPI), and how big-data or DL frameworks(e.g. Apache Spark or PyTorch) can be adapted in order to fit the predominant HPC scheduling environments. The demands for computer resources from the two AI use cases are relatively low, however, the merits that the HPC infrastructures can bring are clearly demonstrated. The benefits will become more evident when the AI systems of the project become operational and when a large number of farmers, researchers and members of extension services start adopting them in a large scale. The hybrid architecture that bridges cloud and HPC systems demonstrated herein will be ready to serve this purpose.

The hybrid architecture elaborated in this chapter contains a login node that connects both the cloud and HPC installations, where the login node is located in two network domains: the domains of HPC cluster and cloud cluster. On an HPC production system, a more portable and secure approach would be restraining the login node within the cloud network domain and connecting it remotely to the HPC cluster via ssh, although this could cause performance degradation in case of a large volume of data transmission.

Lustre [50], a distributed file system that can provide fast data access, can be utilized as the intermediate storage for cloud and HPC cluster instead.

13.7 Conclusion Remarks and Future Works

This chapter has exhibited the necessity of implementing AI technologies in the domain of agriculture. It described the challenges that AI applications are facing. Containerization, which may have paved the way for AI applications on cloud clusters, may also be applied to HPC clusters to ensure compatibility and portability. This chapter has presented a novel hybrid architecture and described necessary methods of parallelizing and deploying AI applications on HPC systems. This CYBELE hybrid architecture unifies the interface for job submission on cloud and HPC systems. Via the same user interface, long-running service programs can be scheduled to cloud clusters, while compute-intensive and/or data-intensive applications can be scheduled onto HPC clusters where their performance can be significantly improved. Two relevant AI applications are showcased to demonstrate the efficiency of incorporating HPC and AI technologies into the field of agriculture.

Continuous increase of data volume pushes the usage of cloud for AI applications as cloud supplies users with on-demand computer resources. Traditional HPC centers are evolving to support AI applications, not only in terms of powerful hardware resources, such as many GPU nodes, large-size memory and SSD, but also with regard to the provision of software environments, such as Singularity, Shifter [51], or Charliecloud [52]. Meanwhile, AI applications are being adapted to HPC environment, for example, the DL Framework Horovod [53], which adopts MPI concept, allows to perform efficiently distributed training on HPC systems.

Acknowledgments

The CYBELE project received funding from the European Union's Horizon 2020 Research and Innovation Programme under grant agreement No. 825355.

Yield monitor data from Pilot Soybean Farming was generously granted by Donau Soya association from farms across the Lower Austria region, through the CYBELE project and are subject to NDA.

The authors would like to express their gratitude to Ms. Aikaterini Papapostolou and Dr. Sophia Karagiorgou for proof-reading the contents.

Many thanks to Dr. Yiannis Georgiou for supporting this work. The authors also appreciate datasets support for Pilot Wheat Ear from Željana Grbović and Marko Panić.

Notes

1 CYBELE: Fostering Precision Agriculture and Livestock Farming through Secure Access to Large-Scale HPC-Enabled Virtual Industrial Experimentation Environment Empowering Scalable Big Data Analytics. www.cybele-project.eu/
2 An ML pipeline is a workflow running a sequence of algorithms to process and learn from data.
3 HLRS Germany, PSNC Poland, CINECA Italy and ATOS France.
4 https://github.com/flannel-io/flannel

References

[1] W. Felter, A. Ferreira, R. Rajamony, and J. Rubio, "An updated performance comparison of virtual machines and Linux containers," in *2015 IEEE International Symposium on Performance Analysis of Systems and Software (ISPASS)*, 2015, pp. 171–172.

[2] D. Bernstein, "Containers and Cloud: From LXC to Docker to Kubernetes," *IEEE Cloud Computing*, vol. 1, no. 3, pp. 81–84, 2014.

[3] J. P. Martin, A. Kandasamy, and K. Chandrasekaran, "Exploring the Support for High Performance Applications in the Container Runtime Environment," *Hum.-Centric Comput. Inf. Sci.*, vol. 8, no. 1, 2018, doi: 10.1186/s13673-017-0124-3.

[4] M. Plauth, L. Feinbube, and A. Polze, "A Performance Survey of Lightweight Virtualization Techniques," in *Service-Oriented and Cloud Computing*, 2017, pp. 34–48.

[5] J. Zhang, X. Lu, and D. K. Panda, "Is Singularity-Based Container Technology Ready for Running MPI Applications on HPC Clouds?," in *Proceedings of The10th International Conference on Utility and Cloud Computing*, 2017.

[6] G. Hu, Y. Zhang, and W. Chen, "Exploring the Performance of Singularity for High Performance Computing Scenarios," in *2019 IEEE 21st International Conference on High Performance Computing and Communications; IEEE 17th International Conference on Smart City; IEEE 5th International Conference on Data Science and Systems (HPCC/SmartCity/DSS)*, 2019, pp. 2587–2593.

[7] A. J. Younge, K. Pedretti, R. E. Grant, and R. Brightwell, "A Tale of Two Systems: Using Containers to Deploy HPC Applications on Supercomputers and Clouds," in *2017 IEEE International Conference on Cloud Computing Technology and Science (CloudCom)*, 2017, pp. 74–81. [Online]. Available: https://doi.org/10.1109/CloudCom.2017.40

[8] L. Abdollahi Vayghan, M. A. Saied, M. Toeroe, and F. Khendek, "Deploying Microservice Based Applications with Kubernetes: Experiments and Lessons Learned," in *2018 IEEE 11th International Conference on Cloud Computing (CLOUD)*, 2018, pp. 970–973.

[9] K. Hightower, B. Burns, and J. Beda, *Kubernetes: Up and Running Dive into the Future of Infrastructure*, 1st ed. Sebastopol, California, USA: O'Reilly Media, Inc, 2017.

[10] G. Staples, "TORQUE Resource Manager," in *Proceedings of the 2006 ACM/IEEE Conference on Supercomputing*, 2006, p. 8.

[11] Morris A. Jette, Andy B. Yoo, and Mark Grondona, "SLURM: Simple Linux Utility for Resource Management," in *In Lecture Notes in Computer Science: Proceedings of Job Scheduling Strategies for Parallel Processing (JSSPP) 2003*, 2002, pp. 44–60.

[12] Vijai Singh, Namita Sharma, and Shikha Singh, "A review of imaging techniques for plant disease detection," *Artificial Intelligence in Agriculture*, vol. 4, pp. 229–242, 2020, doi: 10.1016/j.aiia.2020.10.002.

[13] Tanha Talaviya, Dhara Shah, Nivedita Patel, Hiteshri Yagnik, and Manan Shah, "Implementation of artificial intelligence in agriculture for optimisation of irrigation and application of pesticides and herbicides," *Artificial Intelligence in Agriculture*, vol. 4, pp. 58–73, 2020, doi: 10.1016/j.aiia.2020.04.002.

[14] D. Budaev et al., "Conceptual design of smart farming solution for precise agriculture," *International Journal of Design & Nature and Ecodynamics*, vol. 13, pp. 307–314, 2018, doi: 10.2495/DNE-V13-N3-307-314.

[15] O. G. Selfridge, "Pattern Recognition and Modern Computers," in *Proceedings of the March 1–3, 1955, Western Joint Computer Conference*, 1955, pp. 91–93. [Online]. Available: https:/ / doi.org / 10.1145/ 1455292.1455310

[16] T. Ben-Nun and T. Hoefler, "Demystifying Parallel and Distributed Deep Learning: An In-Depth Concurrency Analysis," *ACM Comput. Surv.*, vol. 52, no. 4, 2019, doi: 10.1145/3320060.

[17] Z. Jia, M. Zaharia, and A. Aiken, "Beyond Data and Model Parallelism for Deep Neural Networks," in *Proceedings of Machine Learning and Systems*, 2019, pp. 1–13. [Online]. Available: https:/ / proceedings.mlsys.org / paper/ 2019/ file/ c74d97b01eae257e44aa9d5bade97baf-Paper.pdf

[18] R. E. Uhrig, "Introduction to artificial neural networks," in *Proceedings of IECON '95 – 21st Annual Conference on IEEE Industrial Electronics*, 1995, 33–37 vol.1.

[19] M.'ı. Abadi et al., "TensorFlow: A System for Large-scale Machine Learning," in *Proceedings of the 12th USENIX Conference on Operating Systems Design and Implementation*, 2016, pp. 265–283. [Online]. Available: http:/ / dl.acm.org / citation.cfm? id= 3026877.3026899

[20] Adam Paszke et al., "PyTorch: An Imperative Style, High-Performance Deep Learning Library," in *Advances in Neural Information Processing Systems 32: Annual Conference on Neural Information Processing Systems 2019, NeurIPS 2019, 8–14 December 2019, Vancouver, BC, Canada*, 2019, pp. 8024–8035.

[21] M. Zaharia et al., "Apache Spark: A Unified Engine for Big Data Processing," *Commun. ACM*, vol. 59, no. 11, pp. 56–65, 2016, doi: 10.1145/2934664.

[22] S. Pandey and V. Tokekar, "Prominence of MapReduce in Big Data Processing," in *2014 Fourth International Conference on Communication Systems and Network*

Technologies, 2014, pp. 555–560. [Online]. Available: https:/ / doi.org / 10.1109/ CSNT.2014.117

[23] N. Narkhede, G. Shapira, and T. Palino, *Kafka: The Definitive Guide Real-Time Data and Stream Processing at Scale*, 1st ed. Sebastopol, California, US: O'Reilly Media, Inc, 2017.

[24] G. Sammons, *Exploring Ansible 2: Fast and Easy Guide*. North Charleston, SC, USA: CreateSpace Independent Publishing Platform, 2016.

[25] G. Mateescu, W. Gentzsch, and C. J. Ribbens, "Hybrid Computing-Where HPC Meets Grid and Cloud Computing," *Future Gener. Comput. Syst.*, vol. 27, no. 5, pp. 440–453, 2011, doi: 10.1016/j.future.2010.11.003.

[26] R. Mayer and H.-A. Jacobsen, "Scalable Deep Learning on Distributed Infrastructures: Challenges, Techniques, and Tools," *ACM Comput. Surv.*, vol. 53, no. 1, 2020, doi: 10.1145/3363554.

[27] D. Brayford, S. Vallecorsa, A. Atanasov, F. Baruffa, and W. Riviera, "Deploying AI Frameworks on Secure HPC Systems with Containers," in *2019 IEEE High Performance Extreme Computing Conference (HPEC)*, 2019, pp. 1–6.

[28] W. Gropp, E. Lusk, and A. Skjellum, *Using MPI: Portable Parallel Programming with the Message-passing Interface*. Cambridge, MA, USA: MIT Press, 1994.

[29] Message Passing Interface Forum and University of Tennessee, Knoxville, Tennessee, *MPI: A Message-Passing Interface Standard*. [Online]. Available: www. mpi-forum.org / docs/ mpi- 3.1/ mpi31-report.pdf

[30] N. Zhou et al., "Container Orchestration on HPC Systems through Kubernetes," *Journal of Cloud Computing: Advances, Systems and Applications*, 2021, doi: 10.1186/s13677-021-00231-z.

[31] Naweiluo Zhou, Yiannis Georgiou, Li Zhong, Huan Zhou, and Marcin Pospieszny, "Container Orchestration on HPC Systems," in *2020 IEEE International Conference on Cloud Computing (CLOUD)*, 2020.

[32] Y. Georgiou et al., "Converging HPC, Big Data and Cloud Technologies for Precision Agriculture Data Analytics on Supercomputers," in *High Performance Computing*, 2020, pp. 368–379. [Online]. Available: https:/ / doi.org / 10.1007/ 978- 3- 030- 59851- 8_ 25

[33] Naweiluo Zhou, "Containerization and Orchestration on HPC Systems," in *Sustained Simulation Performance 2019 and 2020*, 2021.

[34] Vadim Pisaruk and Sasha Yakovtseva, *WLM-operator. Gitlab:* https://github. com/sylabs/wlm-operator (accessed 13/02/2020).

[35] Gregory M. Kurtzer, Vanessa V. Sochat, and Michael Bauer, "Singularity: Scientific containers for mobility of compute," in *PloS one*, 2017. [Online]. Available: https:/ / doi.org / 10.1371 / journal.pone.0177459

[36] D. C. Cireşan, U. Meier, L. M. Gambardella, and J. Schmidhuber, "Deep, Big, Simple Neural Nets for Handwritten Digit Recognition," *Neural Computation*, vol. 22, no. 12, pp. 3207–3220, 2010, doi: 10.1162/NECO_a_00052.

[37] Alon Halevy, Peter Norvig, and Fernando Pereira, "The Unreasonable Effectiveness of Data," *IEEE Intelligent Systems*, vol. 24, pp. 8–12, 2009. [Online]. Available: www.computer.org / portal/ cms_ docs_ intelligent/ intelligent/ homepage/ 2009/ x2exp.pdf

[38] M. Drusch et al., "Sentinel-2: ESA's Optical High-Resolution Mission for GMES Operational Services," *Remote Sensing of Environment*, vol. 120, pp. 25–36, 2012, doi: 10.1016/j.rse.2011.11.026.

[39] X. Yan and X. G. Su, *Linear Regression Analysis: Theory and Computing*. USA: World Scientific Publishing Co., Inc, 2009.
[40] J. R. Quinlan, "Induction of Decision Trees," *Mach. Learn.*, vol. 1, no. 1, pp. 81–106, 1986, doi: 10.1023/A:1022643204877.
[41] Jerome H. Friedman, "Greedy function approximation: A gradient boosting machine," *The Annals of Statistics*, vol. 29, no. 5, pp. 1189–1232, 2001, doi: 10.1214/aos/1013203451.
[42] L. Breiman, "Random Forests," *Mach. Learn.*, vol. 45, no. 1, pp. 5–32, 2001, doi: 10.1023/A:1010933404324.
[43] V. K. Vavilapalli et al., "Apache Hadoop YARN: Yet Another Resource Negotiator," in *Proceedings of the 4th Annual Symposium on Cloud Computing*, 2013.
[44] J. Dean and S. Ghemawat, "MapReduce: Simplified Data Processing on Large Clusters," *Commun. ACM*, vol. 51, no. 1, pp. 107–113, 2008, doi: 10.1145/1327452.1327492.
[45] B. Hindman et al., "Mesos: A Platform for Fine-Grained Resource Sharing in the Data Center," in *Proceedings of the 8th USENIX Conference on Networked Systems Design and Implementation*, 2011, pp. 295–308.
[46] Zeljana Grbovic, Marko Panic, Oskar Marko, Sanja Brdar, and Vladimir S. Crnojevic, "Wheat Ear Detection in RGB: and Thermal Images Using Deep Neural Networks," in *Machine Learning and Data Mining in Pattern Recognition, 15th International Conference on Machine Learning and Data Mining, MLDM 2019, New York, NY, USA, July 20–25, 2019, Proceedings, Volume II*, 2019, pp. 875–889.
[47] O. Ronneberger, P. Fischer, and T. Brox, "U-Net: Convolutional Networks for Biomedical Image Segmentation," in *Medical Image Computing and Computer-Assisted Intervention (MICCAI)*, 2015, pp. 234–241. [Online]. Available: http://lmb.informatik.uni-freiburg.de / Publications/ 2015/ RFB15a
[48] Jeremy Howard and Sylvain Gugger, "Fastai: A Layered API for Deep Learning," *Information*, vol. 11, no. 2, p. 108, 2020, doi: 10.3390/info11020108.
[49] D. Merkel, "Docker: Lightweight Linux Containers for Consistent Development and Deployment," *Linux J*, vol. 2014, no. 239, pp. 76–90, 2014.
[50] R. Salunkhe, A. D. Kadam, N. Jayakumar, and S. Joshi, "Luster a scalable architecture file system: A research implementation on active storage array framework with Luster file system," in *2016 International Conference on Electrical, Electronics, and Optimization Techniques (ICEEOT)*, 2016, pp. 1073–1081.
[51] Lisa Gerhardt et al., "Shifter: Containers for HPC," *Journal of Physics: Conference Series*, vol. 898, p. 82021, 2017, doi: 10.1088/1742-6596/898/8/082021.
[52] R. Priedhorsky and T. Randles, "Charliecloud: Unprivileged Containers for User-Defined Software Stacks in HPC," in *Proceedings of the International Conference for High Performance Computing, Networking, Storage and Analysis*, 2017.
[53] Alexander Sergeev and Mike Del Balso, "Horovod: fast and easy distributed deep learning in TensorFlow," *CoRR*, abs/1802.05799, 2018.

14

European Processor Initiative: Europe's Approach to Exascale Computing

Mario Kovač, Jean-Marc Denis, Philippe Notton, Etienne Walter, Denis Dutoit, Frank Badstuebner, Stephan Stilkerich, Christian Feldmann, Benoît Dinechin, Renaud Stevens, Fabrizio Magugliani, Ricardo Chaves, Josip Knezoviw, Daniel Hofman, Mario Brčić, Katarina Vukušić, Agneza Šandić, Leon Dragić, Igor Piljić, Mate Kovač, Branimir Malnar, and Alen Duspara

CONTENTS

14.1 Introduction

Microprocessor technology and computing, in general, have proliferated into almost every aspect of our individual lives and society. It provides the base for sustainable development in almost every application domain and in almost every aspect of our everyday lives such as healthy aging, climate change, and smart industry. In addition, there are several application areas that play critical roles for sustainable development. These key application areas, such as autonomous driving, health and drug discovery, climate modeling, intelligent transport, smart materials can be seen as vertical pillars

DOI: 10.1201/9781003176664-14

built on top of the horizontally laid computing technologies serving as bases: high-performance computing (HPC), data storage and cloud technologies, embedded high-performance/throughput computing, and, recently, data science, and artificial intelligence. The trends in vertical application areas or pillars have caused the convergence in horizontal bases of computing technologies with the critical workloads and kernels being addressed and carefully designed across all these base computing pools. For example, artificial intelligence workloads and kernels need to be crafted, optimized, and run in each base at different scale: embedded computing systems, data centers, clouds, and HPC. Besides pure performance valued in a number of operations per second, energy-efficiency given as performance per consumed power becomes equally important in the trade-off analysis and design of the computing system. In fact, energy efficiency is recognized as the main challenge to reaching exascale computing of 10^{18} calculations per second within the affordable power budget. Moreover, due to the convergence of the workloads, portability is also taken into the consideration in the design of the base building components (processors, interconnects, accelerators, networks, operating systems, libraries, and APIs), which was correctly recognized by numerous research initiatives within Europe [1]–[5].

The importance of sustainable HPC has been recognized by the European Commission that has strategically initiated efforts to achieve and support activities toward the implementation of European exascale computing systems and related technologies. These efforts have been synchronized in the establishment of the EuroHPC Joint Undertaking, a legal funding entity that will enable pooling of national and European Union (EU)-wide resources in HPC to acquire, build and deploy the most powerful supercomputers in the world within Europe [6].

EuroHPC has a short-term goal with two main ongoing activities:

- **Development of the pan-European supercomputing infrastructure** with three pre-exascale supercomputers and five petascale supercomputers. Pre-exascale supercomputers are meant to position the EU within the world's top 5 supercomputers [6, 7], while petascale machines will become the computing infrastructure available to European private and public entities, academia, and industry to perform R&D activities.
- **Support for the research and innovation activities** with the goal of developing a supercomputing-fed ecosystem for the technology supply industry throughout the whole stack: microprocessor and hardware components, middleware, operating systems, programming languages and models, and applications.

The European Processor Initiative (EPI) project is one of the cornerstones of the EU's strategic plan for future exascale HPC. The vision of EPI is to develop a complete "Made in EU" high-end microprocessor technology capable of addressing a wide range of key application areas for sustainable development in the vast digital future. In the short term, EPI aims to provide the microprocessor technology and the whole management and programming stack to empower new EU exascale HPC machines. From the long-term perspective, EPI will bring forward Europe's goal of reaching sovereignty in the high-performance [8], low-power microprocessor technology, components, intellectual property, and know-how with the potential to create an ecosystem of industrial products and scientific initiatives across the whole computing stack: hardware components, interconnects, NoCs, SoCs, runtimes, operating systems, libraries, frameworks, and applications. Thanks to such new European technologies, European scientists and industry will be able to access exceptional levels of energy-efficient computing performance. As recognized by high-level EU officials, EPI will benefit Europe's scientific leadership, industrial competitiveness, engineering skills, and know-how in vastly important areas critical for sustainable development.

14.2 European Processor Initiative

This proposal for the EPI was the result of European strategic plans to support the next generation of (super)computing infrastructures, as EU efforts have been synchronized in establishment of the EuroHPC Joint Undertaking.

The project gathers 28 partners from 10 European countries with the objective to develop the processor and ensure that the key competence of high-end chip design remains in Europe, and to provide the European scientists and industry with the access to exceptional levels of energy-efficient computing performance.

EPI is currently running under the first stage of the Framework Partnership Agreement signed by the Consortium with the European Commission, whose aim is to design and implement a roadmap for a new family of low-power European microprocessors for extreme-scale computing, HPC, big data, and a range of emerging applications in various domains which require exceptional computing capabilities within the limited power budget.

The specific challenge of the topic was to support the creation of a world class European HPC and big-data ecosystem built on two exascale computing machines, which would rank in the top three in the world. Such a processor would help "foster an HPC ecosystem capable of developing new European technology such as low-power HPC chips."[9]

The consortium has chosen the holistic approach to define the EPI system architecture and all the components of the stack to address the main challenge of energy-efficient computing for diverse application areas that require supercomputing capabilities. The project will enable scientific leadership

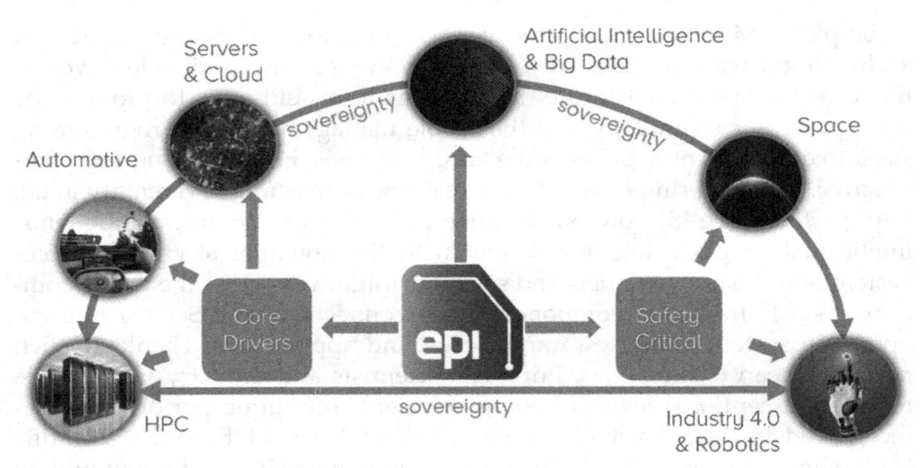

FIGURE 14.1
EPI technology targets.

and industrial competitiveness in both microprocessor technology, and in key application areas exploiting the technology.

The design of a novel HPC processor family cannot be sustainable without thinking about possible additional markets that could support such long-term activities. Therefore, EPI covers a complete range of expertise, skills, and competencies needed to design and execute a sustainable roadmap for research and innovation in processor and computing technology fostering future exascale HPC and emerging applications. In addition, EPI decided to address other converging areas identified as critical for ensuring the overall economic viability of the initiative. This is especially highlighted for the automotive industry because of Europe's strong position in the sector. The project aims to provide value-adding components for key research and industrial stakeholders in the field which will benefit both: the project goals and positioning the European automotive industry in both, short-term automotive development, and long-term autonomous driving sectors. Figure 14.1 illustrates the EPI technology targets and their diversity.

EPI activities are organized into high-level technology domains which will focus on a specific segment or work packages of the project. In the context of EPI, these domains and associated activities are depicted as streams. Streams work together to integrate their results by exercising the interplay of activities.

14.2.1 Global Technical Panstream

Global Technical Panstream manages the interplay of activities within other technological streams and, together with those streams, aims to define the global EPI architecture that will be first realized in a so-called EPI Common Platform (CP) (illustrated in Figure 14.2.) This (pan)stream is responsible for codesign, making, and validating the design choices toward the ECP and future realizations of EPI processors. The stream finally consolidates the results and deliverables of other streams into the EPI Global System Architecture consisting of management-related components (system software, power management, compilation, and simulation tools) in addition to the processor and other hardware components.

Figure 14.2 illustrates the EPI CP which serves as the common basis for the development of future exascale computing systems within the EPI and after the formal end of the SGA project. The platform relies on the global architecture framework that will be followed in the prototyping and production phases. The global architecture includes hardware and software components specifications, common design methodology, and the unified approach to system management with an emphasis on security and power management. The CP will also provide the basis for the development of future commercial processor families named Rhea and Cronos. These processor families will be provided by SiPearl, the industrial spin-off of the EPI project, with the indicative roadmap shown in Figure 14.3.

As shown in Figure 14.2, the EPI CP is organized as the flexible network-on-chip (NoC) which connects a wide range of components developed under the EPI consortium or delivered by external entities:

1. Number of general-purpose Arm Zeus cores with architectural features targeting high-performance and low-power applications (number of cores and architecture specifics depending on the EPI processor family and architecture version) – the development falls within the general-purpose processor (GPP) stream.
2. EPI accelerator tile EPAC based on the RISC-V architecture with vector floating point unit (RISC-V Vector Extension Specification [10]), with other specialized accelerators targeting stencil, neural networks, and variable precision computations, all of them in the focus of the accelerator stream.
3. Accelerator blocks tailored for further specialization into specific domains such as a massively parallel processor array (MPPA) from Kalray, embedded FPGA (eFPGA) from Menta, technology-agnostic cryptographic IP, and fully specialized ASIC blocks, mainly the focus of automotive stream.
4. On- and off-chip/die memory interfaces (HBM, DDR5) and high-speed links (PCIe gen 5, CCIX, CXL), specifics depending on the EPI processor family and architecture.

Figure 14.3 shows the indicative EPI roadmap with major milestones highlighted. The first-generation chip family following the EPI CP, named

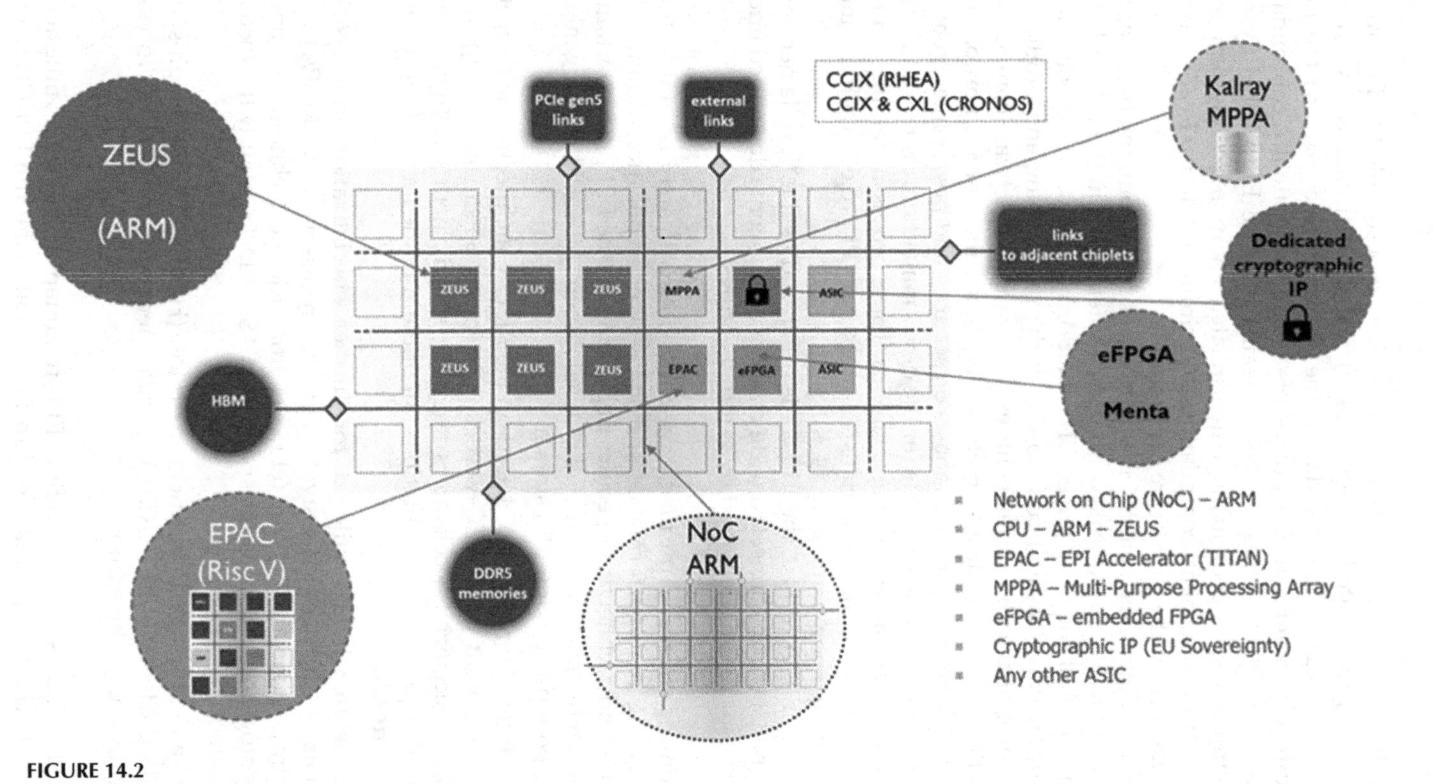

FIGURE 14.2
EPI common platform architecture.

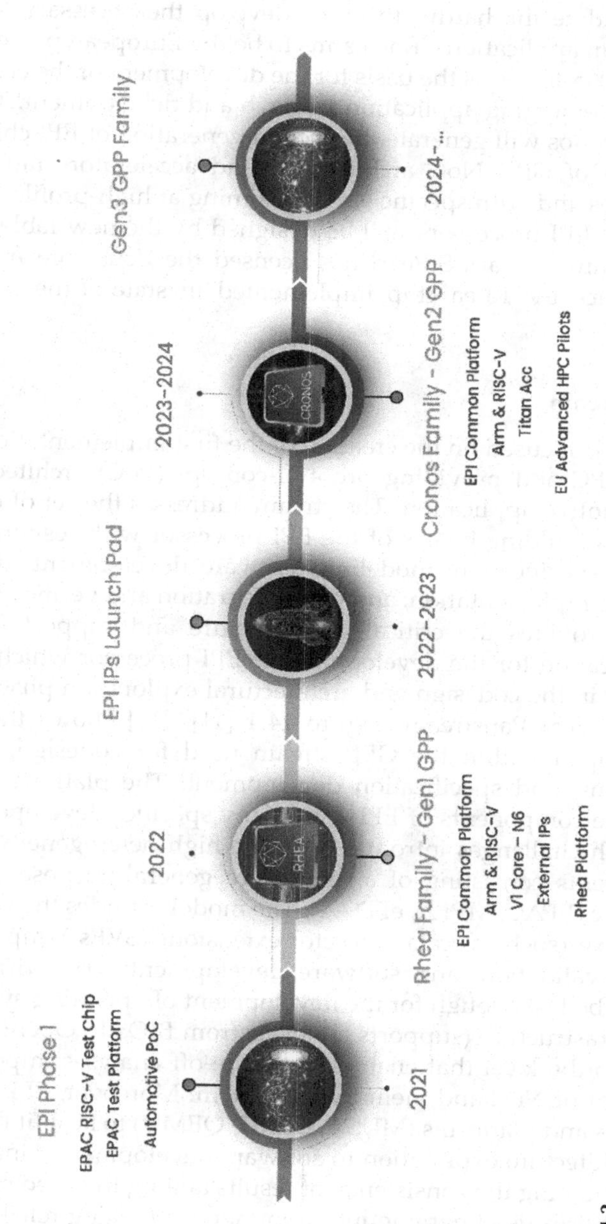

FIGURE 14.3
EPI roadmap.

Rhea, will include Arm architecture general-purpose cores, EPAC accelerator, MPPA, and eFPGA blocks. The Rhea chips will be integrated into test platforms to validate the hardware units, develop the necessary software interfaces, and run applications. Rhea aims to be the European processor for several EU HPC platforms as the basis for the development of the ecosystem for HPC, AI, and emerging application research and development. The next family named Cronos will generate the second generation of EPI chips with the new version of GPP, NoC architecture, and accelerators tailored for exascale machines and with specific variants aiming at high-profile automotive applications. EPI processors will be designed by the new fabless semiconductor company SiPearl. SiPearl has licensed the Zeus core from Arm and will introduce the Rhea chip implemented in state-of-the-art silicon technology.

14.2.2 GPP Stream

The GPP stream is focused on the creation of the first implementation of EPI processor for HPC and providing proof-of-concept (PoC) architecture for emerging automotive application. The stream addresses the set of common technologies and building blocks of the EPI processor with responsibilities for architecture specification, modeling, software development, IP design and verification, implementation, and chip integration and verification.

GPP stream provides the critical infrastructure and support for modeling and verification for the development of EPI processor which is especially important in the codesign and architectural exploration phases of the Global Technological Panstream. Figure 14.4 [11], [12] shows the virtual platform developed within the GPP stream used for codesign, architectural explorations, and specification development. The platform includes models of all the components of EPI with many specifics developed within the projects with challenges introduced by the high heterogeneity of computing components consisting of a many-core general-purpose CPU and accelerator fabric (EPAC, MPPA, eFPGA). The model includes the latest Arm processor features (such as scalable vector extensions (SVEs)) important for benchmarking, validation, and software developments. The virtual platform needed to be fast enough for the development of operating system and compilation infrastructure (supports all levels from BIOS to OpenMPI), but, also, accurate to the level that enables the trade-off analyses important for the development of NoC and memory subsystem. Moreover, EPI leverages other simulators and platforms (MUSA, GEM5, QEMU) to best fit the design needs from architecture exploration to software development. Finally, EPI's approach to validating the consistency of results among the used simulators and virtual platforms is of paramount importance in making reliable design decisions. Therefore, the consortium conducted a thorough covalidation of design decisions and developments with this approach.

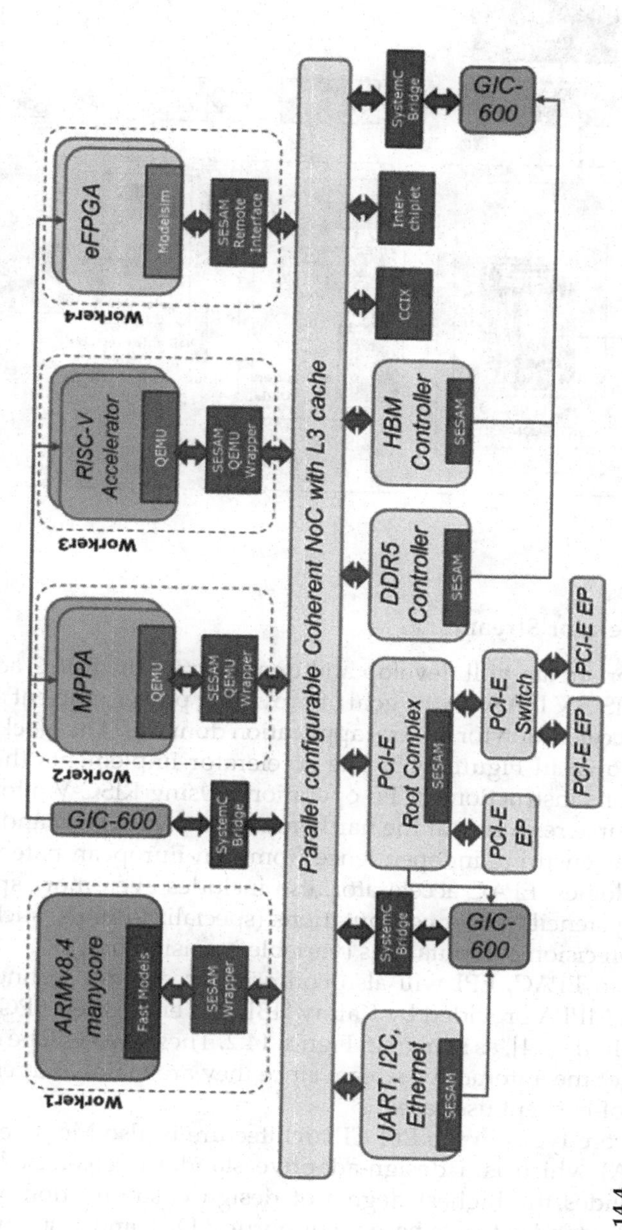

FIGURE 14.4
EPI virtual platform based on SESAM/VPSIM.

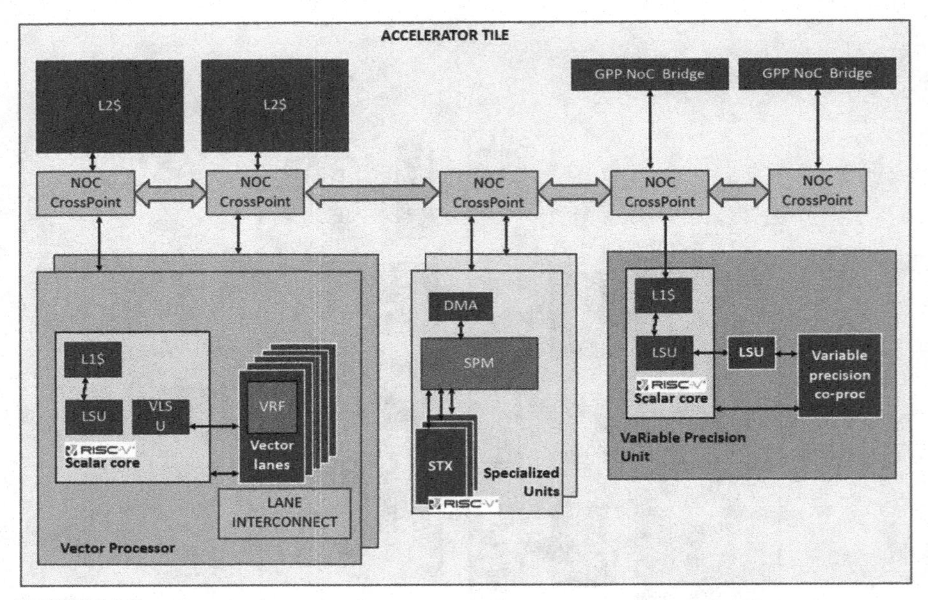

FIGURE 14.5
EPAC accelerator.

14.2.3 Accelerator Stream

The Accelerator stream will develop and demonstrate the EPAC accelerator based on the RISC-V ISA with the goal of providing power-efficient and high-performance acceleration for several application domains. The block diagram of EPAC is shown in Figure 14.5. The accelerator implements the RISC-V vector extensions instruction for FP operations. Using RISC-V allows leveraging open-source resources at the hardware architecture level and software level, as well as ensuring independence from non-European patented computing technologies. EPAC accelerator also includes even more specialized units targeting stencil/tensor computations (specialized units, such as STX) and variable precision computations (variable precision unit).

In addition to EPAC, EPI will also comprise accelerator technologies in the form of an MPPA provided by Kalray [13], and embedded FPGA eFPGA provided by Menta [14], as shown in Figure 14.2. These two will be described in the context of the automotive stream since they are mainly targeting automotive proof-of-concept use case.

One of the core types in the EPI CP architecture is also Menta embedded FPGA (eFPGA) which is a design-adaptive standard cells-based architecture that provides the highest degree of design customization. eFPGA IP cores are 100% standard cells based, supporting DFT and test coverage up to 99,8%. The embedded FPGA cores will be supplied with a proven EDA tool, Origami Programmer, which supports the design from HDL design to bitstream with synthesis, mapping, place, and route. The eFPGA integrated into the EPI CP GPP chip is optimized for not just general-purpose HPC, but also for automotive applications such as signal and image processing. The unique features added by eFPGA are customer specialization (customization) in post-production and deployment phases, and reconfigurability of the EPI system for specific use cases and security improvements. Tight integration of eFPGA blocks with other processing elements of the EPI chip removes the overhead of off-chip communication.

14.2.4 Automotive Stream

The automotive stream activities are targeting main trends driving innovation in the automotive industry including the introduction of autonomous driving (levels 4/5) and the connected car infrastructure. Within this stream, HPC GPPs and HPC accelerators will be integrated into the architectural solutions for a novel embedded HPC (eHPC) platform to demonstrate the approach to be technologically, functionally, and economically successful. The stream activities focus on a concept for an embedded HPC platform (eHPC platform) based on EPI CP. The platform includes two main components: an automotive safety and security microcontroller (eHPC MCU) and a GPP addressing a variety of automotive purposes and applications. The automotive safety and security capable MCU complements the GPP and accelerators and acts as a trusted safety and security backbone supplementing and controlling the safety and security features of the performance microprocessors (GPPs), while the latter will act as number crunchers for high-performance applications (e.g. sensor data fusion). The eHPC MCU provides sufficient performance to serve as a fallback unit to run safety critical applications at least with reduced functionality in order to enable a fail operational system.

As the automotive industry moves to intelligent and interacting vehicles, the need for secure and reliable communications and environments also increases. Therefore, cars also become more susceptible to various security attacks. These attacks can lead to information theft, loss, or corruption of critical data, precluding the correct working of the vehicle. Unsecured systems can also be used as backdoor into secure network. As such, secure protocols need to be used with the intent of securing the communication and the system itself, such as TLS or IPSEC.

The safety and security requirements ensure high functionality of vehicles and protection against unwanted manipulation of built-in devices, which concern more and more embedded systems. The protection against attacks was considered during the development of the hardware and software components for the eHPC MCU. Both requirements – safety as well as security – assumed a much greater effort during the development processes compared with the QM (quality management) processes of standard devices and led to fourfold efforts. This fact requires in-depth knowledge of microcontroller development to achieve the highest safety and security standards for the eHPC MCU.

The requirements for comprehensive safety and security in vehicles present tremendous opportunities for EPI partners like Infineon Technologies AG to deliver products and services to make the breakthrough innovation in the automotive industry. But with great good, there also can be a great risk. Intelligent monitoring and control for cars can increase efficiency and convenience. However, these same powerful tools can be misused by bad actors to disrupt critical embedded systems with dangerous and expensive results. The autonomous driving potentially exposes cars and personal information to malicious or criminal acts.

Fortunately, safety and security techniques developed by Infineon have been applied to the EPI's eHPC MCU to address autonomous driving. These techniques have a proven track record of providing effective, cost-efficient protection while enabling continuing innovation. Many developers working in the fields of safety and security are also experts in car manufacturing. They include their safety and security expertise in products, which meet their domain requirements. That is why experts in Infineon's safety and security solutions are engaging with the many domains affected by autonomous driving to ensure that strong and appropriate safety and security features are built in from the start, as exercised at the EPI project.

Autonomous driving may be the most important technology trend of the 21st century. By connecting numerous electronic sensors and control elements to each other and to interconnected networks, autonomous driving will contribute to increased efficiency, greater convenience, and improved lifestyle for every vehicle user. Autonomous driving consists of a complex combination of connected devices and intelligent car services and applications, where connectivity plays an important role. The number of connected devices is projected to grow at a rate of 15%–20% per year for the next five years with incremental annual revenue in billions of US dollars. The benefits are tremendous, but not without risk. The most dramatic example of safety and security impacts can be seen in connected car technology. It is estimated that if 90% of all vehicles in the USA were fully autonomous (self-driving), "as many as 4.2 million accidents could be avoided each year, saving 21,700 lives and $450 billion in related costs." It is believed that improved safety and reduced stress from the transition to autonomous driving outweighs the risk of successful attacks on connected cars. Infineon agrees that the already realized and potential future benefits make the move to e-vehicles and interconnected cars inevitable. It is vital, that the car industry and policymakers will recognize and address the related safety and security risks, respectively.

A successful attack on a vehicle device can have a significant impact on users, device manufacturers, and service providers by affecting the physical as well as the cyber world. It may expose confidential information such as private user data as well as know-how, intellectual property, and process intelligence. In addition, it can lead to interruption of operations and even danger to a company's brand image, success, and very existence. For policymakers, the principal concerns related to connected cars and e-vehicles is risk mitigation and the protection of public safety and privacy. It is critical that networked systems are protected from both accidental and malicious attacks. Personal information about individuals that are monitored by car devices must also be protected both from accidental exposure and deliberate theft with intent to misuse.

The types of risk associated with connected cars vary. However, the methods of attacks are common across the range of systems. Eavesdropping attacks are aimed at discovering information, which may then be used in future attacks. Other attacks involve subverting or impersonating the server to send bad commands or injecting false information from devices with the intent to cause an unwanted response or hide a physical attack. For an individual, invasion of privacy could escalate to personally catastrophic consequences. An attack may lead to theft of intellectual property or be a precursor to an attack on critical car components or connected cars. The infiltration of counterfeit commands can cause economic damage to a car manufacturer as well as to the producer of car electronics and at the same time lead to loss of confidence in the quality of their products.

While there is no single solution to safety and security, there are commonalities in the approach across the many different application scenarios. In every case, the goal is to prevent unauthorized reading, copying, and analysis of digital information, and to avoid direct manipulation of the protected systems. This can be accomplished with a spectrum of techniques that range from a software-only approach to the use of robust eHPC MCU-based safety and security solutions that are specifically designed to resist even determined attacks by bad actors with access to sophisticated resources.

Safety and security solutions for autonomous driving have been evaluated within EPI using a risk-based approach, in which increasing levels of protection are applied. A scalable safety and security implementation can be designed to protect each car device by isolating it from other devices. In this way, a higher safety level can be built up at critical points. Risk analysis also considers the entire networked system and the many devices that are connected to that system. When devices are linked in a communication network, every linked device represents an attack surface. Even a simple smart entertainment device in a vehicle that is controlled via wireless link can be an entry point for either a malfunction or a more serious attack.

Based on the longstanding experience in the field of safety and security solutions for embedded systems, Infineon and its customers have learned that a software-only approach to protecting systems from malicious attacks leaves both the individual device and larger networked system at risk. eHPC MCU-based hardware security provides a critical layer of protection appropriate to the risk level of the many different devices that make up the whole infrastructure for e-vehicles and autonomous driving. The critical layer revolves around the concept of a "Root of Trust," which is a secured area that resides on a computer chip and provides a memory and processing environment that is isolated from the rest of the system. The "Root of Trust" is shielded from malicious attacks and thus provides security for the other operating layers of the computing system that it protects.

Safety and security for e-vehicles and autonomous driving revolve around three main concepts: confidentiality, identity, and integrity. Is the transfer and storage of sensitive data protected? Are the components of the vehicle infrastructure what they claim to be or are they digitally disguised? Have the components been compromised or infected? These are questions that need to be answered. Powerful eHPC MCU-based hardware security is hardened against attacks and is integrated into the vehicles, networks, or servers.

The lowest level of risk may be a non-programmable end node that simply relays sensor data to some type of gateway or local server, which verifies the source and includes the input in its operating data. Even at this level, the solution provides a way to confirm identity throughout the device life cycle. This also helps to prevent the proliferation of cloned devices at the edge of the network. If there is a requirement that the transmitted data be encrypted or that the device be reconfigured, additional protected storage of keys and certificates are considered. The data and commands that flow between devices and servers should be encrypted sufficiently to resist attempts at eavesdropping and false command injection. This requires cryptographic computation capability at both ends, which are scaled to suit the level of risk.

Even at the lowest level of functionality, hardware-based security uses cryptographic mechanisms to protect secret data. The cryptographic algorithm can be implemented running on the eHPC MCU, but it is advisable for the devices themselves to have at least basic tamper-resistant capability and cryptographic functionality. These chips protect themselves and can even automatically erase their memory if tampering is detected. A holistic approach provides for safety and security throughout the life cycle of every device used in the system. In vehicles that use large numbers of low-cost devices, secure hardware supply chains support shipping devices directly from the manufacturer to the point of assembly. With a preprogrammed identity, the chips can register themselves "over the air" when turned on. It is easier to defend against intrusion and subversion if each device is fitted with a security key at a central point of control.

New autonomous vehicle network architectures require computing platforms to be able to execute complex vehicle perception algorithms that include sensor/imaging processing, data fusion, environment sensing and modeling, low-latency deep machine learning for object classification and behavior prediction with seamless, dependable, and secure interaction between mobile high-performance embedded computing, and stationary server-based HPC. The processing and power requirements in targeted AD levels in automotive use cases can be obtained by accelerating specific computation kernels. EPI's vision is to exploit the Kalray MPPA for the acceleration of automated driving perception functions and other high-performance time-critical functions. MPPA applies the successful principle of GPU many-core architectures: processing elements are regrouped with a multi-banked local memory and a slice of the memory hierarchy into compute units, which share a global interconnect and access to system memory. The key feature of the MPPA many-core architecture is the integration of fully software-programmable cores for the processing elements, and the provision of an RDMA engine in each compute unit. The cores implement a 64-bit, 6-issue VLIW architecture, which is an effective way to design instruction-level parallel cores targeting numerical, signal, and image processing applications. Moreover, the implementation of this VLIW core and its caches ensures that the resulting processing element is fully timing compositional, a critical property with regard to computing accurate bounds on the worst-case response time.

14.3 Conclusion

The EPI project will provide European industry and research with a competitive HPC platform, ecosystem, computing solutions, and capabilities at a world-class level. EPI will harmonize the heterogeneous computing environment by defining an EPI CP that will include the global architecture (hardware and software) specification, common design methodology, and global approach for power management and security. The right balance of computing resources for application matching will be defined through the ratio of the accelerator and general-purpose tiles. We expect to achieve unprecedented levels of performance at very low power.

One of the distinctive, visionary features of the EPI CP approach is the federation of accelerators. This unique feature is enabled by the interposer and multi-chiplet concept of the EPI CP (Figure 14.6). Starting from the introductory SiPearl Rhea family of processors, chiplets will allow building a variety of performance/feature options for members of a family. Besides the obvious fact that the number of chiplets will define processing power provided in a single chip, the CP chiplet approach goes beyond this by also providing the option of HBM configurations and that processing IP can even be integrated within a single chiplet as a core in the 2D mesh.

This design decision has created an opportunity to offer integration of accelerator IP into the Rhea/Cronos platform to those that until now would not have thought it feasible to consider their accelerator IP working in such a powerful processing platform, as illustrated in Figure 14.7. The possibilities with this approach are now wide, globally attractive, and provide many new business opportunities. From the developers' and users' standpoint,

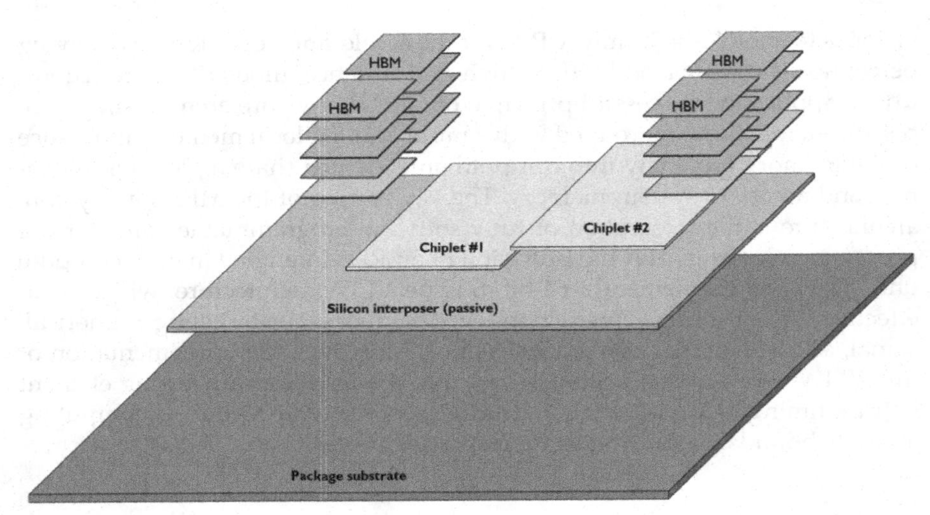

FIGURE 14.6
EPI interposer and multi chiplet modular concept.

the approach presents new opportunities to use specific targets for certain functions/kernels while still relying on widely accepted general computing platform technologies.

Acknowledgment

This work has received funding from the European Union's Horizon 2020 research and innovation programme "European Processor Initiative (EPI)" under grant agreement No 826647.

Bibliography

[1] B. Maitre, G. Massari, A. Cilardo, W. Fornaciari, Y. Hoornenborg, and M. Kovač "Enabling HPC for QoS-sensitive applications: the MANGO approach," *DATE 2016*, 2016, pp. 702–707.
[2] "HiPEAC." [Online]. Available: www.hipeac.net/. [Accessed: 15-Apr-2021].
[3] "OPen TransPREcision COMPuting (OPRECOMP) European project." [Online]. Available: http://oprecomp.eu/. [Accessed: 15-Apr-2021].
[4] "Home | MEEP." [Online]. Available: https://meep-project.eu/. [Accessed: 15-Apr-2021].

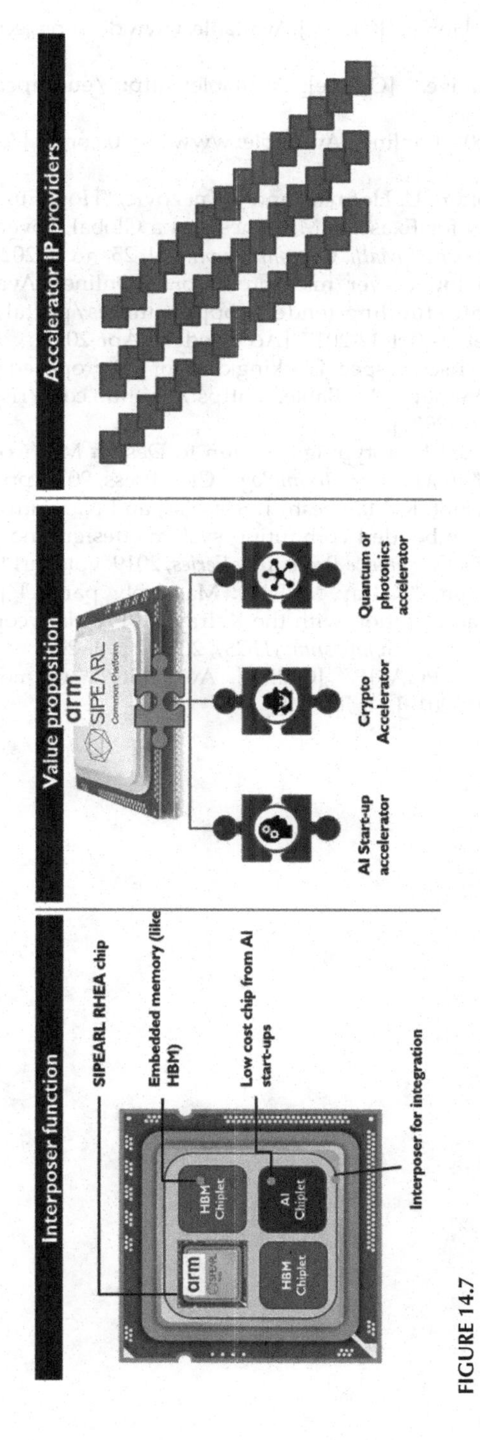

FIGURE 14.7
EPI common platform – federation of accelerators.

[5] "DEEP Projects | Home." [Online]. Available: www.deep-projects.eu/. [Accessed: 15-Apr-2021].
[6] "EuroHPC Initiative." [Online]. Available: http://eurohpc.eu/. [Accessed: 12-Jan-2018].
[7] "Home – TOP500." [Online]. Available: www.top500.org/. [Accessed: 15-Apr-2021].
[8] M. Kovac, P. Notton, D. Hofman, and J. Knezovic, "How Europe Is Preparing Its Core Solution for Exascale Machines and a Global, Sovereign, Advanced Computing Platform," *Math. Comput. Appl.*, vol. 25, no. 3, 2020.
[9] "FPA European low-power microprocessor." [Online]. Available: https://ec.europa.eu/info/funding-tenders/opportunities/portal/screen/opportunities/topic-details/ict-42-2017. [Accessed: 15-Apr-2021].
[10] "GitHub – riscv/riscv-v-spec: Working draft of the proposed RISC-V V vector extension." [Online]. Available: https://github.com/riscv/riscv-v-spec. [Accessed: 19-Apr-2021].
[11] "SESAM: A Virtual Prototyping Solution to Design Multicore Architectures Andriamisaina," in *Multicore Technology*, CRC Press, 2021, pp. 87–130.
[12] A. Charif, G. Busnot, R. Mameesh, T. Sassolas, and N. Ventroux, "Fast virtual prototyping for embedded computing systems design and exploration," in *ACM International Conference Proceeding Series*, 2019, vol. Part F148382, pp. 1–8.
[13] B. D. de Dinechin, "Kalray MPPA®: Massively parallel processor array: Revisiting DSP acceleration with the Kalray MPPA Manycore processor," in *2015 IEEE Hot Chips 27 Symposium (HCS)*, 2015, pp. 1–27.
[14] Menta, "Menta eFPGA IP." [Online]. Available: www.menta-efpga.com/. [Accessed: 02-Oct-2019].

Index

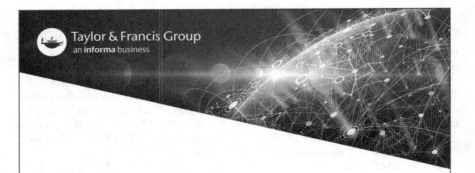

Printed in the United States
by Baker & Taylor Publisher Services